Large-Scale Structures in Acoustics and Electromagnetics

Proceedings of a Symposium

Board on Mathematical Sciences
Commission on Physical Sciences, Mathematics, and Applications
National Research Council

National Academy Press
Washington, D.C. 1996

NOTICE: The project that is the subject of this report was approved by the Governing Board of the National Research Council, whose members are drawn from the councils of the National Academy of Sciences, the National Academy of Engineering, and the Institute of Medicine.

The National Academy of Sciences is a private, nonprofit, self-perpetuating society of distinguished scholars engaged in scientific and engineering research, dedicated to the furtherance of science and technology and to their use for the general welfare. Upon the authority of the charter granted to it by the Congress in 1863, the Academy has a mandate that requires it to advise the federal government on scientific and technical matters. Dr. Bruce Alberts is president of the National Academy of Sciences.

The National Academy of Engineering was established in 1964, under the charter of the National Academy of Sciences, as a parallel organization of outstanding engineers. It is autonomous in its administration and in the selection of its members, sharing with the National Academy of Sciences the responsibility for advising the federal government. The National Academy of Engineering also sponsors engineering programs aimed at meeting national needs, encourages education and research, and recognizes the superior achievement of engineers. Dr. Harold Liebowitz is president of the National Academy of Engineering.

The Institute of Medicine was established in 1970 by the National Academy of Sciences to secure the services of eminent members of appropriate professions in the examination of policy matters pertaining to the health of the public. The Institute acts under the responsibility given to the National Academy of Sciences by its congressional charter to be an adviser to the federal government and, upon its own initiative, to identify issues of medical care, research, and education. Dr. Kenneth I. Shine is president of the Institute of Medicine.

The National Research Council was organized by the National Academy of Sciences in 1916 to associate the broad community of science and technology with the Academy's purposes of furthering knowledge and advising the federal government. Functioning in accordance with general policies determined by the Academy, the Council has become the principal operating agency of both the National Academy of Sciences and the National Academy of Engineering in providing services to the government, the public, and the scientific and engineering communities. The Council is administered jointly by both Academies and the Institute of Medicine. Dr. Bruce Alberts and Dr. Harold Liebowitz are chairman and vice-chairman, respectively, of the National Research Council.

The National Research Council established the Board on Mathematical Sciences in 1984. The objectives of the Board are to maintain awareness and active concern for the health of the mathematical sciences and to serve as the focal point in the National Research Council for issues connected with the mathematical sciences. In addition, the Board conducts studies for federal agencies and maintains liaison with the mathematical sciences communities and academia, professional societies, and industry.

This material is based on work supported by the National Science Foundation under Grant No. DMS-9525898 and relates to Department of Navy Grant N00014-94-1-0571 issued by the Office of Naval Research. Any opinions, findings, and conclusions or recommendations expressed in this material are those of the authors and do not necessarily reflect the views of the sponsors. The United States government has a royalty-free license throughout the world in all copyrightable material contained herein.

Library of Congress Catalog Card Number 95-70716
International Standard Book Number 0-309-05337-4

Additional copies of this report are available from:
National Academy Press
2101 Constitution Avenue, N.W.
Box 285
Washington, D.C. 20055
800-624-6242; 202-334-3313 (in the Washington Metropolitan Area)
B-665

Copyright 1996 by the National Academy of Sciences. All rights reserved.

Printed in the United States of America

BOARD ON MATHEMATICAL SCIENCES

AVNER FRIEDMAN, University of Minnesota, *Chair*
JEROME SACKS, National Institute of Statistical Sciences, *Vice-Chair*
LOUIS AUSLANDER, City University of New York
HYMAN BASS, Columbia University
PETER E. CASTRO, Eastman Kodak Company
FAN R.K. CHUNG, University of Pennsylvania
R. DUNCAN LUCE, University of California, Irvine
PAUL S. MUHLY, University of Iowa
GEORGE NEMHAUSER, Georgia Institute of Technology
ANIL NERODE, Cornell University
INGRAM OLKIN, Stanford University
RONALD F. PEIERLS, Brookhaven National Laboratory
DONALD ST.P. RICHARDS, University of Virginia
MARY F. WHEELER, Rice University
ROBERT ZIMMER, University of Chicago

Ex Officio Member
JON R. KETTENRING, Bell Communications Research
    Chair, Committee on Applied and Theoretical Statistics

Staff
JOHN R. TUCKER, Director
RUTH E. O'BRIEN, Staff Associate
JOHN W. ALEXANDER, Program Officer
BARBARA W. WRIGHT, Administrative Assistant

COMMISSION ON PHYSICAL SCIENCES, MATHEMATICS, AND APPLICATIONS

ROBERT J. HERMANN, United Technologies Corporation, *Chair*
STEPHEN L. ADLER, Institute for Advanced Study
PETER M. BANKS, Environmental Research Institute of Michigan
SYLVIA T. CEYER, Massachusetts Institute of Technology
L. LOUIS HEGEDUS, W.R. Grace and Co.
JOHN E. HOPCROFT, Cornell University
RHONDA J. HUGHES, Bryn Mawr College
SHIRLEY A. JACKSON, U.S. Nuclear Regulatory Commission
KENNETH I. KELLERMANN, National Radio Astronomy Observatory
KEN KENNEDY, Rice University
THOMAS A. PRINCE, California Institute of Technology
JEROME SACKS, National Institute of Statistical Sciences
L.E. SCRIVEN, University of Minnesota
LEON T. SILVER, California Institute of Technology
CHARLES P. SLICHTER, University of Illinois at Urbana-Champaign
ALVIN W. TRIVELPIECE, Oak Ridge National Laboratory
SHMUEL WINOGRAD, IBM T.J. Watson Research Center
CHARLES A. ZRAKET, Mitre Corporation (retired)

NORMAN METZGER, Executive Director

Preface

In response to a request from the Office of Naval Research, the Board on Mathematical Sciences convened a symposium, "Large-Scale Structures in Acoustics and Electromagnetics," on September 26-27, 1994, in Washington, D.C. The symposium's main theme, the dynamics of large-scale structures, refers to structures that are large relative to their operating wavelengths. Large-scale structures typically involve many substructures and are characterized by an extended range of scales. Examples include large man-made objects in the ocean such as naval and maritime vessels, aerospace vehicles, and densely packed microelectronic and optical integrated circuits. Analytical, computational, and experimental procedures for studying large-scale structures entail an extremely large number of degrees of freedom. The excitation of large-scale structures can yield both linear and nonlinear responses, with similar effects in surrounding media. The dynamics of the substructures and their interfaces include time-variant, dispersive, and dissipative aspects.

The symposium focused on computational methods required to determine the dynamics of large-scale electromagnetic, acoustic, and mechanical systems. Over the past two decades, long-dominant frequency-domain methods have been complemented and occasionally supplanted by a growing collection of time-domain techniques. For example, in structural acoustics two recent procedures involve high-order expansions in time, and temporal finite elements. Another noteworthy example is research on integrated microwave and optical circuits that involves electromagnetic and optical scattering and propagation theories, quantum electronics, and solid-state physics. Speakers were advised that one purpose of the symposium was to stimulate discussions of the efficiency, accuracy, and areas of applicability of time- and frequency-domain computational procedures. Another purpose was to address the interplay of time- and frequency-domain computational procedures and experimental procedures with respect to the future goal of comparing them. The symposium emphasized the relationship and synergy between time- and frequency-domain methods rather than their individual advantages, since information that allows a direct comparison between the two types of methods often is not available.

This symposium helped to clarify the roles of and relationships between time-domain and frequency-domain methods in electromagnetics and acoustics. It also inaugurated an exchange of ideas and perspectives between investigators involved mainly with acoustics and others concerned principally with electromagnetics. Chapters 1, 2, 6, and 9 of these Proceedings describe results using mostly time-domain methods in elasto-acoustics problems derived primarily from fluid-solid interactions of the kind that occur in submarine detection. Chapters 4, 5, 7, 8, and 10 similarly describe spatial frequency range results for electromagnetic systems. Chapter 3 describes a unique, successful attempt at generalizing the problems in a mathematically useful way. A number of the papers address the relative applicability of time-domain and fast (or rather, discrete) Fourier transform methods, adaptive grid techniques, error estimates, costs of computation, and so on, to issues arising in other fields.

Regardless of the domain in which the model of a large structure is formulated, a key issue is how computer cost scales with increasing model complexity and model size in wavelengths. For the most part, the papers in this volume emphasize analytical issues and how good a particular approach is for addressing the problem at hand. For constructing models of large-scale, complex structures, hybrid approaches are required and a multidisciplinary approach is de rigueur. The ideal model would simultaneously be efficient enough to be affordable and to permit quantifiable trade-offs between accuracy and efficiency, and would also provide a known, and preferably selectable, amount of accuracy. Several papers address issues such as computer time involved, the number of iterations required to reach acceptable convergence, and similar concerns, and a few describe how the cost, or equivalently the total operation count, depends on spatial frequency.

Chapter 1 describes significant spatial adaptation techniques that are extremely important for ensuring accuracy and efficiency and also addresses the problem of spectral density that haunts the middle frequency range of elasto-acoustic systems. Chapter 2 demonstrates the use of reduced-order models that capture, to acceptable accuracy, a physical behavior of interest—a technique that may substantially improve modeling efficiency for problems amenable to such an approach. The analytically elegant Chapter 3 presents various modeling approaches in an overall analytical framework that may be most relevant as a means for the systematic development of hybrid models. Chapter 4, which tackles some impressively complex problems in ultrafast optical pulses, provides evidence for the computational benefits of the use of hybrid models and points up the challenges posed by a three-dimensional setting with, presumably, its commensurate increased computer costs. Chapter 5 usefully discusses computational electromagnetics from the initial perspective of computational fluid dynamics, and describes some nice results for this computer-intensive approach. Impressive results showing the benefits of adapting a spatial grid to satisfy an error criterion are given in Chapter 6. Chapter 7 discusses the use of reduced-order models in a hierarchical way that seems well suited to that particular physical setting and elsewhere, and that, at least conceptually, offers hope that hybrid models of problems consisting only of wave fields might be similarly, if for different reasons, addressed. An interesting synthesis and analysis problem for large-scale integrated photonic devices and circuits is discussed in Chapter 8. Chapter 9's preconditioners (an approach to improving the model formulation prior to interaction) lead to, among other things, iterative convergence rates that are almost frequency- and discretization-independent; if this independence were to remain true for general problems, a significant advance could result: time-harmonic iterative solutions might be achieved at a cost proportional to the total number of spatial unknowns. Chapter 10 provides a comprehensive discussion of methods being pursued in electromagnetics to reduce the operation count, describes how such methods work, and presents explicit operation-count scaling laws.

Chapter 11, the record of an open discussion in which symposium participants tried to come to grips with the symposium's main theme, is particularly useful. There it is noted, for example, that much of the person-effort needed for numerical modeling is associated with grid generation, so that being able to avoid grid generation would be extremely beneficial. Avoidance of remeshing is particularly important as well when adapting field sampling for error reduction. Furthermore, since error estimation is not in general well developed for computational electromagnetics, the value and reliability of any computed result must be regarded with skepticism until independent confirming data are obtained. Also, knowledge of spatial or temporal errors is needed to determine whether fields are being undersampled, causing decreased accuracy, or oversampled, causing an expenditure of excessive computer time. Thus the importance of error estimation and validation of computed results in computational electromagnetics can hardly be overstated. Chapter 11 also points out the opportunity to compile the common analytical features of electromagnetics and acoustics and to tabulate the most efficient kinds of codes for modeling various kinds of structures. Also of value would be the development of a modeling handbook that catalogs solved problems (according to type and complexity together with the codes used for their solutions, and so forth), and that includes negative as well as positive results so that others can learn what does and does not work.

The symposium's concluding discussion further raised the crucial question of what is really sought in scientific computing, which all too often is done with no regard for whether all the resulting data are needed. If computing entailed no cost penalty, all would be fine, but that is hardly ever the case. Thus, when computing the input impedance of an antenna, with everything else being equal, it would almost always be better to develop an iterative solution rather than compute a factored—or, even worse, an inverted—solution matrix.

As reflected in the symposium papers and in the discussion, numerous research opportunities exist, including enunciating the scaling laws for some of the techniques discussed and addressing the question, Are electromagnetic systems immune to the spectral density problem (addressed in Chapter 1) because they have lower-order field equations? The presentations collected in this volume point out interrelationships and opportunities for future development of heretofore mostly separate research approaches, and so inspire coordinated progress in understanding the behavior of large-scale structures. The papers and discussion help clarify issues and, in emphasizing scientific bridges and methodological commonalities, indicate new and beneficial research directions. It is hoped that they will stimulate investigations of these related frequency- and time-domain approaches and how to use them together to achieve even greater progress in acoustics and electromagnetics, as well as exploration of cross-cutting fundamental questions whose answers would directly benefit efforts in these areas.

Contents

OPENING REMARKS 1
Fred E. Saalfeld

INTRODUCTION 4
John R. Tucker

1 **HIGH-ORDER, MULTILEVEL, ADAPTIVE TIME-DOMAIN METHODS** 5
 FOR STRUCTURAL ACOUSTICS SIMULATIONS
 J. Tinsley Oden, Andrzej Safjan, Po Geng, Leszek Demkowicz

2 **DISTRIBUTED FEEDBACK RESONATORS** 50
 Hermann A. Haus

3 **ACOUSTIC, ELASTODYNAMIC, AND ELECTROMAGNETIC WAVEFIELD** 72
 COMPUTATION—A STRUCTURED APPROACH BASED ON RECIPROCITY
 Adrianus T. de Hoop, Maarten V. de Hoop

4 **NUMERICAL MODELING OF THE INTERACTIONS OF ULTRAFAST** 89
 OPTICAL PULSES WITH NONRESONANT AND RESONANT MATERIALS
 AND STRUCTURES
 Richard W. Ziolkowski, Justin B. Judkins

5 **ALGORITHMIC ASPECTS AND SUPERCOMPUTING TRENDS IN** 103
 COMPUTATIONAL ELECTROMAGNETICS
 Vijaya Shankar, William F. Hall, Alireza Mohammadian, Chris Rowell

6 **ADAPTIVE FINITE ELEMENT METHODS FOR THE HELMHOLTZ** 122
 EQUATION IN EXTERIOR DOMAINS
 James R. Stewart, Thomas J.R. Hughes

7 **MODELING OF OPTICALLY "ASSISTED" PHASED ARRAY RADAR** 143
 Alan Rolf Mickelson

8 **SYNTHESIS AND ANALYSIS OF LARGE-SCALE INTEGRATED** 162
 PHOTONIC DEVICES AND CIRCUITS
 Lakshman S. Tamil, Arthur K. Jordan

9 **DESIGN AND ANALYSIS OF FINITE ELEMENT METHODS FOR** 183
 TRANSIENT AND TIME-HARMONIC STRUCTURAL ACOUSTICS
 Peter M. Pinsky, M. Malhotra, Lonny L. Thompson

**10   AN OVERVIEW OF THE APPLICATION OF THE METHOD OF MOMENTS............204
         TO LARGE BODIES IN ELECTROMAGNETICS**
Edward H. Newman, I. Tekin

11   DISCUSSION...221

APPENDICES 241

A	**Symposium Agenda**..243
B	**Speakers** 246
C	**Symposium Participants**...250

Large-Scale Structures in Acoustics and Electromagnetics

Proceedings of a Symposium

Opening Remarks

Fred E. Saalfeld
Office of Naval Research

Good morning and greetings to our distinguished guests. I am pleased to represent the Office of Naval Research at this National Research Council Symposium on Large-Scale Structures in Acoustics and Electromagnetics. Large-scale structures are those whose dimensions are larger than the operational acoustic or electromagnetic wavelengths and which are complex, comprising many different systems and components with different length scales.

Large-scale structures are important for many missions of the Navy. A few examples are sonar and radar tracking and identification of ships and airplanes, microwave and optical communications, and geophysical remote sensing. The Office of Naval Research sponsors many diverse research programs that involve large-scale structures in acoustics and electromagnetics. In particular, the use of acoustics to "see" in the ocean is a Department of the Navy issue. In the littoral areas of the world, this problem is especially challenging technically since the acoustic reverberations are very complex. For mine location and mapping in the littoral environment, electromagnetics, especially ocean optics, holds the key to many mission requirements. Here, too, the technical challenges are impressive.

The Department of the Navy and the Office of Naval Research have undertaken support of many scientific and engineering endeavors, such as mathematics and the neural sciences, to meet these challenges. I will illustrate some of the Navy's science and technology work with two examples.

An electromagnetic example of a complex, large-scale structure is a very large scale integrated (VLSI) circuit, shown in Figure 0.1. This silicon chip is about 5 millimeters square and contains 108 photodetectors, which are seen as the horizontal line stretching across the center of the chip. Radar pulses modulate a laser beam that is then incident upon this chip. The circuit is a linear photodetector array that was developed for spectrum analysis in an environment with a high density of short radar pulses. The incident light beam wavelength is 830 nanometers, which is nearly 4 orders of magnitude smaller than the chip size. Information is extracted from the optical beam pixels at the focal plane of the photodetector array. Electromagnetic research problems arise, given that typical radar pulse widths may be less than 100 nanoseconds and a dynamic range greater than 60 dB is desired.

The electromagnetic field analysis of complex structures such as this VLSI chip can be accomplished with a combination of analytic and numerical methods. Asymptotic methods can be used depending upon the ratio of the characteristic dimensions of the complex structure to the operational wavelength. For wavelengths much smaller than the characteristic dimensions, geometric and physical optics approximations provide approximate analytic solutions. If the wavelength is much larger than the complex structure, then the Rayleigh approximation for large-scale volume scattering is valid. When the wavelength is approximately equal to the characteristic dimension, exact analytic solutions are possible only for simple, canonical geometries. Numerical integration directly from Maxwell's equations can be used for complex geometries if sufficient computational power is available.

This region of the spectrum, which is called the resonance region, is important for many applications. It is also important since the scattering characteristics are sensitive to small changes in wavelength, scattering angle, and polarization. Microscopic and quantum effects also need to be considered for very short wavelengths. We need a better understanding of electromagnetic interactions in this and other regions of the spectrum. This understanding can be obtained by the appropriate combination of exact and asymptotic analytic methods and also numerical methods.

Large-scale structures that use acoustic radiation and scattering are much larger in size and have complex internal structures. An example is a submarine shown in Figure 0.2.

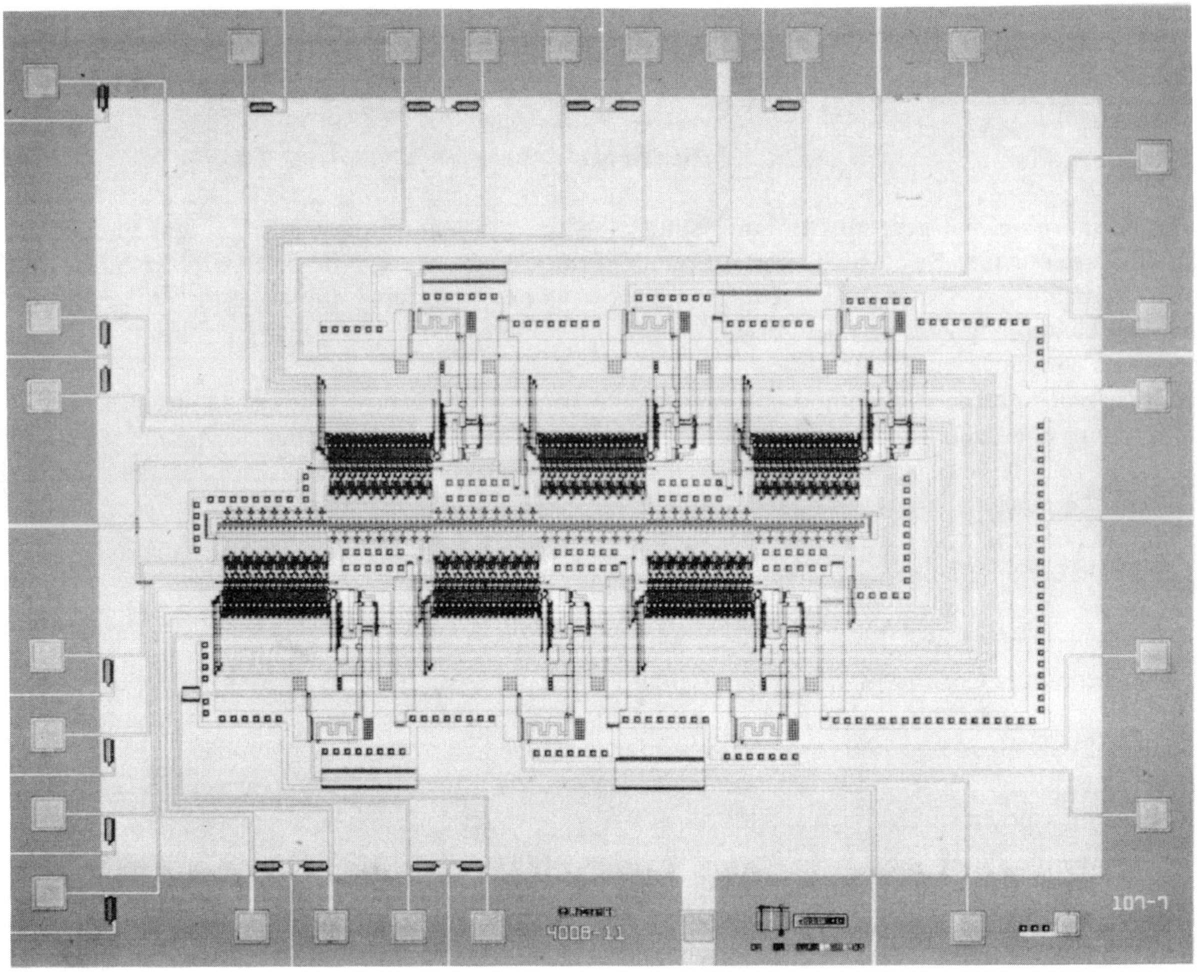

Figure 0.1 Viewgraph of VLSI chip (courtesy of G.W. Anderson, Electronics Science and Technology Division, Naval Research Laboratory).

Figure 0.2 Viewgraph of submarine with cutout.

Acoustic waves that are incident on this submarine will have complex interactions with it. Inside the hull are many smaller structures and equipment. As is the case with electromagnetics, the analysis of the scattering and radiation of sound waves by such complex structures into the surrounding fluid medium or into its passenger cabin can also use asymptotically valid methods. This asymptotic analysis is possible for small or large values of the ratio of characteristic dimensions to acoustic wavelengths.

However, there is a large intermediate range in which this capability is severely limited. The difficulties involved can be illustrated by considering a simple example of an empty submarine hull submerged in water. The surrounding water supports the acoustic wavelength while the steel shell of the hull supports multiple elastic waves with wavelengths ranging from one-fifth of the acoustic wavelength to orders of magnitude larger than the acoustic wavelength. Thus the analysis of this system involves a very wide range of wavelengths that are amenable to widely differing analytic techniques. Due to the strong coupling between wave types, all interactions at different wavelengths must be treated accurately.

The comprehensive analysis of acoustic scattering by large complex structures has long had importance for the Navy. Furthermore, U.S. industry is under increasing pressure to develop analytic and modeling capabilities for complex structural acoustic systems with high strength, light weight, and low noise and vibration. The recognition of these needs has made it imperative to optimize the strength-to-weight and vibration-to-weight ratios.

Whether the interest is noise in submarines, aircraft, or automobiles, engineers and scientists are still unable to predict accurately and reliably the midfrequency acoustic behavior of large complex structures. These needs, coupled with the opportunities offered by broadening computational horizons, have renewed interest in analytical models of systems that in principle are understood, but the analysis of which is limited in practice by complexity.

Large-scale structures in acoustics and electromagnetics exemplify this trend. The VLSI chip and the submarine indicate the general characteristics of large-scale complex structures that need to be better understood:

- First, models of these large-scale complex structures require the coherent combination of asymptotic, computational, empirical, and exact analysis.
- Second, the analysis of large-scale systems generally is multidisciplinary in nature.
- Third, efficiency and accuracy of the analytic techniques are essential for models of realistic large-scale systems.

At this symposium, the ONR looks forward to learning how cooperative research and development in different sciences and technologies will lead to a greater understanding of large-scale systems. I am impressed with the high quality of the abstracts and the breadth of the technical material to be presented. This symposium is indeed appropriate for the National Academy of Sciences because of the excellence of the speakers and their subjects. I am sure you will advance our knowledge of large-scale structures during your time here.

Thank you and best wishes for a successful symposium.

Introduction

John R. Tucker
Board on Mathematical Sciences, National Research Council

Dr. Saalfeld's splendid overview hits just the right chord to open this symposium. Here we seek to identify areas for interplay between different approaches and, wherever possible, to emphasize potential benefits that can arise from synergy between different methods in the analysis of large-scale systems. As director of the Board on Mathematical Sciences, I relish opportunities to spotlight research directions that will help address cutting-edge questions of importance to both government and industry. And so I thank the Office of Naval Research and the Naval Research Laboratory for supporting this symposium, as it is particularly satisfying in that regard. I also want to thank Dr. Philip Abraham of the Office of Naval Research for conceiving the idea of holding this symposium.

The aim is to encourage a great amount of stimulating informal discussion and simultaneously to provide a forum where questions of interest to everyone can be addressed. A panel discussion follows the presentations to summarize the most important points, issues, and ideas in what has been presented and discussed and to indicate what are considered to be the best directions in which to focus research, and what approaches or ideas merit emphasis.

Thank you for coming. I hope you find this event both stimulating and enjoyable, and I welcome your comments on it.

1

High-Order, Multilevel, Adaptive Time-Domain Methods for Structural Acoustics Simulations

J. Tinsley Oden
Andrzej Safjan
Po Geng
Leszek Demkowicz
University of Texas at Austin

> Traditional approaches toward the computer simulation of structural acoustics phenomena have been based primarily on frequency-domain formulations of such problems, and these have dominated both the scientific literature and available analysis software in structural acoustics for many years. These classical approaches have been very successful for a limited class of problems, particularly those involving scales (wavelength) that are large relative to the characteristic dimensions of the structures. In recent times, attempts at large-scale simulations involving high ka-values have revealed limitations of many classical methods and have led to the consideration of new and alternative approaches, including time-domain techniques. This paper describes a new class of high-order methods which employ adaptive hp-version finite-element methods implemented on a special data structure designed to capture multiple-scale response. This is accomplished through the use of multilevel mesh and spectral representations, a posteriori error estimation, adaptive error control, and algorithms for treating non-reflective boundary conditions. Applications to several model problems are presented. In addition, new coupled finite-element and boundary-element methods employing frequency-domain strategies are also discussed, and comparisons with the time-domain strategies are given.

INTRODUCTION

Much of structural acoustics is concerned with the classical problem of interaction of the motion of an elastic structure with that of a perfect inviscid fluid into which the structure is immersed. Small perturbations in the velocity or pressure fields of the fluid create waves that impinge upon the structure and scatter into the fluid-structure domain while small motions of the structure may radiate energy into the surrounding fluid media. The radiation and scattering of waves in such fluid-structure interaction has been the subject of research for many decades.

Mathematically, the problem can be adequately modeled by a system of linear conservation laws, which, with further manipulations, can be reduced to the wave equation for the acoustical pressure coupled to the equations of elastodynamics for the structure. Traditionally, approaches for the computer simulations of structural acoustics phenomena have been primarily based on frequency-domain formulations of such problems, in which a Fourier transformation of the dynamic equations is made, and this results in a system of partial-differential equations in complex-valued field variables with frequency-dependent coefficients. The fluid may be modeled with the exterior Helmholtz equation or, as is commonly done, by an equivalent boundary-integral equation. These classical approaches have been very successful for a large and important class of problems, particularly those involving scales which are not too small in relation to characteristic dimensions of the structure. In recent times, attempts at large-scale simulations involving high ka values (k being the wave number and a the characteristic dimension) have encountered formidable computational difficulties and have exposed serious and fundamental limitations of some classical frequency-domain approaches.

In the present work, some new high-order time-domain methods are presented that have proven to be effective for simulating certain classes of problems in structural acoustics and that may have potential for overcoming many of the shortcomings of the frequency-domain methods for high ka. These techniques are built around several special ideas and algorithms:

1) the structural acoustics problem is formulated as a hyperbolic system of conservation laws and this leads to an abstract Cauchy problem in common Hilbert space settings in which the spectral theory of linear operators is applicable;
2) new, high-order, multistage Taylor-Galerkin (TG) approximations of these equations in the time-domain are constructed which deliver high-order temporal accuracy with unconditional stability;
3) spatial approximations are constructed using hp-finite element methods in which local mesh-size h and local spectral order p are varied in a way to adaptively control error, which provides multi-level approximations of the unknowns;
4) a posteriori error estimates are constructed over each element and time step to provide a running account of solution quality and to provide a basis for the adaptive control strategy;
5) applications to representative model problems reveal that the methods are robust and provide remarkably accurate results for a class of two-dimensional cases.

In addition to the time-domain methods, a brief account of a coupled boundary element/finite element method is also given. This approach also employs hp-methods and is based on a new version of the Burton-Miller integral equations for the acoustical fluid. Results obtained by using large-scale parallel computers for such problems are presented.

We also make comparisons of time-domain and frequency-domain approaches for a model class of problems. In these computations, some advantages of the time-domain approach over traditional frequency-domain approaches are observed.

A FLUID-SOLID INTERACTION PROBLEM

Statement of the Problem

Our investigation focuses on the simulation of structural acoustics phenomena using classical models of acoustics and elastic bodies characterized by systems of hyperbolic equations. The elastic structure is modeled by the classical Navier-Lamé equations of linear elasticity, and the fluid is a perfect acoustical medium modeled as a quiescent, inviscid fluid. The mathematical setting is described as follows, and is depicted in Figure 1.1.

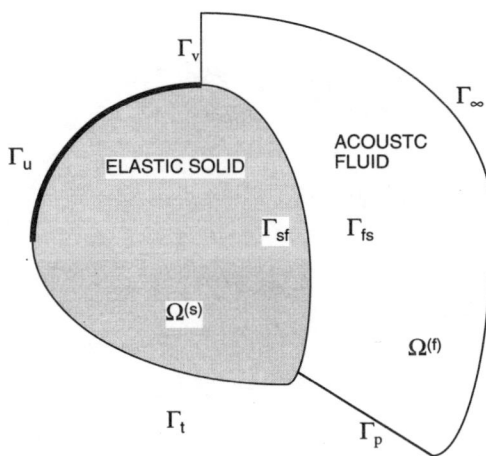

Figure 1.1 The solid-structure interaction problem.

Let $\Omega^{(s)} \subset \mathbb{R}^N, N = 2, 3$, be a region occupied by an elastic body and and let its boundary $\partial \Omega^{(s)}$ consist of three distinct parts, $\partial \Omega^{(s)} = \overline{\Gamma_u \cup \Gamma_t \cup \Gamma_{sf}}, \Gamma_u \cap \Gamma_t \cap \Gamma_{sf} = \emptyset$, where Γ_u and Γ_t are portions of $\partial \Omega^{(s)}$ with prescribed displacements and tractions, respectively, and Γ_{sf} is the the portion of $\partial \Omega^{(s)}$ subjected to the action of the fluid (the contact surface). Similarly, let $\Omega^{(f)} \subset \mathbb{R}^N, N = 2, 3$, be a region occupied by acoustic fluid and let the boundary $\partial \Omega^{(f)}$ consist of four distinct parts, $\partial \Omega^{(f)} = \overline{\Gamma_v \cup \Gamma_p \cup \Gamma_\infty \cup \Gamma_{fs}}, \Gamma_v \cap \Gamma_p \cap \Gamma_\infty \cap \Gamma_{fs} = \emptyset$, where Γ_v and Γ_p are portions of $\partial \Omega^{(f)}$ with prescribed velocities and pressure, respectively, Γ_∞ denotes the truncated exterior boundary, and Γ_{fs} denotes the part of the boundary which is in contact with the elastic body. We further denote $\Gamma_{fs} = \Gamma_{sf} = \Gamma_c$, and we use superscripts s and f to distinguish between variables having $\Omega^{(s)}$ and $\Omega^{(f)}$ as domains.

The fluid-solid interaction problem consists of solving:

- Equations of linear acoustics:

$$p_{,t} + \rho_f c_0^2 \ div\ v^{(f)} = 0 \qquad \text{in } \Omega^{(f)} \times (0, t^*]$$
$$v_{,t}^{(f)} + \rho_f^{-1}\ \mathbf{grad}\ p = 0 \qquad \text{in } \Omega^{(f)} \times (0, t^*]$$
$$v_n \equiv v^{(f)} \bullet n = \hat{v}_n \qquad \text{on } \Gamma_v \times (0, t^*] \qquad (1.1)$$
$$p = \hat{p} \qquad \text{on } \Gamma_p \times (0, t^*]$$
$$\mathcal{B}(v^{(f)}, p) = 0 \qquad \text{on } \Gamma_\infty \times (0, t^*]$$

- Equations of linear elastodynamics:

$$\rho_s u_{,tt}^{(s)} - D^T C D u^{(s)} = 0 \qquad \text{in } \Omega^{(s)} \times (0, t^*]$$
$$u^{(s)} = \hat{u} \qquad \text{on } \Gamma_u \times (0, t^*] \qquad (1.2)$$
$$s(u^{(s)}) \bullet n = \hat{t} \qquad \text{on } \Gamma_t \times (0, t^*]$$

- Contact conditions:

$$v^{(s)} \bullet n = v^{(f)} \bullet n$$
$$(s(u^{(s)}) \bullet n)n = -pn \qquad \text{on } \Gamma_c \times (0, t^*] \qquad (1.3)$$
$$s(u^{(s)}) \bullet n - (s(u^{(s)}) \bullet n)n = 0$$

Here:

$u^{(s)} = (u_i^{(s)})$	=	$u^{(s)}(x,t)$ is the displacement vector at particle $x \in \Omega^{(s)}$ at time t
$s(u^{(s)}) = (s_{ij}) = CDu^{(s)}$	=	the stress tensor evaluated at the displacement $u^{(s)}$
ρ_s	=	mass density of the elastic body
C	=	6×6 symmetric positive definite matrix of elastic constants
$n = (n_i)$	=	the unit outward normal (at the interface, it is the unit outward normal to the elastic body)
p	=	$p(\mathbf{x},t)$ is the acoustic pressure at $x \in \Omega^{(f)}$ at time t
$v^{(f)} = (v_i^{(f)})$	=	$v^{(f)}(x,t)$ is the velocity in the acoustic fluid
ρ_f	=	mass density of the acoustic fluid
c_0	=	small signal sound speed in the acoustic fluid
$\mathcal{B}$	=	boundary operator at the truncation boundary Γ_∞
D	=	generalized gradient operator defined below

$$D \stackrel{def}{=} \begin{pmatrix} \partial_1 & 0 & 0 \\ 0 & \partial_2 & 0 \\ 0 & 0 & \partial_3 \\ 0 & \partial_3 & \partial_2 \\ \partial_3 & 0 & \partial_1 \\ \partial_2 & \partial_1 & 0 \end{pmatrix} \qquad (1.4)$$

Problem (1.1)-(1.2) is completed by specifying initial conditions:

$$v^{(f)}(x,0) = v_0^{(f)} \qquad\qquad p(x,0) = p_0 \quad \text{in } \Omega^{(f)}$$
$$u^{(s)}(x,0) = u_0^{(s)} \qquad\qquad \partial_t u^{(s)}(x,0) = v_0^{(s)} \quad \text{in } \Omega^{(s)} \qquad (1.5)$$

Here $\hat{v}_n, \hat{p}, \hat{u}, \hat{t}, v_0^{(f)}, p_0, u_0^{(s)}, v_0^{(s)}$ are given boundary and initial data.

Equations (1.1) and (1.2) are well known, so we comment only on the interface conditions (1.3). On Γ_c, we have continuity of the normal component of velocity in both media, and continuity of the normal tractions. In addition, since the acoustic fluid is inviscid, we must have zero tangential traction on the elastic body on Γ_c.

We notice that the pressure p in (1.1) is a primitive variable which is coupled with stresses s of (1.2) through contact conditions, and the stress is not a primitive variable in (1.2). These incompatibilities prompt us to seek alternative formulations of the problem.

Velocity-Displacement Formulation

The formulation is similar to that reported in Hubert and Palencia (1989). The principal steps in the proposed methodology are as follows.

1. Introducing an auxiliary variable $u^{(f)}$,

$$u^{(f)}(x,t) \stackrel{def}{=} \int_0^t v^{(f)}(x,\tau) d\tau \qquad (1.6)$$

and assuming the compatibility condition

$$p = -\rho_f c_0^2 \operatorname{div} u^{(f)}, \quad t = 0, \qquad (1.7)$$

equations (1.1) are transformed to the following form:

$$u_{,tt}^{(f)} - c_0^2 \operatorname{\mathbf{grad}} \operatorname{div} u^{(f)} = 0. \qquad (1.8)$$

In addition, we make the usual assumption that $u^{(f)}$ is a potential vector field:

$$u^{(f)} = \operatorname{\mathbf{grad}} \phi \qquad (1.9)$$

for some scalar field ϕ.

Finally, the transition conditions (1.3) take the form

$$u^{(s)} \bullet n = u^{(f)} \bullet n$$
$$(s(u^{(s)}) \bullet n)n = -(\rho_f c_0^2 \operatorname{div} u^{(f)})n \qquad \text{on } \Gamma_c \times (0, t^*) \qquad (1.10)$$

$$s(u^{(s)}) \bullet n - (s(u^{(s)}) \bullet n)n = 0$$

and the fluid-solid interaction problems amount to solving equations (1.2), (1.8), (1.9), (1.10).

2. **Weak formulation.** We now restrict attention to the case in which $\partial\Omega^{(f)} = \overline{\Gamma_v} \cup \overline{\Gamma_c}$ and $\partial\Omega^{(s)} = \overline{\Gamma_c}$, as indicated in Figure 1.2. Then $\Omega^{(f)}$ and $\Omega^{(s)}$ are assumed to be disjoint open connected sets with smooth boundaries which intersect in a C^2 manifold Γ_c of dimension $N-1$. Let u denote the displacement vector in the whole region $\Omega = \Omega^{(f)} \cup \Omega^{(s)} \cup \Gamma_c$, and $u^{(f)}$ and $u^{(s)}$ denote, respectively, its restrictions to $\Omega^{(f)}$ and $\Omega^{(s)}$:

$$\mathbf{u}^{(f)} = \mathbf{u}\big|_{\Omega^{(f)}}, \qquad \mathbf{u}^{(s)} = \mathbf{u}\big|_{\Omega^{(s)}}. \tag{1.11}$$

We start by introducing the spaces of kinematically admissible displacements

$$X^{(s)} = \{w^{(s)} \in (H^1(\Omega^{(s)}))^N\} \tag{1.12}$$

$$\begin{aligned} X^{(f)} = \{w^{(f)} \in (L^2(\Omega^{(f)}))^N, \quad &\text{div}\, w^{(f)} \in L^2(\Omega^{(f)}), \\ \exists \phi \in H^1(\Omega^{(f)})/\mathbb{R}, \quad &w^{(f)} = \text{grad}\,\phi, w^{(f)} \bullet n\big|_{\Gamma_v} = 0\} \end{aligned} \tag{1.13}$$

$$X = \{w : w \in (L^2(\Omega))^N, w^{(s)} \in X^{(s)}, w^{(f)} \in X^{(f)}, w^{(s)} \bullet n\big|_{\Gamma_c} = w^{(f)} \bullet n\big|_{\Gamma_c}\}. \tag{1.14}$$

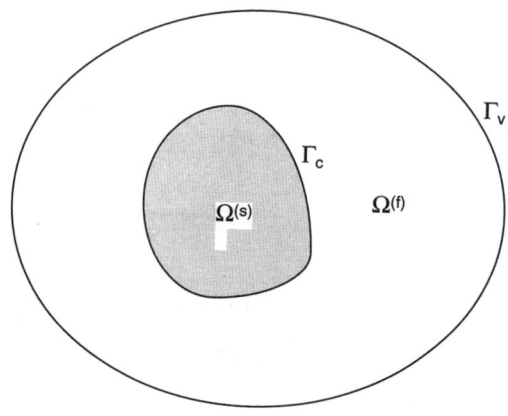

Figure 1.2 A submerged elastic structure.

Next, we define the following Hermitian forms

$$\begin{aligned} a^{(s)}(u,w) &= \rho_s^{-1}(CDu, Dw)_{(L^2(\Omega^{(s)}))^M}, \quad M = N(N+1)/2 \\ a^{(f)}(u,w) &= c_0^2 \int_{\Omega^{(f)}} \text{div}\, u\, \text{div}\,\overline{w}\, dx \end{aligned} \tag{1.15}$$

$$b^{(l)}(u,w) = (u,w)_{(L^2(\Omega^{(l)}))^N}, \quad l = s, f$$

$$(u,w)_{(L^2(\Omega))^N} = \sum_{i=1}^{N} \int_\Omega u_i \overline{w}_i dx$$

and

$$a(u,w) = \rho_s a^{(s)}(u,w) + \rho_f a^{(f)}(u,w)$$
$$(u,w)_H = \rho_s b^{(s)}(u,w) + \rho_f b^{(f)}(u,w). \tag{1.16}$$

The space X, when endowed with the following inner product,

$$(u,w)_X = (u,w)_H + a(u,w) \tag{1.17}$$

is a Hilbert space. In addition, we define space H as the completion of X with inner product $(u,w)_H$, and, finally, we define space $D(\tilde{A})$ as follows

$$D(\tilde{A}) = \{w: w \in X, D^T CDw^{(s)} \in L^2(\Omega^{(s)}), \tag{1.18}$$
$$\mathbf{grad}\ div\ w^{(f)} \in L^2(\Omega^{(f)})\}.$$

Now, we consider the following weak form of the fluid-solid interaction problem (1.2), (1.8), (1.9), (1.10):

Find $u \in C^2([0,t^*], H) \cap C^1([o,t^*], X) \cap C([0,t^*], D(\tilde{A}))$

such that

$$(u_{,tt}, w)_H + a(u,w) = 0 \quad \forall w \in X \tag{1.19}$$

and

$$(u,w)_H = (u_0, w)_H \quad \forall w \in X, \quad t = 0$$
$$(u_{,t}, w)_H = (v_0, w)_H \quad \forall w \in X, \quad t = 0 \tag{1.20}$$
$$u_0 \in D(\tilde{A}), v_0 \in X.$$

It is easily seen that (1.19) is a classical virtual work formulation of (1.2), (1.8), (1.9), (1.10). In fact, (1.19) is valid for any test functions of appropriate regularity, not only for those of the form of gradient.

Let $w \in \mathcal{D}(\Omega^{(f)}) \times \mathcal{D}(\Omega^{(s)})$, where $\mathcal{D}(\Omega^{(l)}), l = f, s$, denotes the space of test functions. Then (1.18) implies that (1.2) and (1.8) are satisfied in a distributional sense. Next, define a as

$$au = \begin{cases} -\rho_f c_0^2\ \mathbf{grad}\ div\ u, & u \in \Omega^{(f)} \\ -D^T CDu, & u \in \Omega^{(s)}. \end{cases} \tag{1.21}$$

Green's formula applied to $\Omega^{(l)}$ and $\Omega^{(s)}$ gives

$$a(u,w) - (au,w)_H = \int_{\Gamma_c} [s_{ij}(u^{(s)})n_i n_j - \rho_f c_0^2 \, div \, u^{(f)})w_k^{(s)} n_k \qquad (1.22)$$
$$+ (s_{ij}(u^{(s)})n_j - s_{kj}(u^{(s)})n_k n_j n_i)w_l^{(s)} \tau_l]ds$$
$$u \in D(\tilde{A}), w \in X,$$

which gives transition conditions (1.10). Here $n = (n_i)$ is the unit outward normal to the elastic body and $\tau = (\tau_i)$ is the corresponding unit outward tangent.

3. Within the above formalism, the initial boundary value problem of structural acoustics can be reinterpreted as an abstract second order Cauchy problem:

$$\begin{cases} \dfrac{d^2 u}{dt^2} + \tilde{A}u = 0 & 0 < t \le t^* \\ u = u_0, \dfrac{du}{dt} = v_0 & t = 0 \end{cases} \qquad (1.23)$$

where $\tilde{A}$ satisfies

$$(\tilde{A}u, w)_H = a(u,w), u \in D(\tilde{A}), w \in X \qquad (1.24)$$

$u = u(t)$ is an H – valued function of time, and u_0 and v_0 are initial conditions.

To proceed further, we reinterpret (1.23) as a first-order system in time. To this end, we introduce:

- the group variable

$$U = \begin{pmatrix} u_1 \\ u_2 \end{pmatrix} = \begin{pmatrix} v \\ u \end{pmatrix}, \quad v = \frac{du}{dt} \qquad (1.25)$$

- the Hilbert space $\mathcal{H}$

$$\mathcal{H} = H \times X \qquad (1.26)$$

$$(U,W)_{\mathcal{H}} = \left(\begin{bmatrix} u_1 \\ u_2 \end{bmatrix}, \begin{bmatrix} w_1 \\ w_2 \end{bmatrix} \right) = (u_1, w_1)_H + (u_2, w_2)_X$$

- the operator A

$$AU \stackrel{def}{=} \begin{pmatrix} 0 & \tilde{A} \\ -I & 0 \end{pmatrix} U \qquad (1.27)$$

$$D(A) = X \times D(\tilde{A}) = \text{ the domain of } A.$$

Then, (1.22) becomes:

$$\begin{cases} \frac{d}{dt}U + AU = 0 & 0 < t \le t^* \\ U = U_0 & t = 0. \end{cases} \tag{1.28}$$

Here U_0 specifies initial conditions, $U_0 = (v_0^T, u_0^T)^T$.

To show the well-posedness of (1.28) we consider an operator $\check{A}$

$$\check{A} = A + I \tag{1.29}$$

and we show that

(i) $\check{A}$ is accretive, i.e.,

$$\mathrm{Re}(\check{A}U, U)_{\mathcal{H}} \ge 0 \tag{1.30}$$

and,

(ii) $\check{A} + I$ is surjective, i.e., the range of $\check{A} + I$ is the whole space $\mathcal{H}$

$$\mathrm{Rg}(\check{A} + I) = \mathcal{H}. \tag{1.31}$$

Indeed,

$$\begin{aligned}(\check{A}U, U)_{\mathcal{H}} &= (\tilde{A}u_2, u_1)_H + (u_1, u_1)_H + (-u_1, u_2)_H + (u_2, u_2)_H \\ &\quad + a(-u_1, u_2) + a(u_2, u_2) \\ &= 2\mathrm{Im}\, a(u_1, u_2) + a(u_2, u_2) + (u_1, u_1)_H + (-u_1, u_2)_H + (u_2, u_2)_H \end{aligned} \tag{1.32}$$

and, hence,

$$\begin{aligned} 2\,\mathrm{Re}(\check{A}U, U)_{\mathcal{H}} &= 2a(u_2, u_2) + (u_1 - u_2, u_1 - u_2)_H \\ &\quad + (u_1, u_2)_H + (u_2, u_2)_H \ge 0. \end{aligned} \tag{1.33}$$

To prove that $\check{A} + I$ is surjective, we need to show that for each $F = \begin{pmatrix} f_1 \\ f_2 \end{pmatrix} \in \mathcal{H}$ there exists a solution of $(A + 2I) = F$, or equivalently, of the following system

$$\begin{aligned} \tilde{A}u_2 + 2u_1 &= f_1 \\ -u_1 + 2u_2 &= f_2 \\ u_1 \in X, u_2 &\in D(\tilde{A}). \end{aligned} \tag{1.34}$$

System (1.34) can be reduced to the following equation

$$\tilde{A}u_2 + 4u_2 = f_1 - 2f_2, \tag{1.35}$$

which, by virtue of the Lax-Milgram Theorem, possesses a unique solution $u_2 \in X$. This, together with (1.35) and the definition of $D(\tilde{A})$, implies that $u_2 \in D(\tilde{A})$. Then $(1.34)_2$ implies that $u_1 \in X$.

Summarizing, $\breve{A}$ is accretive, $\breve{A}+I$ is surjective, and $D(A)$ is dense in $\mathcal{H}$, which implies that the Cauchy problem for operator $\breve{A}$ is well posed (cf., the Lummer-Philips Theorem in Yosida, 1973). Finally, by using the ansatz

$$U(t) := \exp(-t)U(t), \tag{1.36}$$

we get the well posedness of (1.28). Thus, given $U_0 \in D(A)$, there exists a unique $U \in C^1([0,t^*], \mathcal{H}) \cap C([0,t^*], D(A))$ such that (1.28) is satisfied. Moreover,

$$\|U\|_{\mathcal{H}} \leq \exp(t^*)\|U_0\|_{\mathcal{H}}. \tag{1.37}$$

Problem (1.28) is to be solved numerically by using high-order Taylor-Galerkin methods and *hp*-adaptive finite element methods.

Velocity-Stress Formulation

The second approach consists of converting the system of equations of elastodynamics into an equivalent first-order system, and coupling it with (1.1). Thus, we solve the following first-order system. (We label it the "velocity-stress" formulation.)

- equations of linear acoustics:

$$\begin{aligned} p_{,t} + \rho_f c_0^2 \operatorname{div} v^{(f)} &= 0 \\ v_{,t}^{(f)} + \rho_f^{-1} \operatorname{\mathbf{grad}} p &= 0 \end{aligned} \quad \text{in } \Omega^{(f)} \tag{1.38}$$

- equations of elastodynamics:

$$\begin{aligned} v_{,t}^{(s)} - D^T w &= 0 \\ w_t - CDv^{(s)} &= 0 \end{aligned} \quad \text{in } \Omega^{(s)} \tag{1.39}$$

- contact conditions:

$$\begin{aligned} v^{(s)} \bullet n &= v^{(f)} \bullet n \\ (s \bullet n)n &= -p\underline{n} \\ s \bullet n - (s \bullet n)n &= 0 \end{aligned} \quad \text{on } \Gamma_c \tag{1.40}$$

together with appropriate boundary and initial conditions. Here: $w \stackrel{def}{=} (s_{11}, s_{22}, \ldots, s_{23})^T$, and s_{ij} are the components of the stress tensor s.

As before, (1.38), (1.39), (1.40) can be cast in the form of a Cauchy problem

$$\begin{cases} \dfrac{d}{dt} U + iAU = 0 & t > 0 \\ U = U_0 & t = 0, \end{cases} \tag{1.41}$$

where U is the group variable $U = (v^T, w^T)^T$, i is the imaginary unit, and A is an appropriately defined operator. Again, (1.41) is solved numerically by using high-order Taylor-Galerkin methods and hp-adaptive finite element methods.

HIGH-ORDER TAYLOR-GALERKIN (TG) METHODS

The Taylor-Galerkin (TG) schemes represent Galerkin approximations of temporal Taylor expansions of the solution. The first TG scheme was presented by Oden (1974) in connection with a Lax-Wendroff approximation of nonlinear waves by finite element methods. The TG schemes were made popular by a series of important papers by Donea and collaborators (Donea, 1984; Selmin et al., 1985), but were not applied to schemes of higher "order" than three. By a scheme of "order s" we shall understand that the one-step truncation error measured in L^2-norm is bounded by $C\Delta t^{s+1}$, where Δt is a time-step size and C is a constant independent of Δt. The idea of developing multi-stage high-order TG schemes was presented in the dissertation by Safjan (1993) and in subsequent papers by Safjan and Oden (1993, 1994, 1995a,b). A brief summary of the algorithm is presented below.

Let H be a Hilbert space with inner product $(\cdot,\cdot)$ and corresponding norm $\|\cdot\|$, and $A: H \supset D(A) \to H$ be a self-adjoint operator in H. Being self-adjoint, A admits the spectral decomposition of the form (Yosida, 1973)

$$AU = \int_{-\infty}^{\infty} \lambda dE_\lambda U \tag{1.42}$$

$$D(A) = \left\{ U \in H : \int_{-\infty}^{\infty} \lambda^2 d\|E_\lambda U\|^2 < \infty \right\}, \tag{1.43}$$

where E_λ is a uniquely defined spectral family. In this section we introduce TG schemes for solving abstract Cauchy problems of the form

$$\begin{cases} \dfrac{d}{dt} U + iAU = 0 & 0 < t \leq t^* \\ U = U_0 & t = 0. \end{cases} \tag{1.44}$$

Any approximation to (1.44) must involve discretization both in space and time variables. We will adopt the assumption here that the final approximation is obtained by using finite differences in time and finite elements in space variables. Then, two different approaches are possible. In the classical *method of lines,* an approximation in space variables converts the original initial-boundary-value problem into a system of ordinary differential equations (ODEs), which next is discretized in time using one of many time integration schemes for ODE.

The TG schemes belong to a different class of methods, known as *the method of discretization in time,* which consists of the same two steps but done in the reverse order. By discretizing in time first, the initial-boundary-value problem (1.4) is converted into a sequence of boundary-value problems (1.45)

$$U_\tau(t_n + \Delta t) = TU_\tau(t_n)$$
$$U_\tau(0) = U_0 \qquad (1.45)$$
$$\Delta t = t^*/N, \ t_n = n\Delta t, \ n = 0, 1, \ldots, N,$$

which, in turn, give a basis for a spatial approximation and, consequently, a fully discretized scheme,

$$U_{\tau h}(t_n + \Delta t) = T_h U_{\tau h}(t_n)$$
$$U_{\tau h}(0) = U_{0h} \qquad (1.46)$$
$$\Delta t = t^*/N, \ t_n = n\Delta t, \ n = 0, 1, \ldots, N.$$

Here U_τ and $U_{\tau h}$ are the solutions of the semi-discrete problem (1.45) and the fully discrete problem (1.46), respectively, T is an appropriate transient operator, T_h denotes its finite-dimensional realization, and U_{0h} is a suitable approximation of U_0.

Approximation in Time

Let N be a positive integer. We define a partition in time as $\Delta t = t^*/N$, $t_n = n\Delta t$, $n = 0, 1, \ldots, N$ and consider a typical time-step $t_n \to t_n + \Delta t$. Given the solution $U_\tau(t_n)$, we seek the next time step solution $U_\tau(t_n + \Delta t)$ through an *s*-stage scheme of the following form

$$Z_i - \eta \Delta t^2 Z_{i,tt} = Z_0 + \mu_{i0}\Delta t\, Z_{0,t} + \nu_{i0}\Delta t^2 Z_{0,tt} \qquad (1.47)$$
$$+\Delta t \sum_{j=1}^{i-1} \mu_{ij} Z_{j,t} = \Delta t^2 \sum_{j=1}^{i-1} \nu_{ij} Z_{j,tt}, \quad i = 1, 2, \ldots, s,$$

where

$$Z_j = U_\tau(t_n + c_j \Delta t) \qquad j = 0, 1, \ldots, s \qquad (1.48)$$
$$\mu_{ij}, \nu_{ij}, \mu_{i0}, \nu_{i0}, c_i \in \mathbb{R}, \ i = 1, 2, \ldots, s\ ; j = 1, 2, \ldots, i-1$$
$$\eta \in \mathbb{R}_+ \text{ is a stability parameter}$$
$$s = \text{number of stages}.$$

Coefficients $\mu_{ij}, \nu_{ij}, \mu_{i0}, \nu_{i0}, c_i$, are to be chosen so as to obtain the highest possible order of accuracy, subject to stability or other constraints. A free parameter η is to be chosen from stability considerations.

It is convenient to rewrite (1.47) in the following compact form:

$$\begin{pmatrix} Z_1 \\ Z_2 \\ \cdot \\ \cdot \\ Z_s \end{pmatrix} - \Delta t^2 N \otimes \begin{pmatrix} Z_{1,tt} \\ Z_{2,tt} \\ \cdot \\ \cdot \\ Z_{s,tt} \end{pmatrix} - \Delta t M \otimes \begin{pmatrix} Z_{1,t} \\ Z_{2,t} \\ \cdot \\ \cdot \\ Z_{s,t} \end{pmatrix} = K \otimes \begin{pmatrix} Z_0 \\ \Delta t \, Z_{0t} \\ \Delta t^2 Z_{0,tt} \end{pmatrix}, \quad (1.49)$$

where

$$N = (\nu_{ij}), M = (\mu_{ij}) \in \mathbb{R}^{s \times s}, K = (\kappa_{ij}) \in \mathbb{R}^{s \times 3}, c = (c_i) \in \mathbb{R}^{s \times 1}$$

$$N = \begin{pmatrix} \eta & 0 & 0 & 0 & 0 & 0 \\ \nu_{21} & \eta & 0 & 0 & 0 & 0 \\ \nu_{31} & \nu_{32} & \eta & 0 & 0 & 0 \\ \cdot & \cdot & \cdot & \cdot & \cdot & \cdot \\ \nu_{s1} & \nu_{s2} & \nu_{s3} & \cdot & \nu_{s,s-1} & \eta \end{pmatrix} \qquad c = (c_1 c_2 \cdots c_{s-1} \, 1)$$

$$M = \begin{pmatrix} 0 & 0 & 0 & 0 & 0 & 0 \\ \mu_{21} & 0 & 0 & 0 & 0 & 0 \\ \mu_{31} & \mu_{32} & 0 & 0 & 0 & 0 \\ \cdot & \cdot & \cdot & \cdot & \cdot & \cdot \\ \mu_{s1} & \mu_{s2} & \mu_{s3} & \cdot & \mu_{s,s-1} & 0 \end{pmatrix} \qquad K = \begin{pmatrix} 1 & \mu_{10} & \nu_{10} \\ 1 & \mu_{20} & \nu_{20} \\ 1 & \mu_{30} & \nu_{30} \\ \cdot & \cdot & \cdot \\ 1 & \mu_{s0} & \nu_{s0} \end{pmatrix}.$$

(1.50)

Matrices N, M, and K together with vector c completely characterize the difference scheme (1.49).

A distinct feature of (1.49) is that coefficient matrices N and M are lower triangular which makes the resulting scheme semi-implicit (i.e., to compute Z_i it is necessary to know $Z_{i-1}, Z_{i-2}, \ldots, Z_1$, but it is not necessary to know Z_{i+1}). Moreover, all diagonal elements of N are equal ($\nu_{ii} = \eta, i = 1, 2, \ldots, s$) and so are those of M ($\mu_{ii} = 0, i = 1, 2, \ldots, s$). This makes the operator defining the left-hand side of each stage of (1.46) identical for linear (or linearized) problems, and, hence, significantly reduces the cost of applying the method. A particular choice of zero diagonal elements of M is made with an eye on well-posedness of a typical one-stage problem and a possible splitting of the operator defining the left-hand side of each stage.

To make the i-th stage solution Z_i m-th order accurate, it is necessary to satisfy the *order conditions for* Z_i and to make the previous stage solutions $Z_{i-1}, Z_{i-2}, \ldots, Z_1$, at least of the order m-1, (otherwise, some coefficients have to be set to 0; e.g., if Z_{i-1} is of the order m-2, then, necessarily $\mu_{i,i-1} = 0$). The order conditions for Z_i are obtained by expanding it in Taylor series about Z_0:

$$Z_i = Z_0 + (c_i \Delta t) Z_{0,t} + \frac{1}{2!}(c_i \Delta t)^2 Z_{0,tt} + \ldots + \frac{1}{m!}(c_i \Delta t)^m \frac{\partial^m}{\partial t^m} Z_0 + O(\Delta t^{m+1}). \quad (1.51)$$

Introducing this expression into the left-hand side of the i-th stage equation,

$$Z_i - \sum_{j=1}^{s}(\mu_{ij}\Delta t Z_{j,t} + v_{ij}\Delta t^2 Z_{j,tt}) = Z_0 + \kappa_{i2}\Delta t Z_{0,t} + \kappa_{i3}\Delta t^2 Z_{0,tt}, \tag{1.52}$$

and equating coefficients of like powers of Δt to zero, leads to the following system of nonlinear algebraic equations:

$$c_i^k - k\sum_{j=1}^{s} c_j^{k-1}\mu_{ij} - k(k-1)\sum_{j=1}^{s} c_j^{k-2}v_{ij} = \begin{cases} \mu_{i0}, & k=1 \\ 2v_{i0}, & k=2 \\ 0, & \text{otherwise} \end{cases} \tag{1.53}$$

$$k = 1, 2, \ldots, m.$$

Equations (1.53) are referred to herein as the *order conditions*.

It is important to realize that the particular forms of the coefficient matrices M and N limit the attainable order of (direct) TG schemes. For example, it is not possible to attain order 6 in 3 stages. To remedy this situation, a more general class of transformed Taylor-Galerkin schemes has been introduced in Safjan and Oden (1994, 1995a).

Approximation in Space

First, using the original equations (1.44), we calculate the time derivatives in terms of spatial derivatives as follows:

$$U_{,t} = -iAU \tag{1.54}$$

$$U_{,tt} = -A^2 U. \tag{1.55}$$

Next, replacing the time derivatives in (1.49) by formulas (1.54) and (1.55), we arrive at the following system of equations,

$$\begin{bmatrix} Z_i - \sum_{j=1}^{s}(\mu_{ij}\Delta t[-iAZ_j] + v_{ij}\Delta t^2[-A^2 Z_j]) \\ = Z_0 + \kappa_{i2}\Delta t[-iAZ_0] + \kappa_{i3}\Delta t^2[-A^2 Z_0] \\ i = 1, 2, \ldots, s. \end{bmatrix} \tag{1.56}$$

where the indices i and j are used to denote a particular stage. Finally, multiplying (1.56) by a vector-valued test function W, integrating over Ω and integrating by parts, we arrive at a variational formulation of the form:

$$\begin{bmatrix} \text{Given } \mathbf{Z}_0 \in X \\ \text{Find } \mathbf{Z}_i \in X, i = 1, 2, \ldots, s, \text{ such that} \\ A(\mathbf{Z}_i, \mathbf{W}) - \Delta t \sum_{j=1}^{s} \mu_{ij} C(\mathbf{Z}_j, \mathbf{W}) + \Delta t^2 \sum_{j=1}^{s} \nu_{ij} B(\mathbf{Z}_j, \mathbf{W}) \\ = A(\mathbf{Z}_0, \mathbf{W}) + \kappa_{i2} \Delta t^2 C(\mathbf{Z}_0, \mathbf{W}) - \kappa_{i3} \Delta t^2 B(\mathbf{Z}_0, \mathbf{W}) \\ \text{for all test functions } \mathbf{W} \in X, \end{bmatrix} \quad (1.57)$$

where $X = D(A)$, and the bilinear (sesquilinear) forms A, B, and C are defined by

$$\begin{aligned} A, B, C &: X \times X \to \mathbb{C} \\ A(U, W) &= (U, W) \\ B(U, W) &= -(AU, AW) \\ C(U, W) &= -i(AU, W). \end{aligned} \quad (1.58)$$

Weak form (1.57) is the basis for finite element approximations.

The properties of TG schemes are summarized as follows:

1. the TG methods lead to a series of well-posed problems that can be solved using *hp*-adaptive finite element methods with very high accuracy and high rates of convergence,
2. the TG methods have excellent accuracy; some of them possess a built-in temporal error estimate,
3. with a proper choice of parameters, TG methods can be unconditionally stable for certain classes of problems,
4. the TG methods can be classified as semi-implicit schemes, so that they are generally more efficient than fully implicit schemes, and provide a natural splitting of the operator that preserves accuracy, and
5. in most cases, the resulting system of linear algebraic equations is symmetric and positive definite, which is a comfortable setting for most iterative solvers.

For more details on TG schemes, we refer to Safjan and Oden (1993, 1994, 1995a,b).

HIGHLY ABSORBING BOUNDARY CONDITIONS

In many situations, we wish to solve a transient problem in an infinite domain. However, any computational domain is necessarily bounded, so it is essential to introduce appropriate boundary conditions at the boundaries. In particular, we are interested in radiating boundary conditions that have the following features:

(i) the boundary conditions substantially reduce the (unphysical) reflections from the artificial boundaries,

(ii) the boundary conditions are local,

(iii) the boundary conditions together with governing differential equation result in a well-posed problem,

(iv) the boundary conditions fit into the framework of our Taylor-Galerkin schemes.

We first develop radiating boundary conditions for the scalar wave equation and then deduce boundary conditions suitable for the equivalent 1st-order hyperbolic system. Our development follows Enquist and Majda (1977, 1979).

Model Problem

Let p denote the acoustical pressure and let us consider the following initial boundary-value problem for the wave equation

$$\begin{aligned}&p_{,tt} - (p_{,xx} + p_{,yy}) = 0, \quad x > 0, t > 0 \\ &p(0,y,t) = g(y,t)H(t) \\ &p(x,y,0) = p_{,t}(x,y,0) = 0.\end{aligned} \quad (1.59)$$

Here $g(y, t)$ is an arbitrary square-integrable source function and $H(t)$ is the Heaviside step function. Let δ be an acceptable error and suppose that we are interested in the solution of (1.59) in the strip $[0, a^*] \times \mathbb{R}, a^* > 0$, rather than in the whole right half-plane. Our objective is to find a boundary operator, $\mathcal{B}$, on the wall $x = a$ with $a \geq a^*$, but as close to a^* as possible, so that if $\tilde{p}$ is the solution of

$$\begin{aligned}&\tilde{p}_{,tt} - (\tilde{p}_{,xx} + \tilde{p}_{,yy}) = 0, \quad x > 0, t > 0 \\ &\tilde{p}(0,y,t) = g(y,t)H(t) \\ &\mathcal{B}\,\tilde{p}\big|_{x=a} = 0 \\ &\tilde{p}(x,y,0) = \tilde{p}_{,t}(x,y,0) = 0,\end{aligned} \quad (1.60)$$

then

$$\int_0^T \int_{-\infty}^{\infty} \int_0^{a^*} \left| p - \tilde{p} \right|^2 dxdydt \leq \delta^2, \quad (1.61)$$

where time T is so large that M reflections from the artificial boundary at $x = a$ might have occurred.

A Family of Highly Absorbing Boundary Conditions

We first derive a theoretically exact nonreflecting boundary condition. By taking the Fourier transform with respect to y and the Laplace transform with respect to t, we arrive at the following solution of (1.59):

$$p(x,y,t) = \iint \exp\left\{-i\omega\left[(1 - \xi^2/\omega^2)^{1/2} x - t\right] + i\xi y\right\} \hat{g}(\xi,\omega)\, d\xi d\omega \quad (1.62)$$

where

$$\hat{g}(\xi,\omega) = \iint \exp\{-i\omega t - i\xi y\} g(y,t) H(t)\, dy dt \qquad (1.63)$$

ξ and ω are dual variables to y and t, respectively, and $\wedge$ denotes the Fourier transform. We notice the correspondences $i\omega \leftrightarrow \partial/\partial t$ and $i\xi \leftrightarrow \partial/\partial y$.

Let us introduce the operator $\Re$ which is non-local in space and time, defined by

$$\Re p\big|_{x=a} = \iint \exp\{i\omega t + i\xi y\} i\omega(1-\xi^2/\omega^2)^{1/2}\, \hat{p}(a,\xi,\omega)\, d\xi d\omega. \qquad (1.64)$$

Then, a simple computation reveals that the solution p satisfies the following non-local theoretical radiation condition at $x = a$:

$$(p_{,x} + \Re p)\big|_{x=a} = 0. \qquad (1.65)$$

By taking the Fourier transform on (1.65) we arrive at

$$\left[\hat{p}_{,x} + i\omega(1-\xi^2/\omega^2)^{1/2}\hat{p}\right]\big|_{x=a} = 0. \qquad (1.66)$$

Introducing the approximation,

$$(1-\xi^2/\omega^2)^{1/2} = 1 + O(|\xi/\omega|^2), \qquad (1.67)$$

(1.66) reads

$$\left[\hat{p}_{,x} + i\omega\hat{p} + i\omega O(|\xi/\omega|^2)\hat{p}\right]\big|_{x=a} = 0. \qquad (1.68)$$

By taking the inverse Fourier transform on (1.68), we get the first radiating boundary condition $\mathcal{B}_1$:

$$\mathcal{B}_1 p\big|_{x=a} = \left[p_{,x} + p_{,t}\right]\big|_{x=a} = 0. \qquad (1.69)$$

Using the next approximation to $(1-\xi^2/\omega^2)^{1/2}$:

$$(1-\xi^2/\omega^2)^{1/2} = 1 - \tfrac{1}{2}\xi^2/\omega^2 + O(|\xi/\omega|^4),$$

we get

$$\left[\hat{p}_{,x} + i\omega\hat{p} - \tfrac{1}{2}i\xi^2/\omega\hat{p} + i\omega O(|\xi/\omega|^4)\hat{p}\right]\big|_{x=a} = 0, \qquad (1.70)$$

which, after taking the inverse Fourier transform, yields the boundary operator $\mathcal{B}_2$:

$$\mathcal{B}_2 p\big|_{x=a} = \left[p_{,tt} + p_{,xt} - \tfrac{1}{2}p_{,yy}\right]\big|_{x=a} = 0. \qquad (1.71)$$

To get *stable* boundary operators of arbitrary order, we consider the Padé approximation to $(1-\xi^2/\omega^2)^{1/2}$ of the form

$$(1-\xi^2/\omega^2)^{1/2} = P_N(\xi^2/\omega^2) + O(|\xi/\omega|^{2N}). \tag{1.72}$$

Here $P_N(z)$ are recursively defined by

$$\begin{aligned} P_1(z) &= 1 \\ P_2(z) &= 1 - \tfrac{1}{2}z^2 \\ &\vdots \\ P_{N+1}(z) &= 1 - \frac{z^2}{1+P_N(z)} \equiv 1 + \frac{T_{N+1}}{S_{N+1}} \end{aligned} \tag{1.73}$$

and T_{N+1} and S_{N+1} are even polynomials in z, satisfying the following relations:

$$\begin{aligned} T_{N+1} &= 2T_N - z^2 T_{N-1}, \quad T_1 = 0, \quad T_2 = -\tfrac{1}{2}z^2 \\ S_{N+1} &= 2S_N - z^2 S_{N-1}, \quad S_1 = 0, \quad S_2(z) = 0. \end{aligned} \tag{1.74}$$

Introducing (1.72) into (1.66) we get

$$\left[\hat{p}_{,x} + i\omega P_N(\xi^2/\omega^2)\hat{p} + i\omega O(|\xi/\omega|^{2N})\hat{p}\right]\big|_{x=a} = 0. \tag{1.75}$$

By clearing the denominator we obtain

$$S_N\left(\frac{i\xi}{i\omega}\right)\hat{p}_{,x}\big|_{x=a} + i\omega\left(S_N\left(\frac{i\xi}{i\omega}\right) + T_N\left(\frac{i\xi}{i\omega}\right)\right)\hat{p}\big|_{x=a} = 0, \tag{1.76}$$

and, observing from (1.74) that $\deg S_N\left(\frac{i\xi}{i\omega}\right) \leq N-1$ and $\deg T_N\left(\frac{i\xi}{i\omega}\right) \leq N$, we get

$$(i\omega)^{N-1} S_N\left(\frac{i\xi}{i\omega}\right)\frac{\partial}{\partial x} + (i\omega)^N S_N\left(\frac{i\xi}{i\omega}\right) + (i\omega)^N T_N\left(\frac{i\xi}{i\omega}\right) \equiv \mathcal{B}_N(i\omega, i\xi, \frac{\partial}{\partial x}). \tag{1.77}$$

Using identities (1.74), it may be verified that $\mathcal{B}_N$ satisfies

$$\mathcal{B}_{N+1}(i\omega, i\xi, \frac{\partial}{\partial x}) = i\omega \mathcal{B}_N(i\omega, i\xi, \frac{\partial}{\partial x}) + \frac{1}{4}\omega^2 \mathcal{B}_{N-1}(i\omega, i\xi, \frac{\partial}{\partial x}). \tag{1.78}$$

Finally, defining $\mathcal{B}_N(\frac{\partial}{\partial t}, \frac{\partial}{\partial y}, \frac{\partial}{\partial x})$ via the correspondence $i\omega \leftrightarrow \partial/\partial t$ and $i\xi \leftrightarrow \partial/\partial y$, we get

$$\mathcal{B}_{N+1}(\frac{\partial}{\partial t}, \frac{\partial}{\partial y}, \frac{\partial}{\partial x}) = \frac{\partial}{\partial t}\mathcal{B}_N(\frac{\partial}{\partial t}, \frac{\partial}{\partial y}, \frac{\partial}{\partial x}) - \frac{1}{4}\frac{\partial^2}{\partial y^2}\mathcal{B}_{N-1}(\frac{\partial}{\partial t}, \frac{\partial}{\partial y}, \frac{\partial}{\partial x}) \qquad (1.79)$$

or

$$\mathcal{B}_{N+1}p\big|_{x=a} = \left(\frac{\partial}{\partial t}\mathcal{B}_N - \frac{1}{4}\frac{\partial^2}{\partial y^2}\mathcal{B}_{N-1}\right)p\big|_{x=a}. \qquad (1.80)$$

We now record the following fundamental result (Enquist and Majda, 1977).

(a)   Every $\mathcal{B}_N$ is a local strongly well-posed boundary condition for the wave equation.
(b)   Given δ, an acceptable error, $a^* > 0$, an integer M, and an arbitrary source function satisfying

$$\int_0^\infty \int_{-\infty}^\infty |g(y,t)|^2 \, dy\, dt < \infty, \qquad (1.81)$$

there is an $a \geq a^*$ and N so that if $\tilde{p}_N$ solves the boundary value problem in (1.60) with $\mathcal{B} = \mathcal{B}_N$ then

$$\int_0^T \int_{-\infty}^\infty \int_0^{a^*} |p - \tilde{p}_N|^2 \, dx\, dy\, dt \leq \delta^2 \qquad (1.82)$$

for any time T with $0 \leq T \leq 2aM$.

Remarks:
(1) Operator $\mathcal{B}_1$ is exact for a single space dimension and can easily be derived from physical considerations.
(2) The distance a dictating the size of computational domain, and the order of approximation N are critical parameters in practical computations.
(3) Using a Taylor expansion of $(1-\xi^2/\omega^2)^{1/2}$ rather than Padé approximations may result in strongly unstable boundary conditions (Enquist and Majda, 1979).

Operators $\mathcal{B}_1$ and $\mathcal{B}_2$ for Taylor-Galerkin Schemes

We first notice that wave equation (1.59) can be derived from the 1st-order system

$$\begin{aligned} u_{,t} + p_{,x} &= 0 \\ v_{,t} + p_{,y} &= 0 \\ p_{,t} + u_{,x} + v_{,y} &= 0 \end{aligned} \qquad (1.83)$$

and vice versa. Here u and v denote the x- and the y-component of velocity vector, respectively, and we notice that (1.83) is nothing but the system of equations of linear acoustics in appropriately defined nondimensional variables. Then the boundary operator $\mathcal{B}_2$,

$$\mathcal{B}_2 p\big|_{x=a} = \left[p_{,tt} + p_{,xt} - \tfrac{1}{2} p_{,yy} \right]\big|_{x=a} = 0, \tag{1.84}$$

can be written as

$$\left[-(u_{,x} + v_{,y})_{,t} + p_{,xt} + \tfrac{1}{2} v_{,yt} \right]\big|_{x=a} = 0. \tag{1.85}$$

Next, assuming that the acoustic field is quiescent in the vicinity of the artificial boundary at time $t = 0$,

$$u\big|_{x=a} = v\big|_{x=a} = p\big|_{x=a} = 0, \quad t = 0, \tag{1.86}$$

we integrate (1.85) with respect to time and obtain the following boundary condition:

$$\left[-u_{,x} + p_{,x} - \tfrac{1}{2} v_{,y} \right]\big|_{x=a} = 0. \tag{1.87}$$

Similarly, boundary operator $\mathcal{B}_1$ for system (1.83) reads

$$\left[-u_{,x} + p_{,x} \right]\big|_{x=a} = 0, \tag{1.88}$$

Finally, we notice that boundary operator $\mathcal{B}_3$ is not easily implementable. Indeed, using recurrence relation (1.80) we compute $\mathcal{B}_3$ as

$$\mathcal{B}_3 p\big|_{x=a} = \left[p_{,xtt} - p_{,ttt} - \tfrac{1}{4} p_{,xyy} + \tfrac{3}{4} p_{,tyy} \right]\big|_{x=a} = 0. \tag{1.89}$$

Next, introducing (1.83) into (1.89) and integrating with respect to time we get

$$\left[p_{,xt} - p_{,tt} - \tfrac{1}{4} u_{,yy} + \tfrac{3}{4} p_{,yy} \right]\big|_{x=a} = 0. \tag{1.90}$$

Since (1.90) contains spatial derivatives of order 2, it is not acceptable for a second order operator. A new class of split boundary operators circumvents this difficulty (Safjan and Oden, 1995b).

ADAPTIVITY

In this section we give a brief outline of the hp-finite element schemes used in construction of approximation spaces $\mathbf{X}_h$ ($\mathbf{X}_h \subset \mathbf{X}, \dim \mathbf{X}_h < \infty$).

First, we introduce a partition of $\overline{\Omega}$ ($\Omega \subset \mathbb{R}^2$) into N_{ELEM} subdomains (elements) with disjoint interiors

$$\overline{\Omega} = \cup\{K \in Q^h\} \; K \cap J = \varnothing$$
$$h_K = \text{dia}(K); h = \sup_{K \in Q^h} h_K \,, \tag{1.91}$$

and each element K is the image of a master element $\hat{K}$ under an invertible map F_K,

$$\mathbf{x} = (x_1, x_2) = F_K(\xi), \; \xi = (\xi_1, \xi_2) \in \hat{K} = [-1,1] \times [-1,1]. \tag{1.92}$$

On $\hat{K}$ we define the following three categories of shape functions:

- *Node functions.* These are bilinear:

$$\chi_\Delta(\xi, \eta) = \tfrac{1}{4}(1 \pm \xi)(1 \pm \eta), \quad \Delta = 1, 2, 3, 4, \quad \xi = \xi_1, \quad \eta = \xi_2. \tag{1.93}$$

- *Edge functions.* Defining nodes at the midpoints of each side of $\hat{K}$ we define the edge functions

$$\begin{aligned}
\chi_{\partial K}^{1,k_1}(\xi, \eta) &= \tfrac{1}{2}(1 + \eta)\phi_{k_1}(\xi) \\
\chi_{\partial K}^{2,k_2}(\xi, \eta) &= \tfrac{1}{2}(1 - \xi)\phi_{k_2}(\eta) \\
\chi_{\partial K}^{3,k_3}(\xi, \eta) &= \tfrac{1}{2}(1 - \eta)\phi_{k_3}(\xi) \\
\chi_{\partial K}^{4,k_4}(\xi, \eta) &= \tfrac{1}{2}(1 + \xi)\phi_{k_4}(\eta) \\
k_\Delta &= 2, 3, \ldots p_\Delta, \Delta = 1, 2, 3, 4.
\end{aligned} \tag{1.94}$$

Here $\chi_{\partial K}^{\Delta, k_\Delta}$ is associated with side $\hat{\Gamma}_\Delta$ of $\hat{K}$, $\Delta = 1, 2, 3, 4$, and ϕ_k are integrated Legendre polynomials; e.g.,

$$\phi_k(\xi) = \sqrt{\frac{2k-1}{2}} \int_{-1}^{\xi} P_{k-1}(s) ds, \tag{1.95}$$

$P_{k-1}(s) =$ the Legendre polynomial of degree $k-1$. Thus, polynomials of differing degree k_Δ can be assigned to each edge $\hat{\Gamma}_\Delta$.

- *Bubble functions.* On the interior of $\hat{K}$, at its centroid, we assign the shape functions,

$$\begin{aligned}
\chi_{jk}^B &= \phi_j(\xi)\phi_k(\eta) \\
2 &\leq j \leq p_\xi, \; 2 \leq k \leq p_\eta.
\end{aligned} \tag{1.96}$$

Again, different orders can be assigned to different directions at $(0, 0)$.

The master element shape functions define a space,

$$\hat{P}(\hat{K}; p_1, p_2, p_3, p_4, p_\xi, p_\eta,) = span\{\chi_\Delta, \chi_{\partial K}^{\Delta,k_\Delta}, \chi_{jk}^B; \quad \Delta = 1,2,3,4 \\ k_\Delta = 2,3,\ldots,p_\Delta \\ j = 2,3,\ldots,p_\xi \\ k = 2,3,\ldots,p_\eta\}, \tag{1.97}$$

which, together with (1.92), define P_K, a space of local shape functions for element K:

$$P_K = \{v = v(x): v \circ F_K^{-1} \in \hat{P}(\hat{K}; p_1, p_2, p_3, p_4, p_\xi, p_\eta)\}. \tag{1.98}$$

We now define $X_{h,p}$, a global finite element approximation space:

$$X_{h,p}(\Omega) \subset H^1(\Omega), \overline{\Omega} = \cup \{K \in Q^h\} \\ X_{h,p}(\Omega) = \{v \in C^0(\Omega): v|_K \in P_K\}. \tag{1.99}$$

The space $X_{h,p}$ possesses standard interpolation properties of hp-finite element methods (see, e.g., Babuška and Suri, 1987). In particular, for element K which is affine equivalent to a master element $\hat{K}$ (i.e., T_K is an affine map), if $h_K = \text{dia}(K)$ and p_K is the highest degree of the complete polynomials contained in the span of shape functions associated with this element, then, for any function $u \in H^s(K)$, an interpolant $w_h \in P_{pK}(K)$ exists, such that

$$\|u - w_h\|_{r,K} \leq C \frac{h_K^{\mu-r}}{p_K^{s-r}} \|u\|_{s,K} \\ \mu = \min(p_K + 1, s), \ 0 \leq r \leq s, \tag{1.100}$$

where $P_{pK}(K) =$ space of polynomials of degree $\leq p_K$ defined on K, constant C is independent of u, p_K or h_k, and $\|\cdot\|_{s,K}$ denotes the usual Sobolev norm. Then, globally, for uniform h and p, and for $u \in H^s(\Omega)$, there exists an interpolant $w_h \in X_{h,p}(\Omega)$, such that

$$\|u - w_h\|_{r,\Omega} \leq C \frac{h^{\mu-r}}{p^{s-r}} \|u\|_{s,\Omega} \\ \mu = \min(p+1, s), \ 0 \leq r \leq s, \tag{1.101}$$

where, again, C is independent of u, p or h.

Finally, we define $X_h \subset X$ as

$$X_h = [X_{h,p}]^N \cap X. \tag{1.102}$$

As mentioned above, (1.94) and (1.95) allow us to use shape functions of different orders for different edges (and different directions), and, hence, to construct a global mesh with a non-uniform distribution of spectral orders p. In addition, we treat mesh size h as a parameter by using h-adaptive techniques for bisecting elements and imposing hp-constraints to maintain continuity across element edges. In particular, we employ *unstructured, anisotropic, 1-irregular meshes*. For $\hat{K}=[-1,1]^2$, the mesh is composed of quadrilateral elements which are refined by dividing appropriate elements into two siblings, as opposed to four, in such a way that only one constrained node exists between any two active vertices. This provides for directional refinement whenever refinement in one direction is needed but none is called for in an orthogonal direction. An h-refined mesh of this type is depicted in Figure 1.3.

Continuity of global basis functions is maintained by imposing constraints at the interface of elements supporting edge functions of different order. Thus, for example, one encounters cases of the type illustrated in Figure 1.4 where an isotropic h-refinement has resulted in two small elements interfacing a large element with p-shape functions of differing order assigned to the edge (and interior) nodes. In the figure, open circles denote active vertex nodes which support bilinear shape functions, open rectangles denote edge and interior nodes which support various high-order shape functions, and darkened circles and rectangles denote constrained nodes. Thus, along the interface, values of the order of p at the p-nodes of the smaller elements (in this example) are constrained to exactly fit the shape function of the larger element to produce full continuity across this interface. For details on hp-constraints, see Demkowicz et al. (1989, 1991a,b).

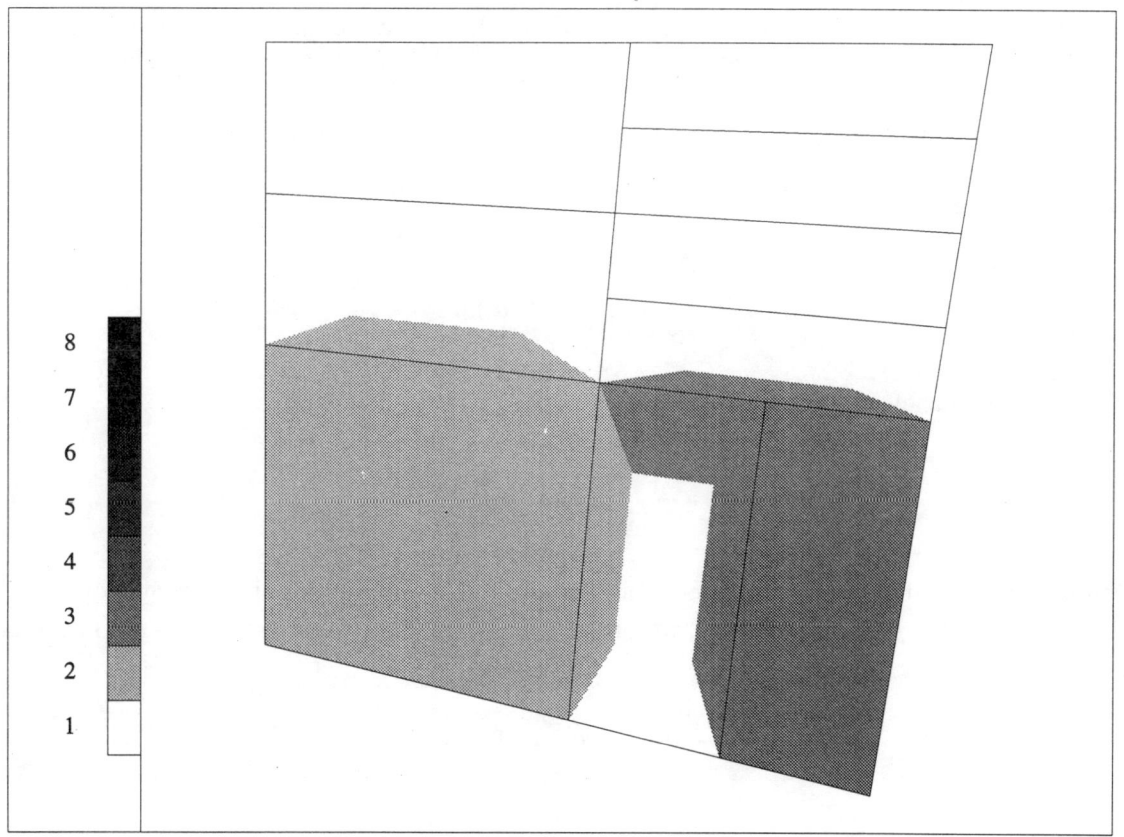

Figure 1.3 Unstructured, 1-irregular mesh of quadrilateral elements with anisotropic h-refinements. Different shadings correspond to different element spectral orders p.

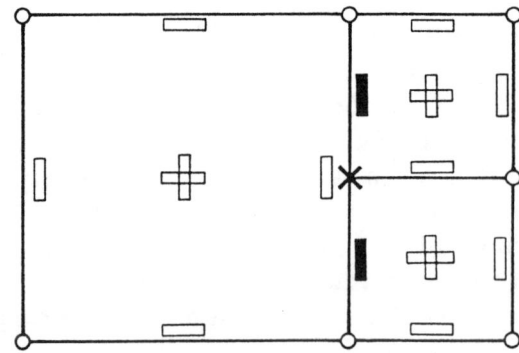

Figure 1.4 Mesh topology in two dimensions showing constrained *h*- and *p*-nodes.

A FREQUENCY-DOMAIN APPROACH

For completeness, we briefly summarize parallel work we have done on the analysis of structural acoustic problems using new coupled boundary-element and finite-element approaches in which the problem is formulated in the frequency domain.

Frequency-domain formulations are obtained by Fourier transformations of the time-domain equations and variables. For a function $u = u(\mathbf{x},t) \in L^2(H^1(\Omega), \mathbb{R})$, its Fourier transform in time is a function $\hat{u}(\mathbf{x}, \omega)$ defined by

$$\hat{u}(\mathbf{x},\omega) = \int_{-\infty}^{\infty} u(\mathbf{x},t) \exp(\iota \omega t) dt \quad \mathbf{x} \in \Omega, \tag{1.103}$$

where ω is the frequency. We recover $u(\mathbf{x},t)$ from $\hat{u}(\mathbf{x},t)$ by the inverse transform

$$u(\mathbf{x},t) = \frac{1}{2\pi} \int_{-\infty}^{\infty} \hat{u}(\mathbf{x},\omega) \exp(-\iota \omega t) d\omega. \tag{1.104}$$

Denoting the transformed displacement and pressure fields (amplitudes at the frequency ω) in (1.38) and (1.39) by $\mathbf{u}(\mathbf{x})$ and $p(\mathbf{x})$, respectively, we observe that a frequency-domain formulation of dynamical problem (1.27) is of the form

$$-\mathbf{D}^T \mathbf{E} \mathbf{D} \mathbf{u} - \rho_s \omega^2 \mathbf{u} = 0 \quad \text{in } \Omega^{(s)} \tag{1.105}$$

and

$$\Delta p + k^2 p = 0 \quad \text{in } \Omega^{(f)}. \tag{1.106}$$

Also, the following boundary conditions must be enforced:

$$\rho_f \omega^2 u_n = -\partial p / \partial n, \quad \sigma_n(\mathbf{u}) = -p, \quad \sigma_\tau(\mathbf{u}) = 0, \quad \text{on } \Gamma_C \tag{1.107}$$

$$p^s = p - p^{inc} = o(|\mathbf{x}|^{-\frac{1}{2}}) \quad \text{at infinity,} \tag{1.108}$$

where $k = \omega / c$ is the wave number, Γ_C is the interface boundary between the interior domain $\Omega^{(s)}$ and the exterior domain $\Omega^{(f)}$, and σ_n, σ_τ and p^{inc} represent amplitudes of normal stress, tangential stress, and incident wave pressure at the frequency ω, respectively.

A major advantage of solving problems in the frequency-domain is that the Helmholtz equation on the exterior domain $\Omega^{(f)}$ can be converted into an integral equation on the boundary Γ_C, i.e.,

$$\tfrac{1}{2} p(\mathbf{x}) - \int_{\Gamma_C} [p(\mathbf{y}) \frac{\partial \Phi(kr)}{\partial n(\mathbf{y})} - \frac{\partial p(\mathbf{y})}{\partial n(\mathbf{y})} \Phi(kr)] ds(\mathbf{y}) = p^{inc}(\mathbf{x}). \tag{1.109}$$

In this way, the dimensionality of the problem on $\Omega^{(f)}$ can be reduced by 1 and the infinite boundary condition (1.108) is satisfied automatically. In (1.107), Φ represents the fundamental solution of the Helmholtz equation and is of the form

$$\Phi(kr) = \tfrac{i}{4} H_0^1(kr) \quad \text{in 2 dimensions}$$

and

$$\Phi(kr) = \frac{e^{ikr}}{4\pi r} \quad \text{in 3 dimensions}$$

where H_0^1 is the Hankel function of the first kind of 0 order and $i = \sqrt{-1}$.

As is well known, a major problem in using integral forms of the Helmholtz equation on exterior domains is that its solutions are nonunique at certain frequencies (forbidden frequencies). The CHIEF (Combined Helmholtz Integral Equation Formulation) method, proposed by Schenck (1967), is one of the earliest and simplest methods to overcome the nonuniqueness problem. In this method, the system of linear equations is combined with a few additional equations generated from the Helmholtz integral equation with the source points in the interior of the closed boundary. However, the CHIEF method is not reliable or effective in problems with complicated boundary geometry and frequencies in the higher range.

Burton and Miller (1971) proved that a complex linear combination of (1.109) and a so-called hypersingular integral equation,

$$\frac{1}{2} \frac{\partial p(\mathbf{x})}{\partial n(\mathbf{x})} - \frac{\partial}{\partial n(\mathbf{x})} \int_{\Gamma_C} [p(\mathbf{y}) \frac{\partial \Phi(kr)}{\partial n(\mathbf{y})} - \frac{\partial p(\mathbf{y})}{\partial n(\mathbf{y})} \Phi(kr)] ds(\mathbf{y}) = \frac{\partial p^{inc}(\mathbf{x})}{\partial n(\mathbf{x})}, \tag{1.110}$$

is equivalent to the original differential equation. Replacing (1.106) and (1.108) by the Burton-Miller formulation leads to the following problem

Find amplitude $\mathbf{u} = \mathbf{u}(\mathbf{x})$ and $p = p(\mathbf{x})$ such that

$$-\mathbf{D}^T\mathbf{E}\mathbf{D}\mathbf{u} - \rho_s\omega^2\mathbf{u} = 0 \quad \text{in } \Omega \tag{1.111}$$

and

$$\tfrac{1}{2}p(x) - \int_{\Gamma_C}[p(y)\frac{\partial\Phi(kr)}{\partial n(y)} - \frac{\partial p(y)}{\partial n(y)}\Phi(kr)]ds(y) +$$
$$\alpha[\tfrac{1}{2}\frac{\partial p(x)}{\partial n(x)} - \frac{\partial}{\partial n(x)} - \int_{\Gamma_C}[p(y)\frac{\partial\Phi(kr)}{\partial n(y)} - \frac{\partial p(y)}{\partial n(y)}\Phi(kr)]ds(y)] \tag{1.112}$$
$$= p^{inc}(x) + \alpha\frac{\partial p^{inc}(x)}{\partial n(x)} \quad \text{on } \Gamma_C$$

$$\left.\begin{array}{rcl}\rho_f\omega^2 u_n &=& \partial p/\partial n \\ \sigma_n(u) &=& -p \\ \sigma_{\tau n}(u) &=& 0\end{array}\right\} \quad \text{on } \Gamma_C, \tag{1.113}$$

where α is a complex constant whose imaginary part cannot be 0.

Equations (1.111) through (1.113) define the mathematical model which we use to solve the acoustic scattering problem at a given frequency ω. It can be proven that the problem defined in (1.111-1.113) is strongly elliptic on the space of $V = (H^1(\Omega) \times H^1(\Omega)) \times H^{\frac{1}{2}}(\Gamma_C)$, and at non-resonant frequency the problem is well defined.

The Galerkin-Burton-Miller Formulation

A special problem that arises in the implementation of the Burton-Miller formulation is that the integral

$$\frac{\partial}{\partial n(x)}\int_{\Gamma_C}\frac{\partial\Phi(kr)}{\partial n(y)} - p(y)ds(y) \tag{1.114}$$

in (1.112) cannot be evaluated easily in a meaningful way. Symbolically, it is customary to write (1.114) in the form

$$\int_{\Gamma_C}\frac{\partial^2\Phi(kr)}{\partial n(x)\partial n(y)}ds(y). \tag{1.115}$$

The integral kernel $\partial^2\Phi(kr)/\partial\mathbf{n}(x)\partial\mathbf{n}(y)$ contains a hypersingular term and it easily verifies that as $r \to 0$ in 2 dimensions

$$\frac{\partial^2\Phi(kr)}{\partial n(x)\partial n(y)} = O\left(\frac{1}{r^2}\right),$$

and in 3 dimensions

$$\frac{\partial^2 \Phi(kr)}{\partial n(x)\partial n(y)} = O\left(\frac{1}{r^3}\right).$$

Although the hypersingular integrals can be computed numerically in the sense of the finite part integral (see Kutt, 1975 and Demkowicz et al., 1991a for details), the implementation for such procedures is complicated and the accuracy of its result is usually very poor.

The method used in this study to avoid hypersingular integral kernels associated with (1.115) was proposed in Karafiat et al. (1993) and Demkowicz et al. (1992). The method requires that the problem be formulated in a weak (Galerkin) approach and is based on the transformation

$$\int_{\Gamma_C} q(y) \int_{\Gamma_C} p(y) \frac{\partial^2 \Phi(kr)}{\partial n(x)\partial n(y)} ds(y) ds(x) =$$
$$- \int_{\Gamma_C} \int_{\Gamma_C} \Phi(kr)(\boldsymbol{n}(y) \times \nabla_y p(y) \cdot (\boldsymbol{n}(x) \times \nabla_x q(x)) ds(y) ds(x) \qquad (1.116)$$
$$+ \int_{\Gamma_C} \int_{\Gamma_C} k^2 \boldsymbol{n}(x) \cdot \boldsymbol{n}(y) \Phi(kr) p(y) ds(y) ds(x)$$

and, in 2 dimensions, this reduces to

$$\int_{\Gamma_C} q(x) \int_{\Gamma_C} p(y) \frac{\partial^2 \Phi(kr)}{\partial n(x)\partial n(y)} ds(y) ds(x) =$$
$$- \int_{\Gamma_C} \int_{\Gamma_C} \frac{\partial q(x) \partial p(y)}{\partial \tau(x) \partial \tau(y)} \Phi(kr) ds(y) ds(x) \qquad (1.117)$$
$$+ \int_{\Gamma_C} \int_{\Gamma_C} [k^2 q(x) \tau(x) \cdot \tau(y) \Phi(kr) p(y) ds(y)] ds(x),$$

where $q(\mathbf{x})$ is the test function and $\tau(\cdot)$ is the positive tangential direction at the corresponding position of $\mathbf{x}$ and $\mathbf{y}$.

With (1.116) or (1.117), the variational formulation of (1.116)-(1.118) can be written as:

Find $(\mathbf{u}, p) \in V$ such that for any $(\mathbf{v}, q) \in V$

$$\int_{\Omega} [D^T \bar{v}(x) E D u(x) - \rho_s \omega^2 u(x) \bar{v}(x)] d\Omega(x)$$
$$+ \int_{\Gamma_C} p(x) \bar{v}(x) \cdot n(x) ds(x) = 0 \qquad (1.118)$$

and

$$\tfrac{1}{2} \int_{\Gamma_C} [p(x) + \alpha \rho_f \omega^2 u_n(x)] \bar{q}(x) ds(x)$$
$$- \int_{\Gamma_C} \int_{\Gamma_C} [p(y) \frac{\partial \Phi(k|y-x|)}{\partial n(y)} - \rho_f \omega^2 u_n(y) \Phi(k|y-x|)] \bar{q}(x) ds(y) ds(x)$$

$$+\alpha \int_{\Gamma_c} \int_{\Gamma_c} \{\frac{\partial \overline{q}(x) \partial p(x)}{\partial \tau(x) \partial \tau(x)} \Phi(k|y-x|) - [k^2 p(y)\tau(x) \cdot \tau(y)\Phi(k|y-x|)$$
$$-\rho_f \omega^2 u_n(y) \frac{\partial \Phi(k|y-x|)}{\partial n(x)}] \overline{q}(x)\} ds(y)ds(x) \qquad (1.119)$$
$$= \int_{\Gamma_c} [p^{inc}(x) + \alpha \frac{\partial p^{inc}(x)}{\partial n(x)}] \overline{q}(x)ds(x) \quad \text{in 2 dimensions}$$

or

$$\tfrac{1}{2} \int_{\Gamma_c} [p(x) + \alpha \rho_f \omega^2 u_n(x)]\overline{q}(x)ds(x)$$
$$- \int_{\Gamma_c} \int_{\Gamma_c} [p(y)\frac{\partial \Phi(k|y-x|)}{\partial n(y)} - \rho_f \omega^2 u_n(y)\Phi(k|y-x|)]\overline{q}(x)ds(y)ds(x)$$
$$-\alpha \int_{\Gamma_c} \int_{\Gamma_c} [-\Phi(kr)(n(y) \times \nabla_y p(y)) \cdot (n(x) \times \nabla_x \overline{q}(x) \qquad (1.120)$$
$$+k^2 p(y)n(x) \cdot n(y)\Phi(kr) - \rho_f \omega^2 u_n(y)\frac{\partial \Phi(kr)}{\partial n(x)}]\overline{q}(x)ds(y)ds(x)$$
$$= \int_{\Gamma_c} [p^{inc}(x) + \alpha \frac{\partial p^{inc}(x)}{\partial n(x)}]\overline{q}(x)ds(x) \quad \text{in 3 dimensions,}$$

where $V = (H^1(\Omega) \times H^1(\Omega)) \times H^{\frac{1}{2}}(\Gamma_c)$.

Equations (1.119) and (1.120) contain only weak singular and regular items, which can be treated easily in numerical implementations; and the whole problem can be solved by the coupled finite element and boundary element methods.

The coupled boundary element and finite element codes in this research are developed in a special *hp*-hierarchical data structure proposed in Demkowicz et al. (1991b), and, as mentioned, this data structure allows the mesh size and order of polynomial interpolation to be varied over the mesh and provides a good basis for the development of an adaptive method. Comparisons with our time-domain approach are given in the next section for the elastic and rigid-scattering problems, respectively.

NUMERICAL EXAMPLES

Example 1: Diffraction of a Plane Wave from an Elastic Cylindrical Shell (2D)

We consider the problem of diffraction of a plane wave consisting of a short train of pulses, by an elastic shell, see Figure 1.5. The problem was solved for the following data:

Acoustic fluid
- Small signal sound speed, $c_0 = 1500 \, m/s$
- Ambient mass density, $\rho_f = 7700 \, kg/m^3$

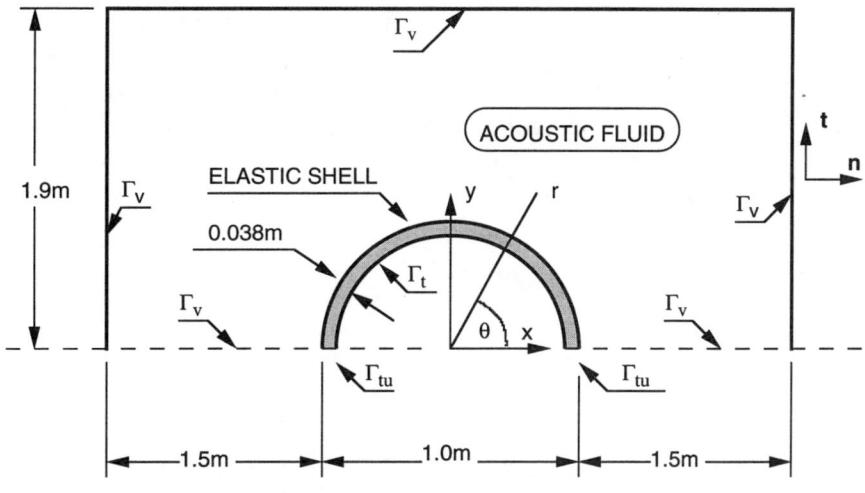

Figure 1.5 Plane progressive wave impinging on a cylindrical elastic shell. Problem definition.

Elastic cylinder
- Young modulus, $E = 19.5 \times 10^{10}\ Pa$,
- Poisson's ratio, $\nu = 0.28$,
- Mass density, $\rho_S = 7700\ kg/m^3$

The initial condition function:

$$g(x) = A[H(x+\tfrac{3}{2}) - H(x+\tfrac{1}{2})] \times \begin{cases} 1 - 32(\tfrac{5}{4}+x)^2 + 256(\tfrac{5}{4}+x)^4, -\tfrac{3}{2} \le x < -\tfrac{5}{4} \\ -\cos[10\pi(x+\tfrac{3}{4})], -\tfrac{5}{4} \le x < -\tfrac{3}{4} \\ -1 + 32(\tfrac{3}{4}+x)^2 - 256(\tfrac{3}{4}+x)^4, -\tfrac{3}{4} \le x \le -\tfrac{1}{2} \end{cases} \quad (1.121)$$

where $u_0 = c_0 g$, $p_0 = \rho_f c_0^2 g$, and $A = 0.005$.

The problem was treated as a 1st-order system, and was solved by a 2-stage 4th-order Taylor-Galerkin method with stability parameter $\eta = 0.471$ and constant time step $\Delta t = 0.01$. The solution algorithm described by Safjan and Oden (1993) was used. As the solution of the problem is fairly smooth, we chose to use *p*-enrichments/unenrichments only. The only exception from this rule was made in elements with contribution to the global error exceeding 50 percent; those elements were first *h*-refined. In addition, only *p*- or anisotropic *h*-refinements were allowed at the interface.

Figures 1.6 through 1.16 show the computed spatial waveforms of various acoustical and elastic variables in nondimensional form. Figure 1.6 shows the initial finite element mesh and Figure 1.7 shows the initial condition function in the form of contour maps. Figures 1.8, 1.9, and 1.10 show pressure distribution at time t = 0.5, 1.0, and 1.5, respectively. A forward radiation from the shell can be observed, as the wave propagation speed in the shell is approximately 4 times greater than that in acoustic fluid. Figures 1.11-1.13 show the propagation of $\theta - \theta$ (bending) stress waves in the elastic shell. Finally, Figures 1.14-1.16 show the evolution of finite element meshes.

Example 2: The Vibrating Cylinder Problem

The problem is defined as follows (see Figure 1.17):

1. Governing equations of linear acoustics are to be solved in the domain

$$\Omega = \mathbb{R}^2 - \{(r,\theta): r \leq a\}. \tag{1.122}$$

2. Boundary condition at $r = a$,

$$u_n = A \sin \omega t [H(t) - H(t-T)], \tag{1.123}$$

where $A = -0.005$, $\omega = 2\pi$, $T = 1$ and $H(\cdot)$ is the Heaviside step function.

3. Sommerfeld radiation condition at $r = \infty$.

4. Initial condition

$$\mathbf{U}_0 = 0. \tag{1.124}$$

5. Unit material constants: $c_0 = 1$ and $\rho_f = 1$.

For computations we accept as the computational domain $[-2.1, 0] \times [0, 1.5] - \{(x,y): x^2 + y^2 \leq a^2\}$ and we use the second highly absorbing boundary condition $\mathcal{B}_2$, on truncation boundary Γ_∞. The problem was solved using a 2-stage 4th-order Taylor-Galerkin method with stability parameter $\eta = 0.471$, constant time step $\Delta t = 0.01$, and a fixed finite element mesh. Figure 1.18 shows the finite element mesh, and the consecutive Figures 1.19, 1.20, 1.21, and 1.22 present the computed pressure distributions at time stages t = 1, 1.5, 2.0, and 2.5, respectively. As can be seen, the wave is leaving cleanly the computational domain. For comparison, the same experiment is performed except that the first highly absorbing boundary condition $\mathcal{B}_1$ is used on truncation boundary Γ_∞. Figure 1.23 presents the computed pressure distribution at time t = 2.5. From a comparison of the performance of $\mathcal{B}_2$ with that of $\mathcal{B}_1$, it may be inferred that $\mathcal{B}_2$ represents a significant improvement over $\mathcal{B}_1$.

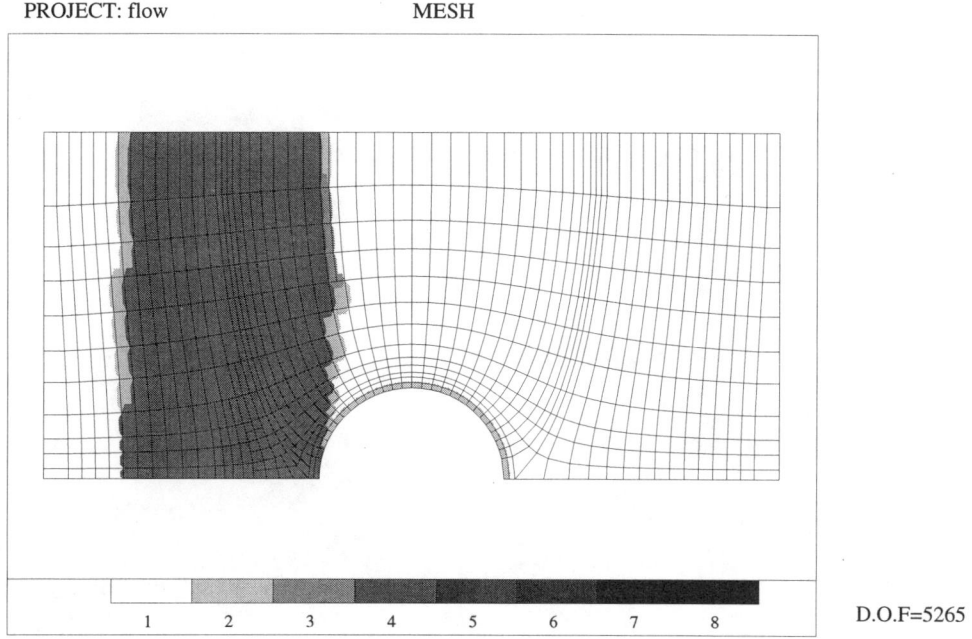

Figure 1.6 Elastic scattering problem, initial FE mesh. Different shadings correspond to different element spectral orders p.

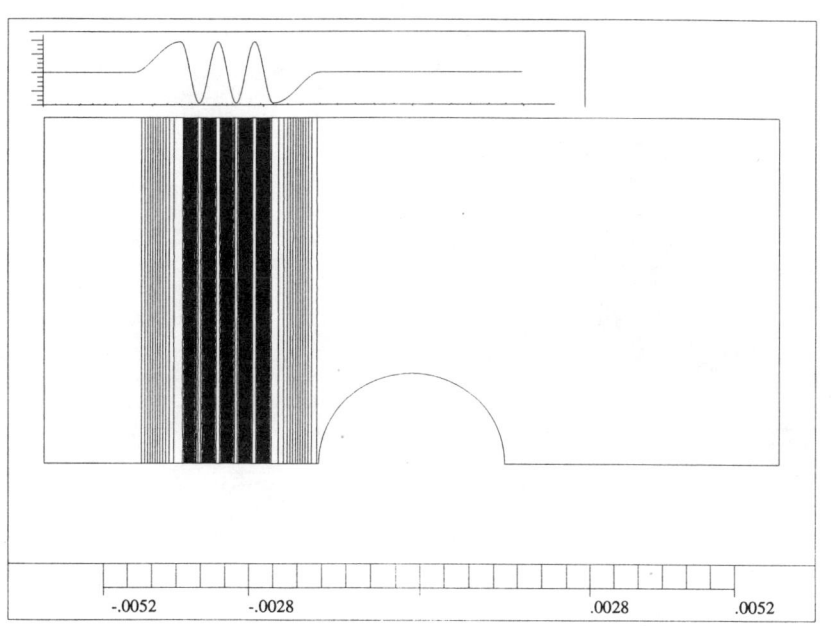

Figure 1.7 Elastic scattering problem. Initial condition, acoustical pressure.

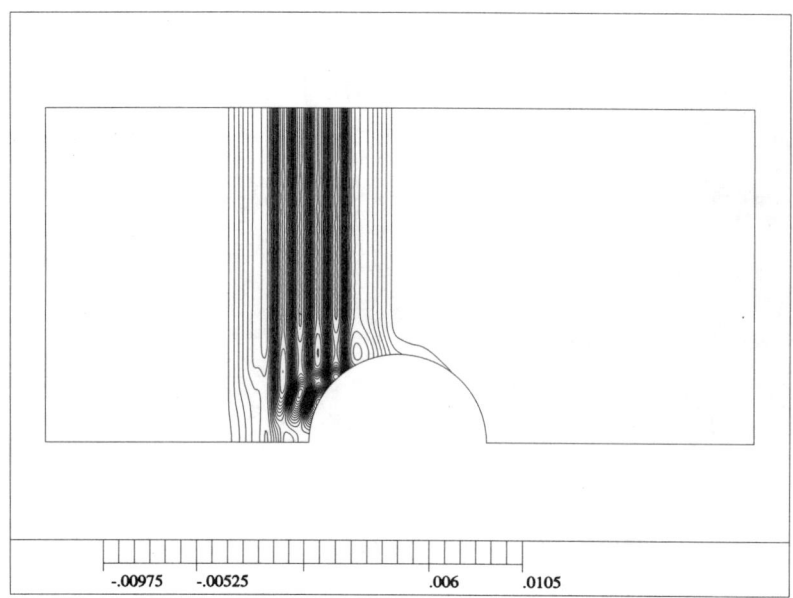

Figure 1.8 Elastic scattering problem. The pressure at time $t = 0.5$.

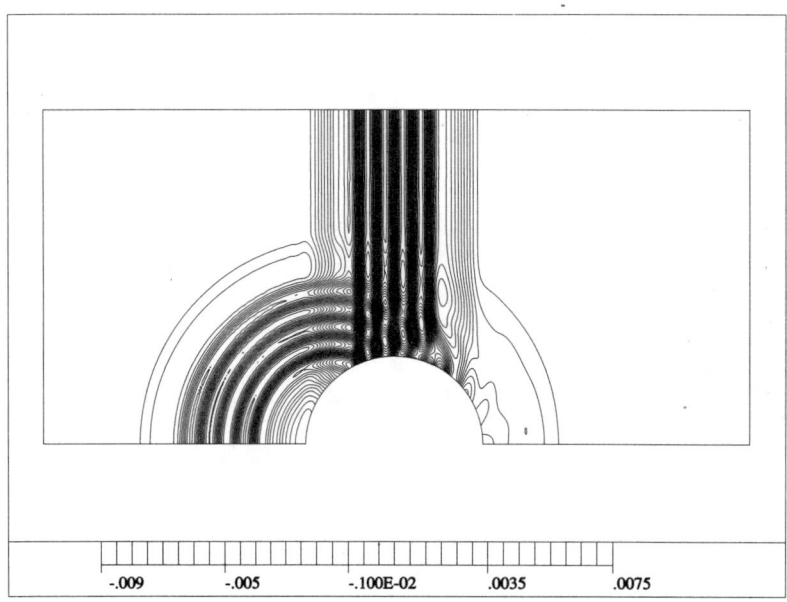

Figure 1.9 Elastic scattering problem. The pressure at time $t = 1.0$.

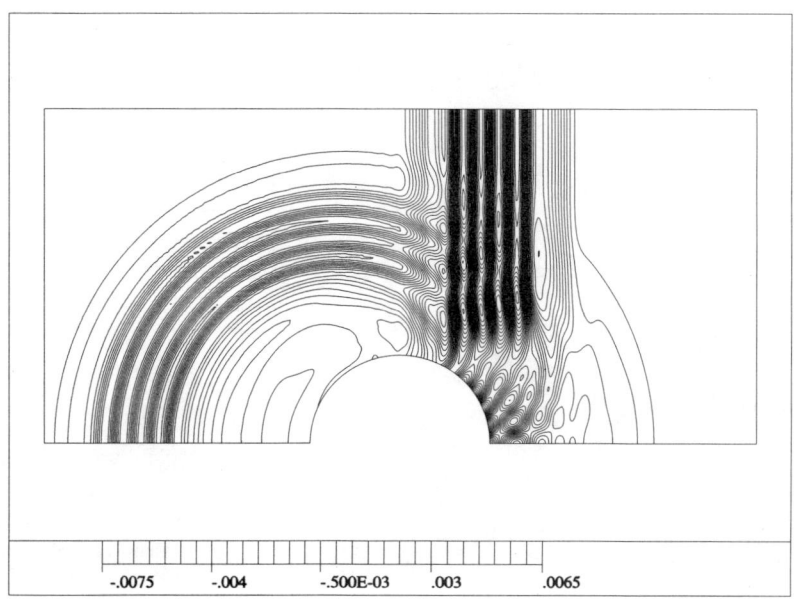

Figure 1.10 Elastic scattering problem. The pressure at time $t = 1.5$.

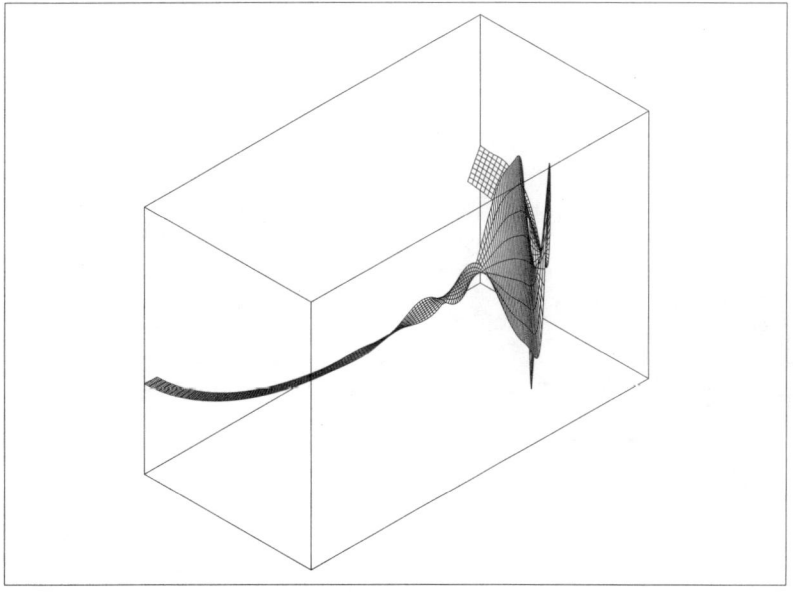

Figure 1.11 Elastic scattering problem. The $\theta - \theta$ component of stress tensor in elastic shell at time $t = 0.5$.

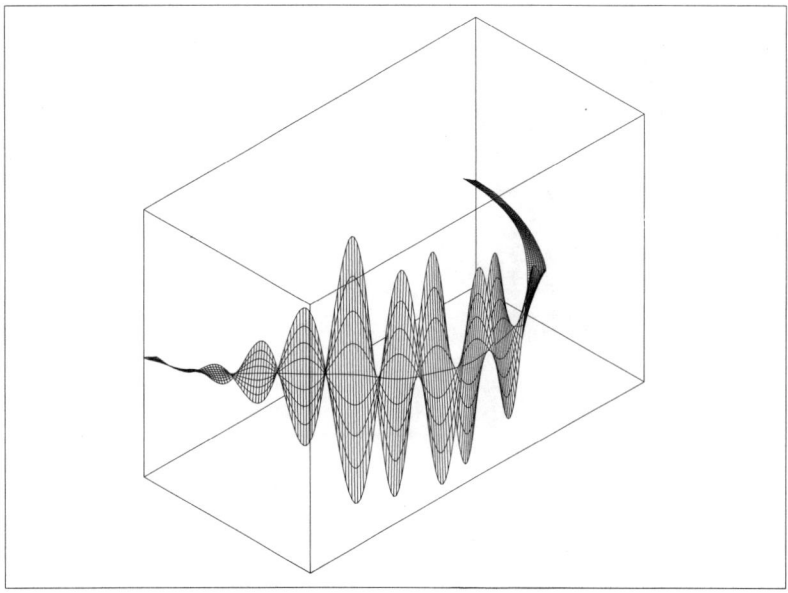

MIN=-.1261909
MAX=.1130277

Figure 1.12 Elastic scattering problem. The $\theta - \theta$ component of stress tensor in elastic shell at time $t = 1.0$.

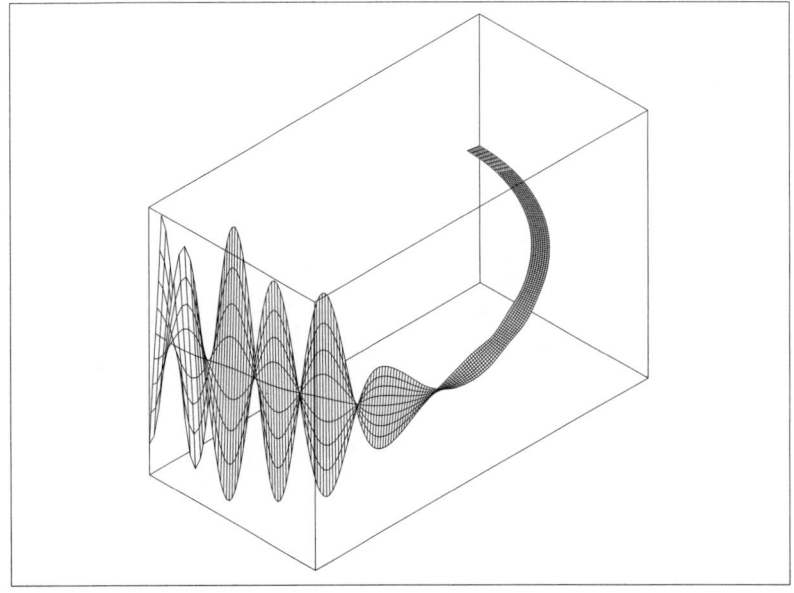

MIN=-.1059902
MAX=.1408646

Figure 1.13 Elastic scattering problem. The $\theta - \theta$ component of stress tensor in elastic shell at time $t = 1.5$.

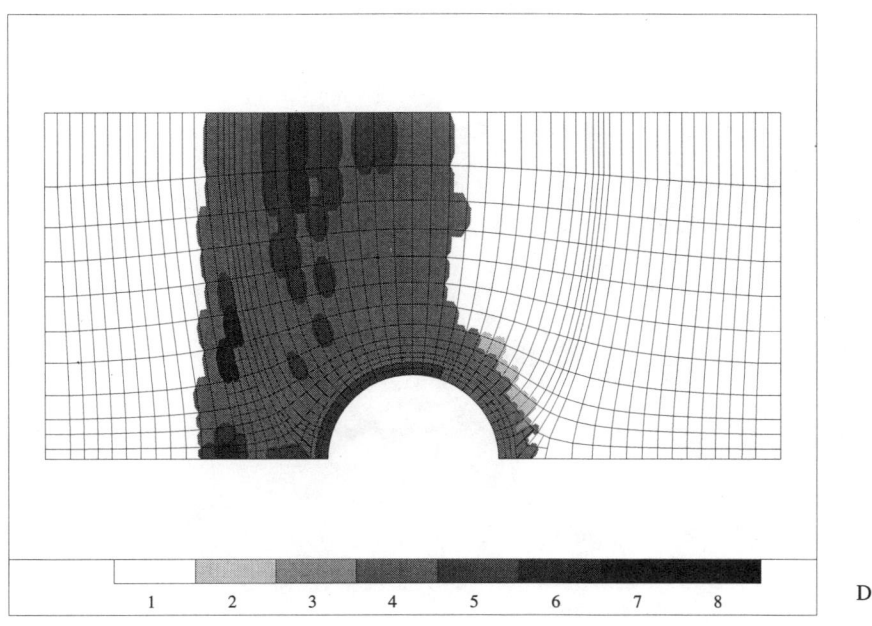

Figure 1.14 Elastic scattering problem. Finite element mesh at time $t = 0.5$. Different shadings correspond to different element spectral orders p.

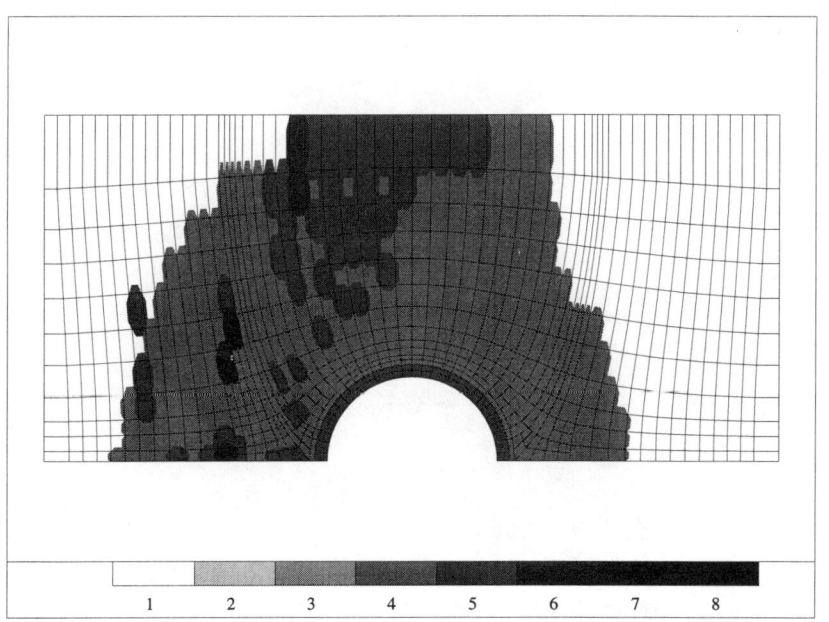

Figure 1.15 Elastic scattering problem. Finite element mesh at time $t = 1.0$. Different shadings correspond to different element spectral orders p.

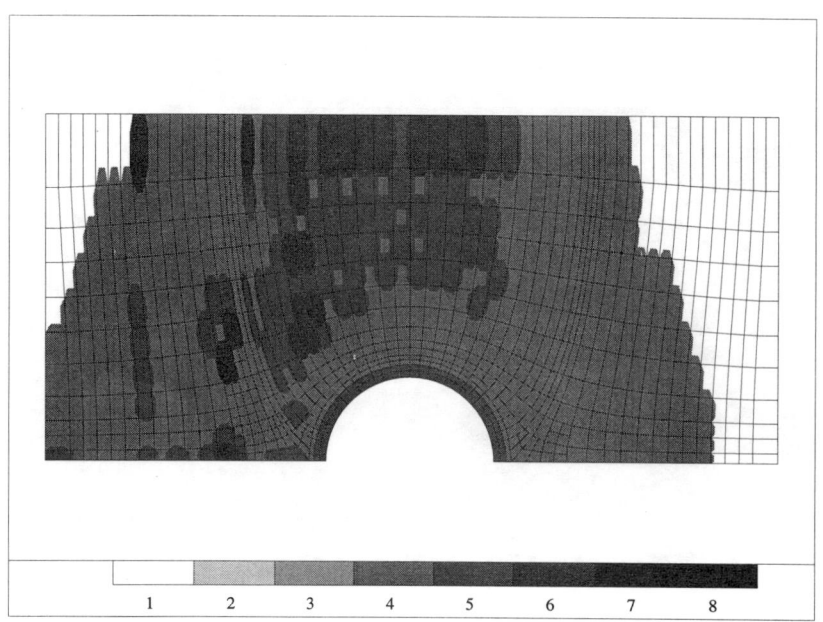

Figure 1.16 Elastic scattering problem. Finite element mesh at time $t = 1.5$. Different shadings correspond to different element spectral orders p.

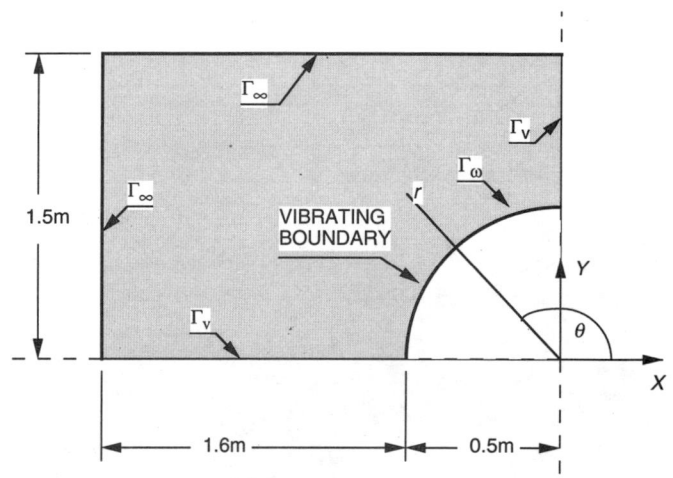

BOUNDARY CONDITIONS

Γ_v : zero normal velocity
Γ_∞ : approximate non-reflexive
Γ_ω : $\mathbf{v}_n = f(t)[H(t) - H(t - 1)]$
   $\mathbf{v}_n$ = normal velocity
   $H(.)$ = the Heaviside step function

Figure 1.17 The vibrating cylinder problem.

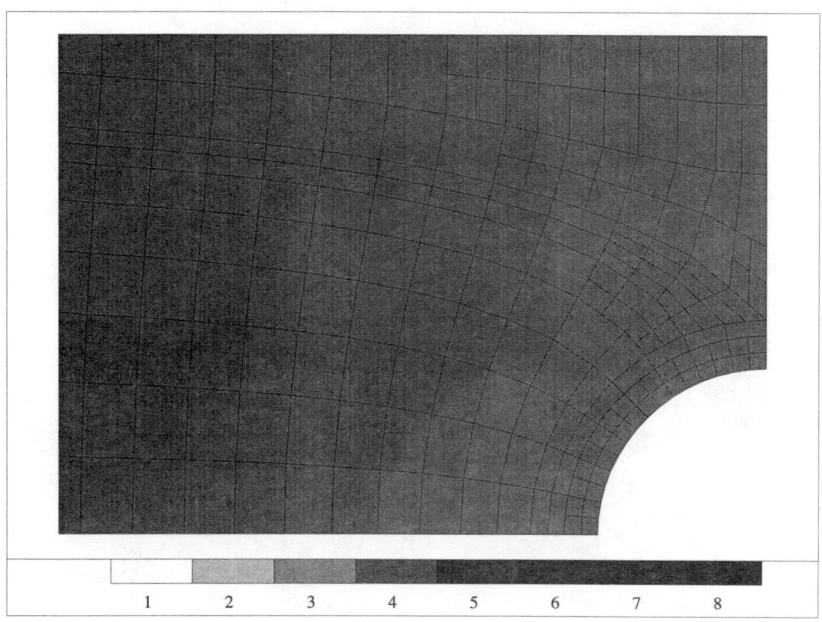

Figure 1.18 The vibrating cylinder problem. Mesh of elements of fourth order.

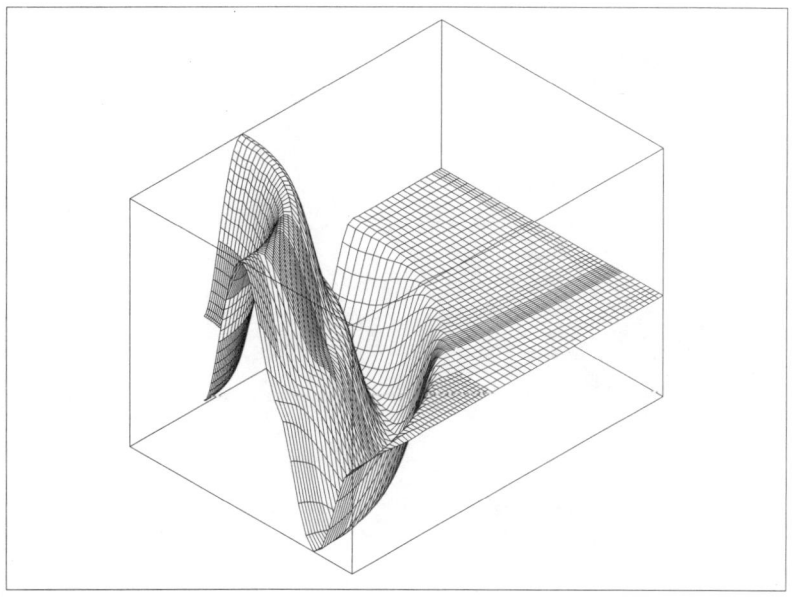

Figure 1.19 The vibrating cylinder problem. Operator $\mathcal{B}_2$. Pressure distribution at time $t = 1.0$.

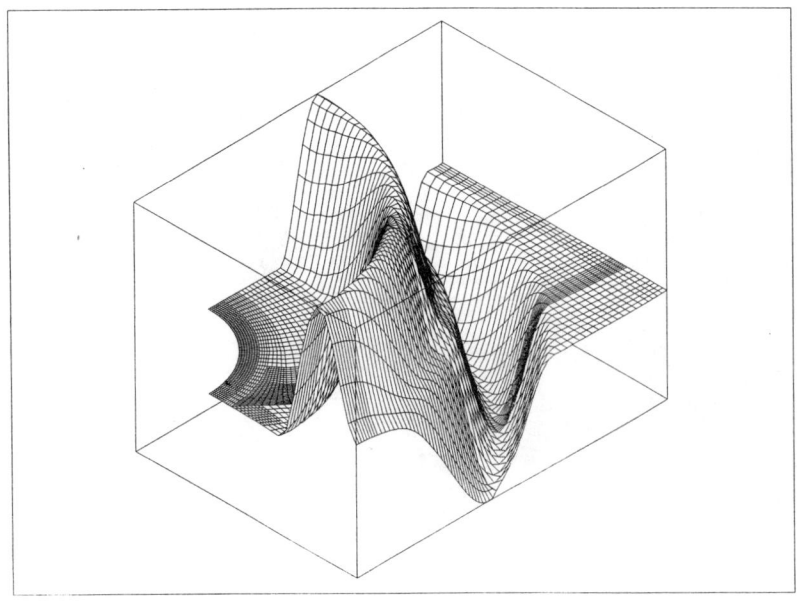

MIN=-.0025104
MAX=.0032418

Figure 1.20 The vibrating cylinder problem. Operator $\mathcal{B}_2$. Pressure distribution at time $t = 1.5$.

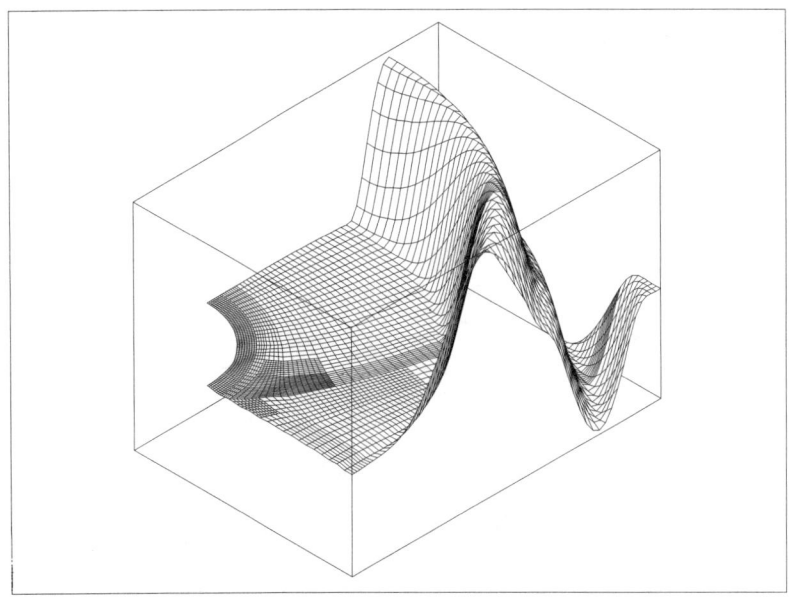

MIN=-.0022478
MAX=.0028677

Figure 1.21 The vibrating cylinder problem. Operator $\mathcal{B}_2$. Pressure distribution at time $t = 2.0$.

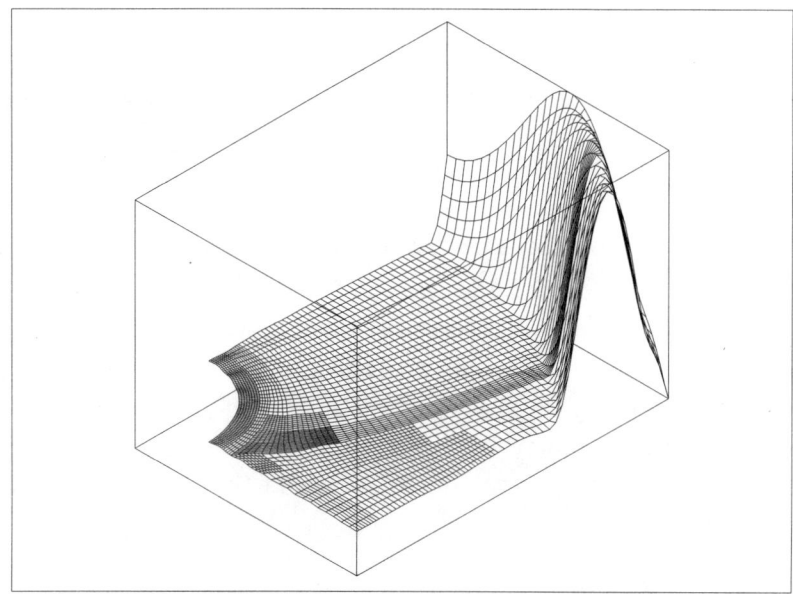

MIN=-.636E-03
MAX=.0025368

Figure 1.22 The vibrating cylinder problem. Operator $\mathcal{B}_2$. Pressure distribution at time $t = 2.5$.

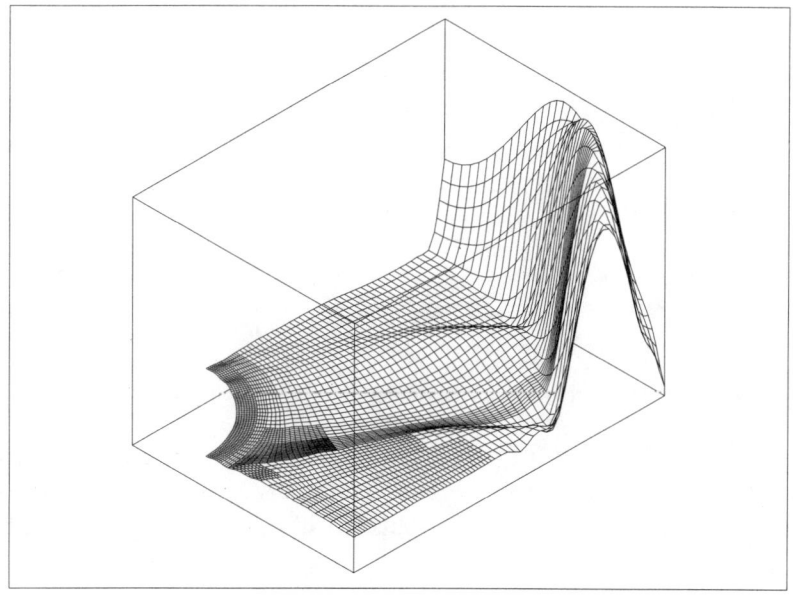

MIN=-.491E-03
MAX=.0026153

Figure 1.23 The vibrating cylinder problem. Operator $\mathcal{B}_1$. Pressure distribution at time $t = 2.5$.

Example 3: The Coupled FE/BE Methods versus the Time-Domain Method

The Galerkin-Burton-Miller formulation developed here is intended to be used to solve problems at very high frequencies. To verify this, we consider an example of a harmonic plane incident wave

$$p^{inc}(\mathbf{x},t) = P_{inc} e^{ik(x-ct)} \qquad (1.125)$$

impinging on a cylinder with a boundary condition $\partial p / \partial n = \beta p$ on the surface of the cylinder (as shown in Figure 1.24). The analytic solution is available in the case of β being a constant (Demkowicz et al., 1991b). Clearly, for $\beta = 0$, we have the problem of acoustical wave scattering on a rigid body. Figure 1.25 compares the numerical solution and corresponding analytic solution for the pressure distribution around the surface of the cylinder with $ka = 400$ and $\partial p / \partial n = p$. With a uniform mesh of 1,024 quadratic elements and 2,048 degrees of freedom, the result illustrated in Figure 1.25 shows a good correlation between the numerical solutions and analytic solution and the L^2-norm of relative error, computed by

$$\left\| \hat{e}^h \right\|_0^2 = \frac{\int_{\Gamma_C} (p^h - p_e)\overline{(p^h - p_e)} ds}{\int_{\Gamma_C} p_e \overline{p_e} ds},$$

is around 2 percent in this example where p_e represents the exact solution and p^h is the numerical solution.

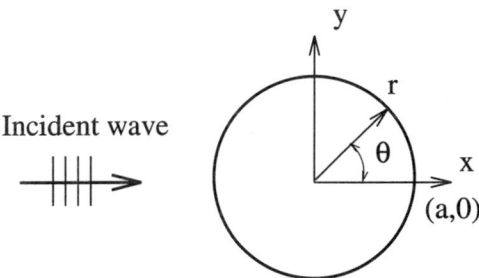

Figure 1.24 The test problem: a plane harmonic wave impinging on a cylinder.

Equations (1.105)-(1.108) define a problem of an elastic structure surrounded by an inviscid fluid field. Because of the absence of damping, the problem is not uniquely solvable at certain frequencies (resonant frequencies) and the stiffness matrix becomes singular or ill-conditioned when the frequency ω is at or close to one of these resonant frequencies.

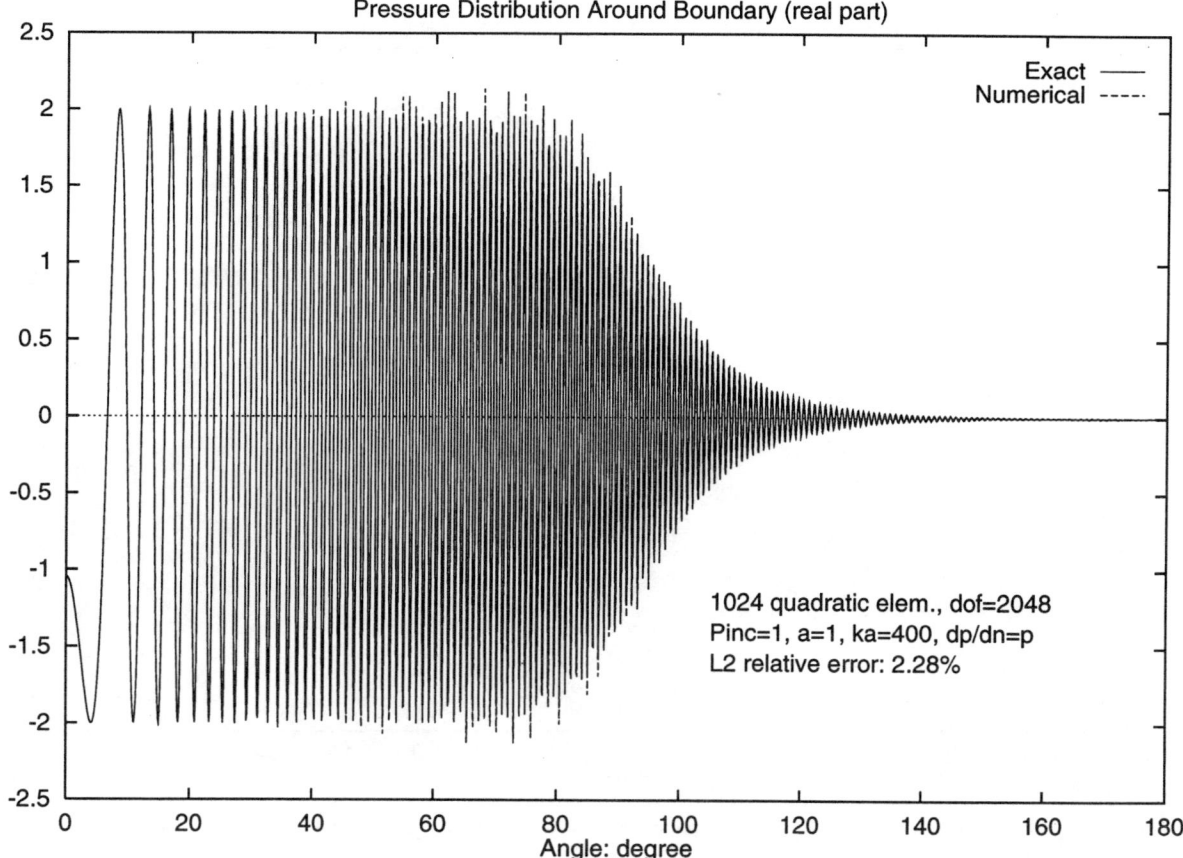

Figure 1.25 The comparison between numerical solution and analytic solution around the cylindrical boundary with $ka = 400$ and $dp/dn = p$. Because of symmetry, only the solution from 0° to 180° around the cylindrical boundary is plotted.

To verify this, the coupled elastic cylinder and fluid model given in Example 1 is solved for the same incident wave given in (1.125) and the wave number here varies from $k = 0.02$ to 14 with $\Delta k = 0.02$. A uniform mesh of 256 quadratic finite elements and 128 quadratic boundary elements is used in the computation. Figure 1.26 plots the relationship between the amplitude of total pressure at a far field point (1,0) and wave number k. The experimental results show that the test problem possesses several resonant frequencies in the range of $0 \le k \le 14$. As a comparison, the analytic solution based on elastic thin shell theory (see Junger and Feit, 1972) is also given in Figure 1.26, and for the small wave number, the elastic solution and thin shell solution are close to each other.

To understand the influence of resonant frequencies on the numerical solution, the test problem given in Example 1 is solved again by the frequency-domain approach. The following incident wave is used

$$p^{inc}(\tau) = \begin{cases} -1 + 32(\tau + 0.25)^2 - 256(\tau + 0.25)^4, & -0.5 \le \tau \le 0 \\ 1 - 32(\tau - 0.25)^2 + 256(\tau - 0.25)^4, & 0 \le \tau \le 0.5 \\ 0, & 0.5 < |\tau| \le T \end{cases} \quad (1.126)$$

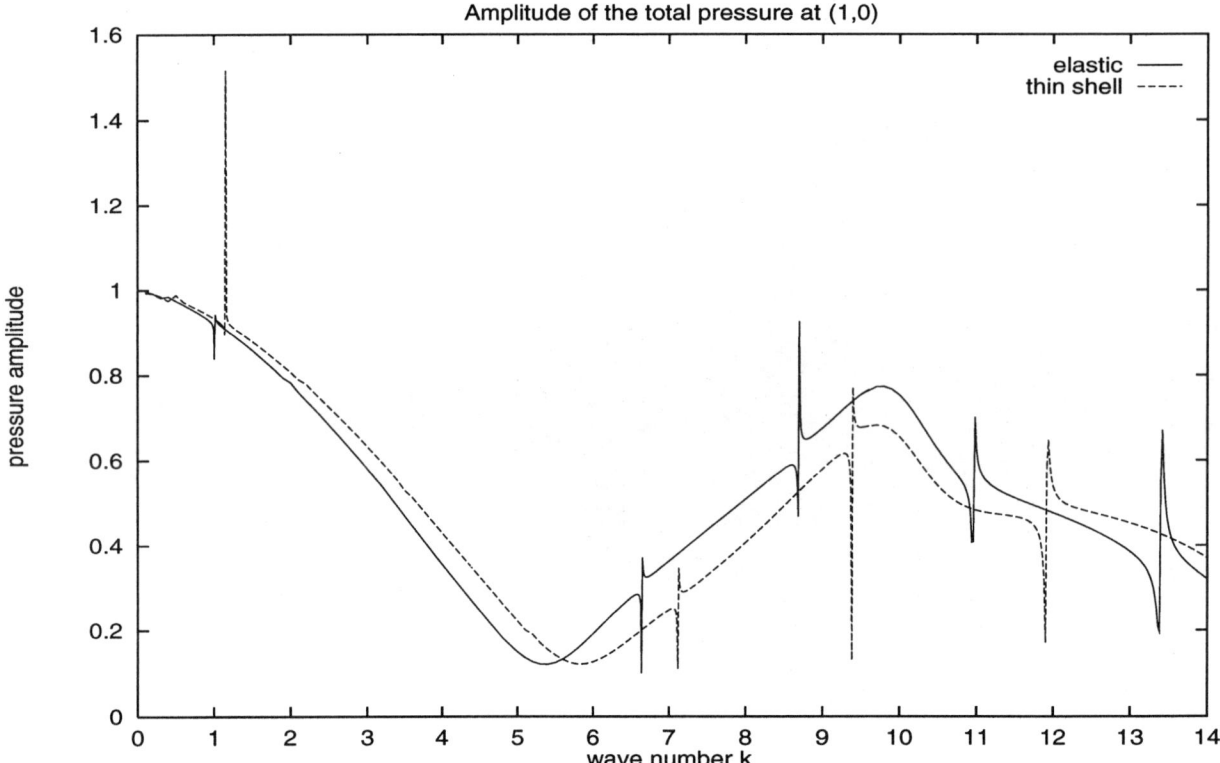

Figure 1.26 The resonant frequencies of the coupled elastic cylinder and fluid model. The plain harmonic incident wave is used and the problem is solved from $k = 0.02$ to $k = 14$ with $\Delta k = 0.02$.

where $\tau = x - ct$ and the time unit is $1/c$. As shown in (1.126), the incident wave pulse must be supposed as a periodical function and T here must be large enough so that the solution at the time period in question is not affected by the solution wave from other periods. In this example, we are interested in the solution in the time period $0 \leq t \leq 2$ and $T = 5.2$ is selected. Then, the incident wave in (1.126) is decomposed into the sum of a series of sine functions, i.e.,

$$p^{inc}(\tau) = \sum_{i=1}^{N} p_i \sin(\frac{i\pi\tau}{T}) \qquad (1.127)$$

and $N = 2{,}048$. At most frequencies, p_i is in fact almost zero; here, we solve the problem only at a frequency when its corresponding amplitude satisfies

$$p_i > -0.00 \ \underset{1 \leq j \leq N}{\boldsymbol{max}} (\boldsymbol{p_j}). \qquad (1.128)$$

Then, instead of 2,048 different frequencies, we need to solve the problem only in 124 different frequencies.

Figures 1.27 and 1.28 present the results at $t = 0$ and $t = 2/c$ for the rigid scattering ($\partial p / \partial n = 0$, upper half part) and elastic scattering (lower half part), respectively. For the rigid scattering, because the

Helmholtz equation with Neumann boundary condition is well defined on the exterior domain (no resonant frequencies), the frequency-domain method produces essentially the same result obtained by the time-domain method. However, the frequency-domain method does not provide a satisfactory result for elastic scattering, because the method fails at the frequencies close to the resonant frequencies of the coupled solid-fluid system. Figure 1.27 shows spurious scattering waves produced by aliasing in the frequency-domain approach. These nonexistent waves are produced ahead of the plane wave that actually excites the structure. It is likely that these types of erroneous excitations could be controlled, if viscous and material damping effects were added to the model.

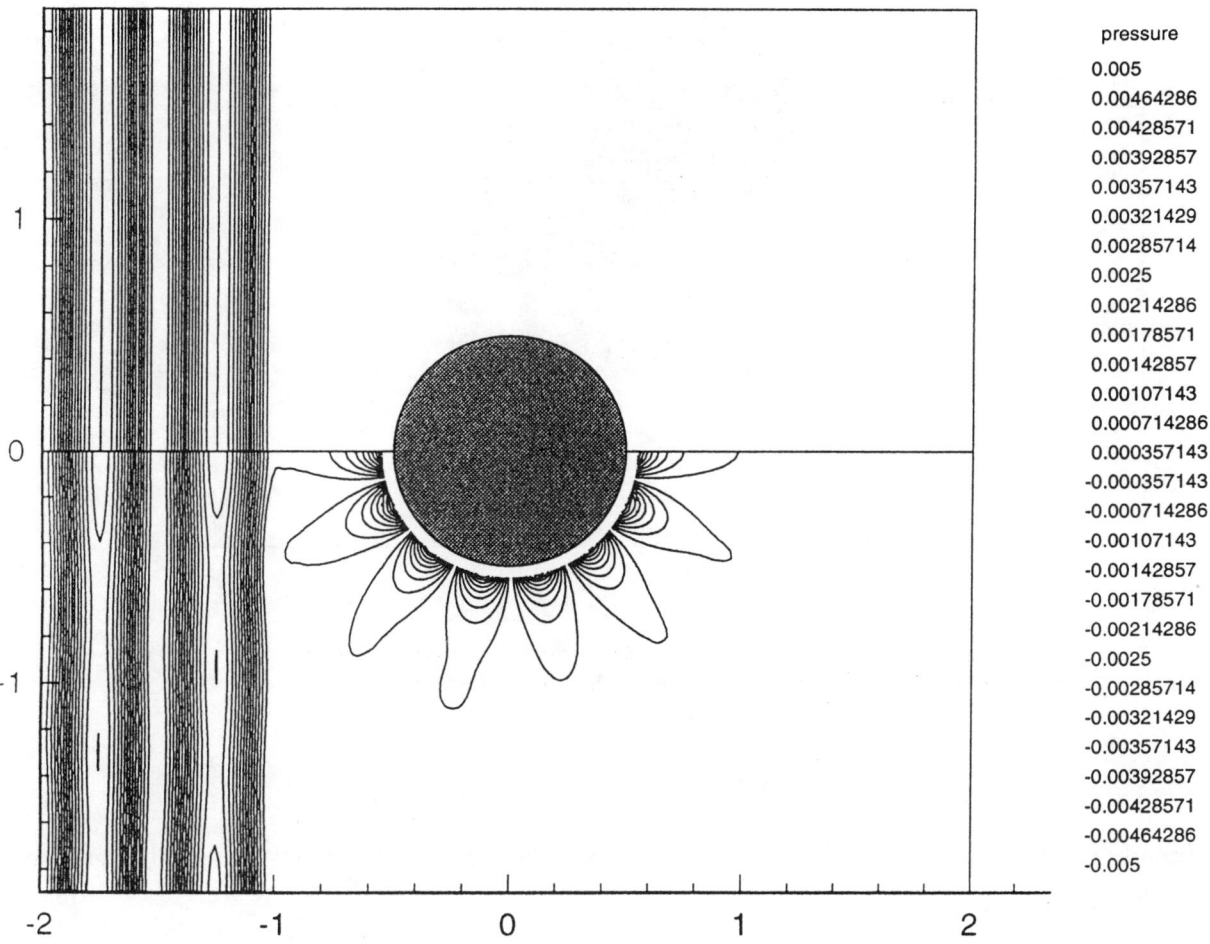

Figure 1.27 The acoustic pressure contours for the rigid scattering (upper half) and elastic scattering (lower half) at $t = 0$. The frequency-domain approach is used and the problem is solved in 124 different frequencies. The elastic cylinder produces spurious waves before the incident plane wave excites it.

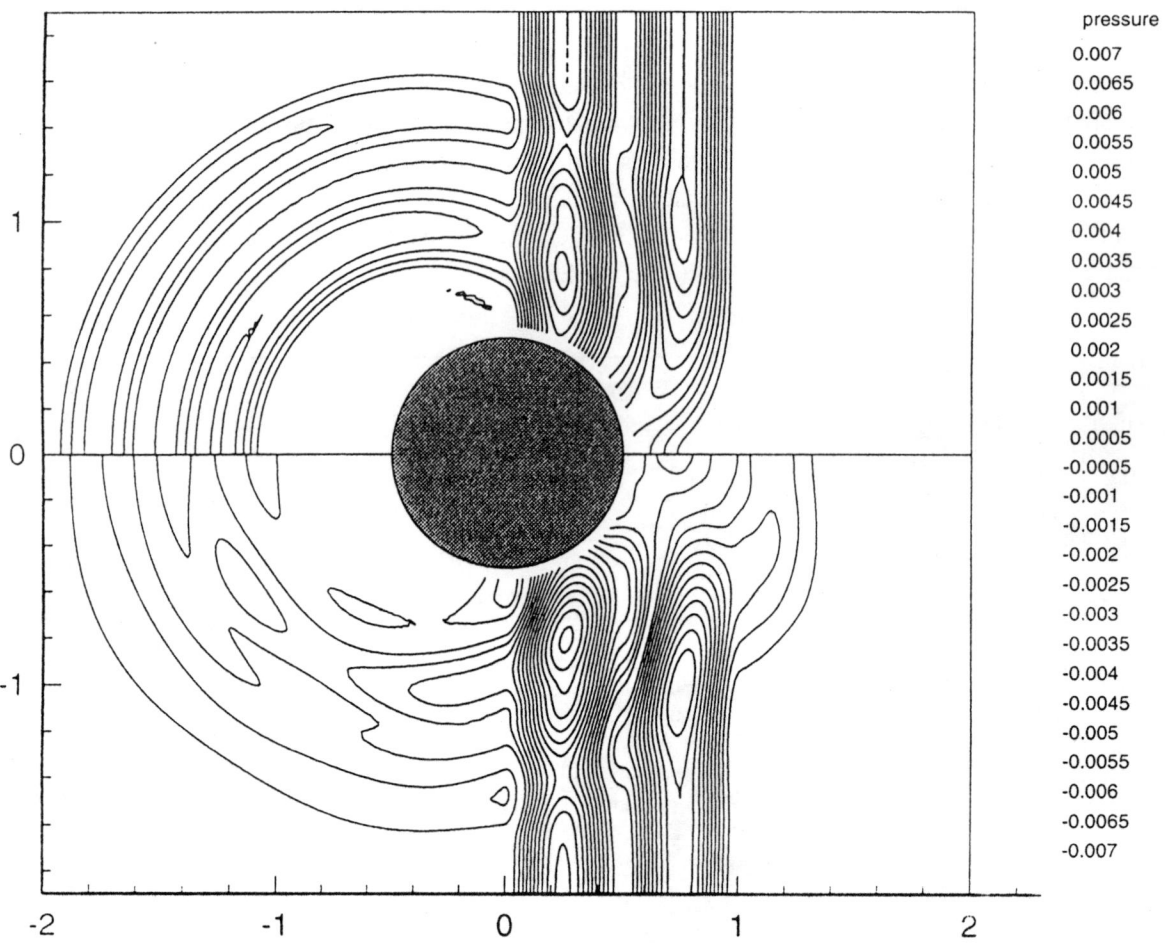

Figure 1.28 The acoustic pressure contours for the rigid scattering (upper half) and elastic scattering (lower half) at $t = 2$. The frequency-domain approach is used and the problem is solved in 124 different frequencies.

ACKNOWLEDGMENT

Support of this work by the Office of Naval Research (ONR) under #N00014-89-J-1451 is gratefully acknowledged.

REFERENCES

Babuška, I., and M. Suri, 1987, "The hp version of the finite element method with quasiuniform meshes," *RAIRO Math. Model. Numer. Anal.* **21**, 199-238.

Burton, A.J., and G.F. Miller, 1971, "The application of integral equation methods to the numerical solution of some exterior boundary-value problems," *Proc. R. Soc. London Ser. A* **323**, 201-210.

Demkowicz, L., J.T. Oden, W. Rachowicz, and O. Hardy, 1989, "Toward a universal hp adaptive finite element strategy. part 1: constrained approximation and data structure," *Comput. Meth. Appl. Mech. Eng.* **77**, 79-112.

Demkowicz, L., J.T. Oden, M. Ainsworth, and P. Geng, 1991a, "Solution of elastic scattering problems in linear acoustics using hp boundary element methods," *J. Comput. Appl. Math.* **36**, 29-63.

Demkowicz, L., J.T. Oden, W. Rachowicz, and O. Hardy, 1991b, "An hp Taylor-Galerkin finite method for compressible Euler equations," *Comput. Meth. Appl. Mech. Eng.* **88**, 363-396.

Demkowicz, L., A. Karafiat, and J.T. Oden, 1992, "Solution of elastic scattering problems in linear acoustics using hp boundary element method," *Comput. Meth. Appl. Mech. Eng.* **101**, 251-282.

Donea, J., 1984, "A Taylor-Galerkin method for convective transport problems," *Int. J. Numer. Meth. Eng.* **88**, 101-120.

Enquist, B., and A. Majda, 1977, "Radiation boundary conditions for the numerical simulation of waves," *Math. Comp.* **31**, 629-651.

Enquist, B., and A. Majda, 1979, "Radiation boundary conditions for acoustic and elastic wave calculations," *Commun. Pure Appl. Math.* **32**, 313-357.

Hubert, J.H., and E.S. Palencia, 1989, *Vibration and Coupling of Continuous Systems*, Berlin, Heidelberg, New York: Springer-Verlag.

Junger, M.C., and D. Feit, 1972, *Sound, Structures, and Their Interaction,* Cambridge, Mass.: MIT Press.

Karafiat, A., J.T. Oden, and P. Geng, 1993, "Variational formulations and hp-boundary element approximation of hypersingular integral equations for Helmholtz-exterior boundary-value problems in two dimensions," *Int. J. Eng. Sci.* **31**, 649-672.

Kutt, H.R., 1975, "On the numerical evaluation of finite part integrals involving an algebraic singularity," *Report WISK* 179, Pretoria: The National Research Institute for Mathematical Sciences.

Oden, J.T., 1974, "Formulation and application of certain primal and mixed finite element models of finite deformations of elastic bodies." In: *Computing Methods in Applied Sciences and Engineering*, R. Glowinski and J.L. Lions (eds.), Berlin, Heidelberg, New York: Springer-Verlag.

Safjan, A., 1993, "High-order Taylor-Galerkin and adaptive hp methods for hyperbolic systems with application to structural acoustics," Ph.D. Dissertation, University of Texas at Austin.

Safjan, A., and J.T. Oden, 1993, "High-order Taylor-Galerkin and adaptive hp methods for second-order hyperbolic systems: application to elastodynamics," *Comput. Meth. Appl. Mech. Eng.* **103**, 187-230.

Safjan, A., and J.T. Oden, 1994, "High-order Taylor-Galerkin and adaptive hp methods for first-order hyperbolic systems," TICAM Report.

Safjan, A., and J.T. Oden, 1995a, "High-order Taylor-Galerkin and adaptive hp methods for hyperbolic systems," submitted to *J. Comput. Phys.*

Safjan, A., and J.T. Oden, 1995b, "Highly accurate nonreflecting boundary conditions for wave propagation problems," to be submitted.

Schenck, A.H., 1967, "Improved integral formation for acoustic radiation problems," *J. Acoust. Soc. Am.* **44**(l), 41-58.

Selmin, V., J. Donea, and L. Quartapelle, 1985, "Finite element methods for nonlinear advection," *Comput. Meth. Appl. Mech. Eng.* **52**, 817-845.

Yosida, K., 1973, *Functional Analysis*, Fourth edition, Berlin, Heidelberg, New York: Springer-Verlag.

2

Distributed Feedback Resonators

Hermann A. Haus
Massachusetts Institute of Technology

> This paper describes the invention of distributed feedback (DFB) structures for integrated optics. The realization of complex structures was carried out in acoustics as cascades of Surface Acoustic Wave (SAW) resonators. I review their theory and describe experiments of acoustic wave resonators, using equivalent circuits as design guides. Finally, I present photonic bandgap materials intended for greater bandwidth along with numerical methods for their analysis.

INTRODUCTION

The thrust of this conference is centered on electromagnetic and acoustic structures that are large compared with a wavelength. In addition there is the constraint of "structures that cannot be simplified because of irregularity in the structures." The following treatise fits into the general theme of structures that are large compared with a wavelength. It deviates from the constraint in its engineering approach to the analysis of the optical and acoustic structures described. Composite structures are formed from simpler components on the basis of simplified models (equivalent circuits) of the individual components. Only at the very end, when introducing photonic bandgap filters, is a full computer solution of the wave equation presented that includes coupling to radiation modes. Even here comparisons are made with the response of simple models. The author hopes that this approach will not be too foreign to the aims of the conference.

Optical resonators are generally large compared with the wavelength. A simple Fabry-Perot transmission resonator as shown in Figure 2.1, with two partially transmitting mirrors, has a resonance for every half-wavelength change of the incident radiation. If such a resonator is to be used as an optical device, the multiplicity of resonances is usually undesirable. A laser diode, which employs a Fabry-Perot formed of the cleavage planes of the semiconductor, selects one resonance frequency via the gain-bandwidth; the mode with the maximum gain oscillates and "robs" all the other modes of their gain.

In order to make resonators and oscillators manufacturable as components in (integrated) waveguide optics, Kogelnik and Shank (1972) introduced the distributed feedback (DFB) resonator. It consists of a uniform grating, a corrugation on top of the optical waveguide, a structure that can be fabricated by photolithography and etching techniques. The waveguide contains the active medium, a heterojunction that confines the carriers which, by recombining, provide the optical gain. The same structure can be used as a transmission filter. It is to this passive realization that we shall devote our attention. The schematic of this simple grating structure is shown in Figure 2.2.

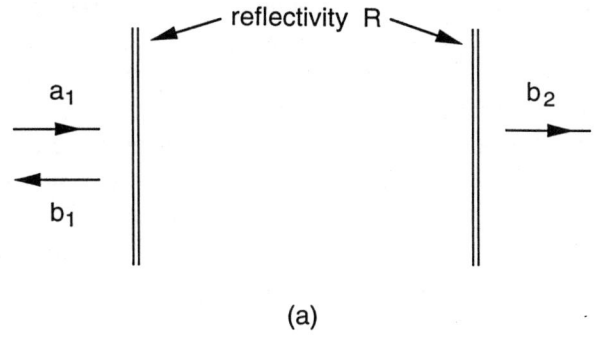

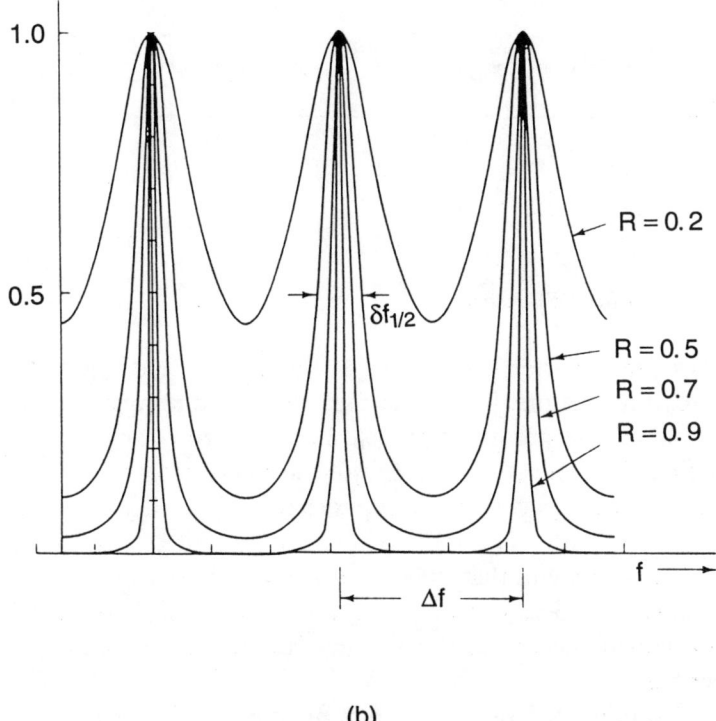

Figure 2.1 A (a) Fabry-Perot transmission resonator and (b) its transmission characteristic.

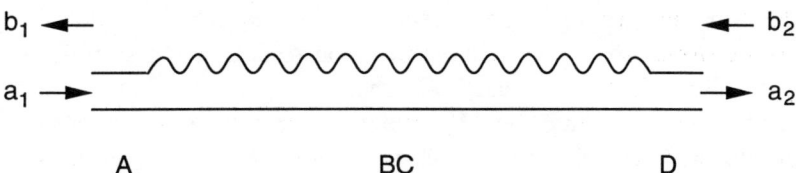

Figure 2.2 Schematic of DFB resonator, and position of reference plane for envelope amplitudes A and B with choice of coupling coefficient in (2.9) and (2.10).

The simple DFB structure has resonances to either side of the stopband as we shall show in detail in the reflection filter section below. If used as a laser resonator, the oscillator may choose either one of the two frequencies depending upon small perturbations that break the gain degeneracy. To overcome this problem and to arrive at a structure with no gain degeneracy, in 1975, Dr. C.V. Shank, then of Bell Laboratories, and the author arrived at a resonator design, the quarter wave shifted DFB resonator with the schematic shown in Figure 2.3 (Haus and Shank, 1976). The grating consists of two sections, and at their juncture they are separated by a quarter wave gap. This gap acts like a local defect in a crystal lattice, generating a local state at the center of the stopband (bandgap). Thus the degeneracy is removed and the resonance at the center of the bandgap experiences the grating reflectivity where it is strongest, not like the uniform structure whose resonances lie outside the stopband.

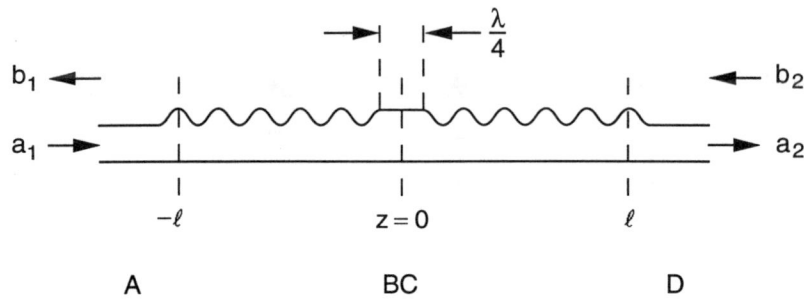

Figure 2.3 Schematic of quarter wave shifted DFB resonator.

The simple resonances of the quarter wave shifted DFB structure suggest that one may construct filter structures from cascaded resonators. The basic idea of cascading quarter wave shifted resonators for the purpose of filter synthesis was presented almost 20 years ago (Haus and Schmidt, 1977a,b). The idea was aimed at optical filter synthesis, but was applied to SAW resonators, since at that time it was not possible to fabricate optical gratings within the required tolerances. The idea was tested with Surface Acoustic Wave (SAW) filters in the 100 MHz regime in which grating groove spacings of the order of 20 microns are required, easily achievable at that time. The idea did not catch on in the SAW field, in part because the design did not use piezoelectric coupling that gives SAW design added flexibility. This advantage does not exist in optical applications, and hence it is likely that the approach will find greater acceptance in waveguide optics.

In wavelength division multiplexed optical communications it is desirable to access one of the channels from a signal bus, without affecting the other channels. Resonant coupling can be used for this purpose. Again, the quarter wave shifted DFB resonators, side-coupled to a signal bus, can accomplish this function. The section on side-coupled resonators discusses the analysis of such structures.

Thus far, we have been describing grating structures with a corrugation etched onto an optical waveguide. The reflection per grating groove tends to be small, the structures tend to be many wavelengths long, and the stopbands are correspondingly narrow. If wider stopbands are to be achieved, the photonic bandgap structures have to be used. An analytic treatment of such structures is inadequate, since they are open and prone to radiate. At MIT we are currently pursuing both the analysis and the fabrication of such structures. In the closing section on photonic bandgap crystals we shall discuss some of the issues.

THE EQUATIONS OF DISTRIBUTED FEEDBACK GRATING

A traveling wave of a single mode in an optical waveguide, of frequency ω and amplitude a, obeys the equation

$$\frac{da}{dz} = -j\beta a, \qquad (2.1)$$

where β is the propagation constant of the mode, in general a function of frequency. For narrow band signals, dispersion can be neglected and expansion around a frequency ω_0 gives

$$\beta(\omega) = \beta_o + (\omega - \omega_o)\beta' = \beta_o + \frac{\omega - \omega_o}{v_g}, \qquad (2.2)$$

where v_g is the group velocity. The wave of amplitude β traveling in the opposite direction obeys the equation

$$\frac{db}{dz} = j\beta b. \qquad (2.3)$$

In the absence of a perturbation, in a uniform waveguide, these two waves are uncoupled. If there is a perturbation, like a local change of index, the forward wave couples to the backward wave. If the perturbation is cosinusoidal, as produced by a cosinusoidal corrugation on top of the waveguide (see Figure 2.2), the equation for a is modified. Aside from coupling to the backward mode, radiation may be produced:

$$\frac{da}{dz} = -j\beta_o a - j\frac{\omega - \omega_o}{v_g}a - j\kappa[e^{jk_g z} + e^{-jk_g z}]b + radiation. \qquad (2.4)$$

For a weak perturbation, the coupling to b will be strong only when the grating is close to the Bragg condition

$$\beta_o - k_g = -\beta_o \quad \text{or} \quad \beta_o = \frac{k_g}{2}. \qquad (2.5)$$

Figure 2.4 indicates this with a vector diagram of the wave vector of the forward wave, of the backward wave, and of the grating. Only the $\exp(+jk_g z)$ component of the grating couples the forward wave to the backward wave. The coupling to radiation fields can be made negligible if

(a) The vector $\beta_o - k_g$ lies outside the surface of wave vectors of the radiation field. In this case there is no phase match to the radiation modes.

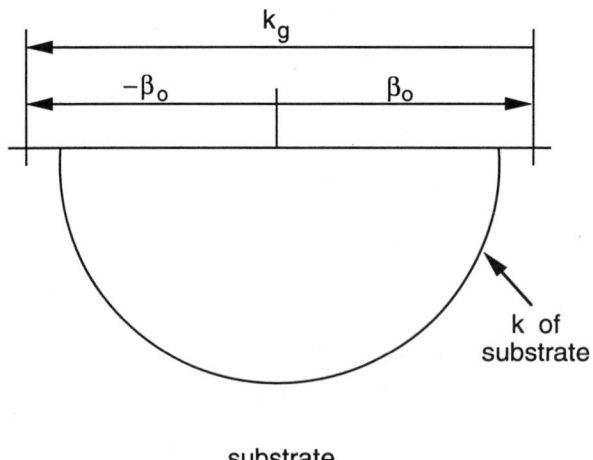

Figure 2.4 The radiation suppression condition for isotropic substrate.

(b) The grating is weak, so that the spatial dependence of a deviates little from its natural $\exp(-j\beta_o z)$ dependence.

The conditions are generally met in integrated optics with gratings on the surface of the optical waveguide. Under these conditions (2.4) simplifies to

$$\frac{da}{dz} = -j\beta_o a - j\frac{\omega - \omega_o}{v_g}a - j\kappa e^{-jk_g z}b. \qquad (2.6)$$

The coefficients in the equation for the backward wave are the complex conjugate of those for the forward wave, since it is the time reversed version of the equation for the forward wave

$$\frac{db}{dz} = j\beta_o a + j\left(\frac{\omega - \omega_o}{v_g}\right)a + j\kappa e^{jk_g z}a. \qquad (2.7)$$

With the introduction of envelope variables A and B

$$a(z,\omega) = A(z,\omega)e^{-j\frac{k_g}{2}z} \qquad (2.8a)$$

$$b(z,\omega) = B(z,\omega)e^{j\frac{k_g}{2}z}, \qquad (2.8b)$$

one gets the simple coupled mode equations

$$\frac{dA}{dz} = -j\delta A - j\kappa B \qquad (2.9)$$

$$\frac{dB}{dz} = j\delta B + j\kappa A, \qquad (2.10)$$

where

$$\delta \equiv \frac{\omega - \omega_o}{v_g}.$$

With κ positive, one can show (Haus, 1984) that the reference planes for evaluation of A and B are picked at the maximum of the index corrugation as shown in Figure 2.2.

THE REFLECTION FILTER

A DFB structure of length ℓ acts as a filter. In this section we derive the filter response of such a structure. Consider the reflection from a structure of length ℓ, with one end matched, $B = 0$ at $z = 0$ (Figure 2.5). For $|\delta| < |\kappa|$, the solutions of (2.9) and (2.10) are of the form $\exp(\mp \gamma z)$, where

$$\gamma = \sqrt{\kappa^2 - \delta^2}. \qquad (2.11)$$

The solutions are growing and decaying exponentials, whereas they are periodic functions in the range $|\delta| > |\kappa|$.

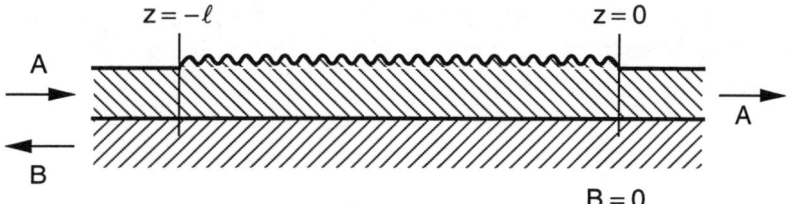

Figure 2.5 The boundary conditions on reflection filter.

The general solutions with arbitrary constants are

$$A = A_+ e^{-\gamma z} + A_- e^{\gamma z} \qquad (2.12)$$
$$B = B_+ e^{-\gamma z} + B_- e^{\gamma z}, \qquad (2.13)$$

where only two of the four constants are independent. From (2.9) we find a relation between $B_\pm$ and $A_\pm$:

$$B_\pm = \frac{\mp g + jd}{-jk} A_\pm. \qquad (2.14)$$

At $z = 0$ there is no reflected wave, $B_+ = -B_-$, and thus

$$B = -2B_+ \sinh \gamma z. \tag{2.15}$$

Using (2.10), we obtain for $A(z)$

$$A = -2B_+ \left(\frac{\gamma}{j\kappa} \cosh \gamma z - \frac{\delta}{\kappa} \sinh \gamma z\right). \tag{2.16}$$

The reflection coefficient $\Gamma = B/A$ at $z = -\ell$ is

$$\Gamma(-\ell) = -\frac{\sinh \gamma \ell}{(\gamma/j\kappa)\cosh \gamma \ell + (\delta/\kappa)\sinh \gamma \ell}. \tag{2.17}$$

The analysis can be generalized to cover an arbitrary reflection at $z = 0$, $\Gamma(0)$. One then finds after some simple manipulation

$$\Gamma(z) = -\frac{1 + \Gamma(0)[(\gamma/j\kappa)\coth \gamma z + (\delta/\kappa)]}{\Gamma(0) - [(\gamma/j\kappa)\coth \gamma z - \delta/\kappa]}. \tag{2.18}$$

Equation (2.17) is a special case of (2.18) with $\Gamma(0) = 0$ and $z = -\ell$. Figure 2.6 shows $\Gamma(-\ell)$ of (2.17) as a function of $\delta-$, the frequency parameter. The reflection coefficient is zero at a set of frequencies within the passband. At these frequencies the backward wave has a sinusoidal distribution within the structure, according to (2.15), with γ pure imaginary, and vanishes at the two end planes of the structure. These frequencies are the resonance frequencies of the periodic structure acting as a "distributed" Fabry-Perot transmission resonator. Instead of a single pair of mirrors, the periodic structure has a large number of reflecting "mirrors," one for each pair of periods of the structure lying symmetrically with respect to the center plane of the structure.

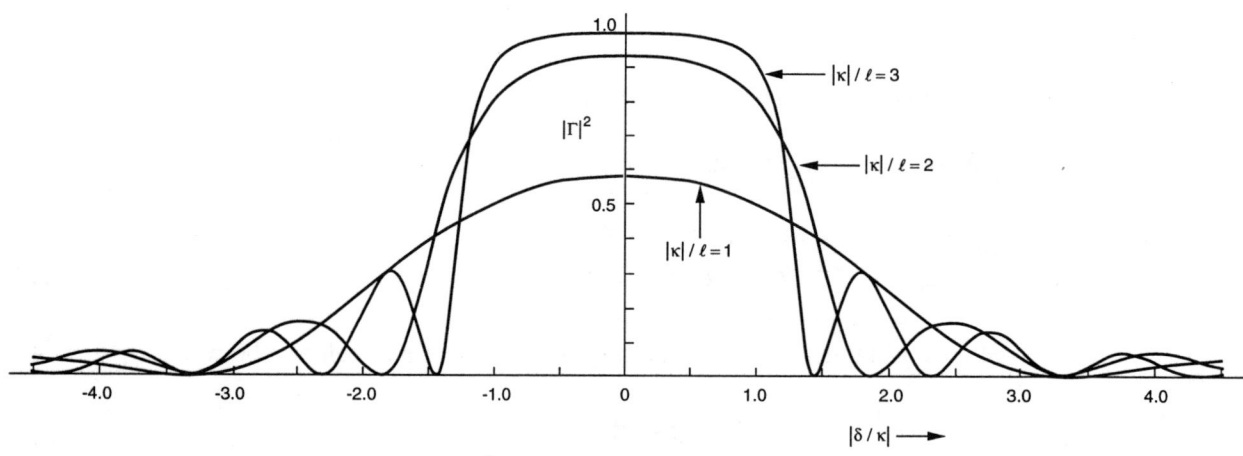

Figure 2.6 The reflection of the filter of Figure 2.5.

In Figure 2.6, $|\kappa|$ is kept constant at its value for $\delta = 0$; in other words, it is treated as if it were frequency independent (δ independent). This approximation is consistent with the approximations inherent in coupled mode theory. The two modes a and b couple within a normalized frequency range $\pm\delta$ of the order of $\pm|\kappa|$. A frequency dependence of $|\kappa|$ can be ignored as it is of second order (i.e., equal to the product of δ and $d\kappa/d\delta$).

THE QUARTER WAVE SHIFTED DFB RESONATOR

The transmission resonances of the uniform DFB structure lie to either side of the grating stopband. If an active medium in the resonator provides sufficient gain, the system oscillates at one of the transmission resonances. The resonance with highest Q is most prone to lasing (if the gain peak is close to the resonance). Thus the uniform DFB structure may lase on either of the two transmission resonances to either side of the stopband, since they possess the same Q, a Q that is larger than those of the resonances further away (as evidenced by their transmission bandwidths). Random reflections from the end of the structure favor either one or the other of the two resonances.

In order to remove this uncertainty in lasing frequency, C.V. Shank and the author proposed the quarter wave shifted resonator of Figure 2.3. This resonator has a high Q-resonance at the center of the stopband. If symmetric, it is a filter that transmits all the power at resonance. This can best be seen by a graphical construction in the complex reflection coefficient Γ plane, the Smith chart familiar to microwave engineers. Figure 2.7 shows the loci of the transformation of Γ with varying lengths of grating with a detuning such as to remain inside the stopband (Haus, 1975). There are two fixed points: these are the eigensolutions of the equations; their characteristic Γ does not change as one moves along the grating.

Figure 2.8 looks at the transformation at the center-frequency of the stopband. The fixed points are at $\pm j$. If the right-hand side of the structure of Figure 2.3 sees a match, then the length ℓ of grating transforms the reflection coefficient of point A into that of point B in Figure 2.8a. The quarter wave shift transforms it into point C. Another section of grating of the same length brings the reflection coefficient back to the origin: the structure appears matched at the input plane.

When the frequency is slightly detuned, the fixed points move as shown in Figure 2.8b. The transformation proceeds as indicated, and the reflection coefficient at the input moves. If the grating structure is long, $\kappa\ell > 1$, then the Q of the structure is high and the transmission characteristic approaches that of a simple resonator; it is Lorentzian, a circle in the Γ plane which closes at $\Gamma = -j$ as shown. This is the resonance of a parallel L-C circuit as viewed through a segment of transmission line of length $\lambda/8$. Because the structure is symmetric, such a segment must be added to the other side as well. In the construction of Figure 2.8 this segment has no effect because we assumed a match on the right-hand end. Thus, with reference planes chosen at the peak of the index corrugation, the equivalent circuit for the center resonance of the quarter wave shifted DFB resonator is a parallel LC circuit with $\lambda/8$ segments of transmission line added to both sides. The resonance is described by two parameters: (a) the resonance frequency (in the equivalent circuit $1/(LC)^{0.5}$, and (b) the external Q (related to Y_0/C). The Q can be evaluated by simple perturbation theory.

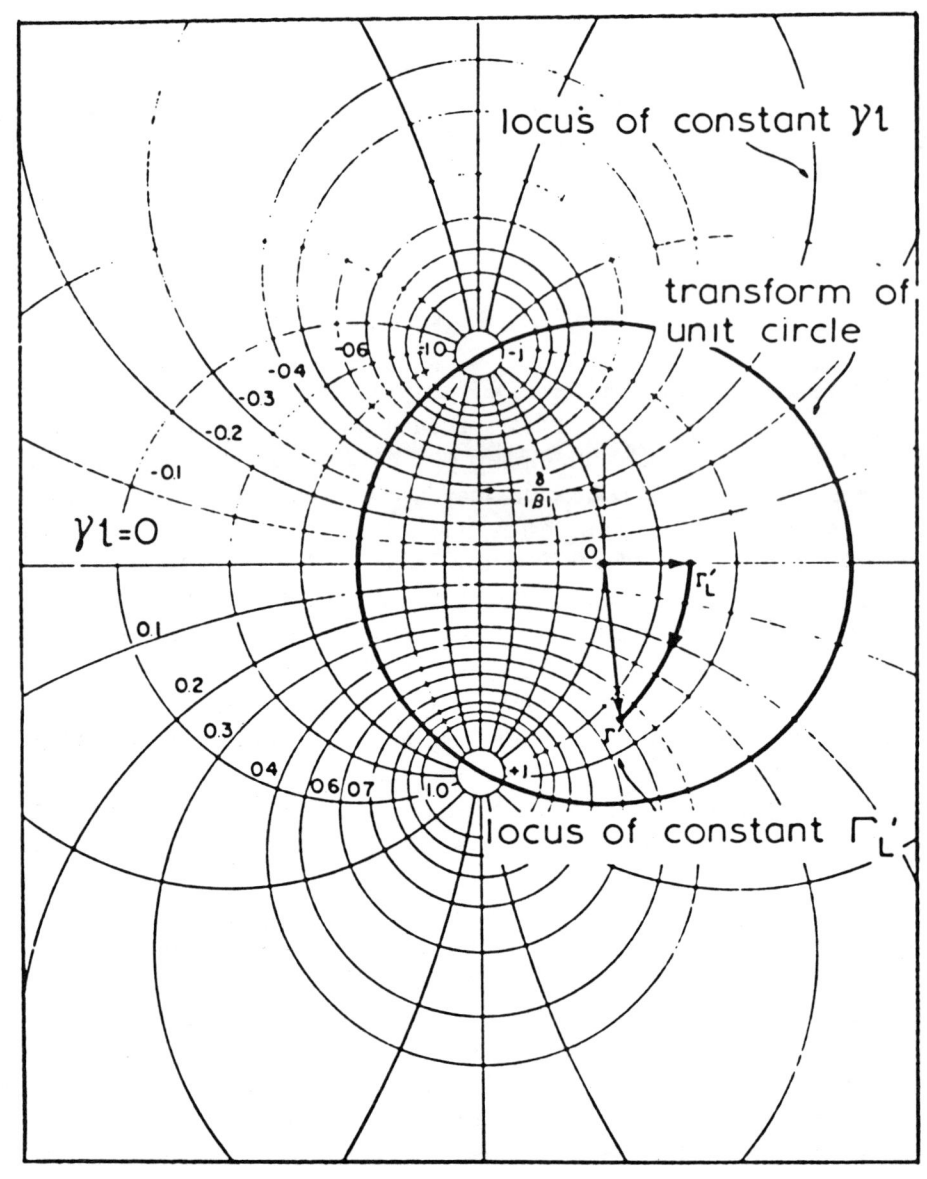

Figure 2.7 Grating transformation graph in complex reflection coefficient plane (Smith chart) at the center of the stopband (Haus, 1975).

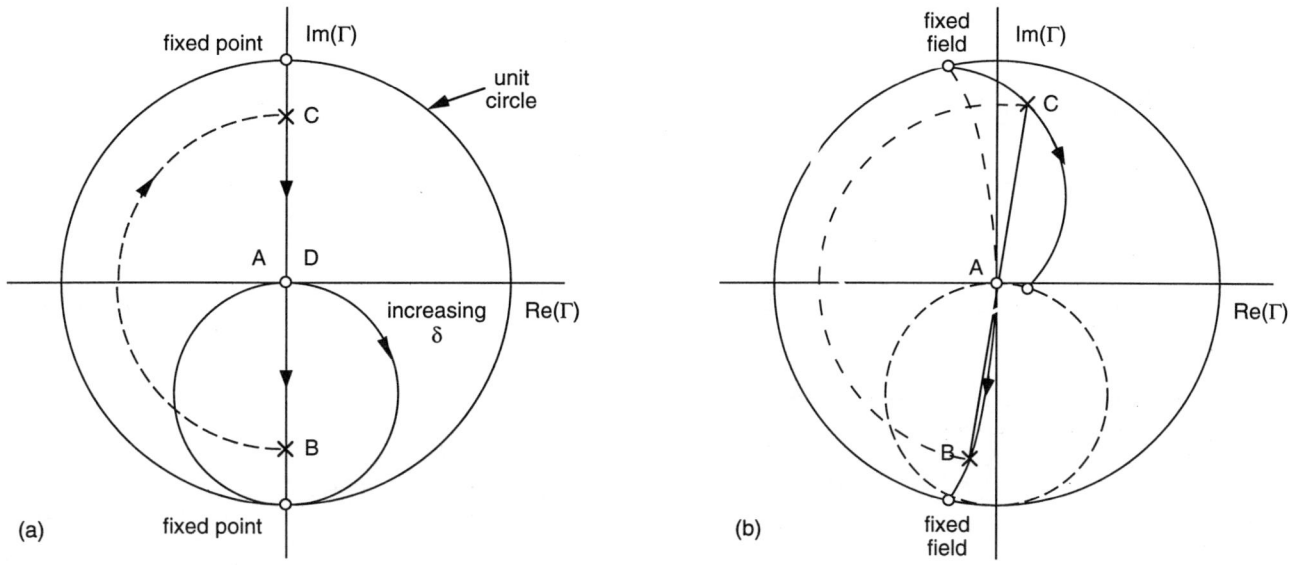

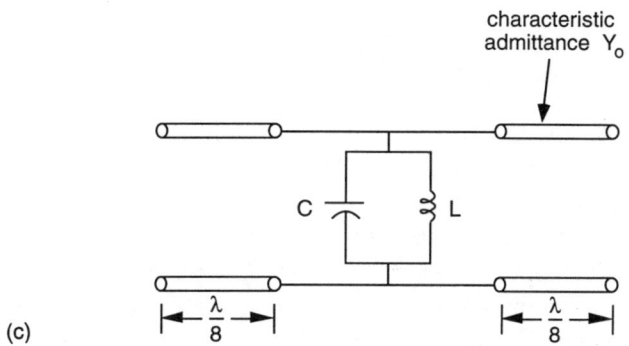

Figure 2.8 The graphical construction for quarter wave shifted resonator and its equivalent circuit: (a) at resonance, (b) slightly off resonance, and (c) equivalent circuit.

One performs the thought experiment in which one considers the resonance excited at $t = 0$ and one determines the power P_{ei} escaping from either end of the structure, $i = 1,2$. The external Q of the resonator follows from the definition

$$\frac{1}{Q_{ei}} = \frac{P_{ei}}{\omega_o W}, \qquad (2.19)$$

where W is the stored energy. The dependence upon z of A (and B) at the gap-center-frequency is exponential, $A = A_o \exp(\pm \kappa z)$, except near the ends where the structure must see a match, where either $A = 0$ on the left hand side, or $B = 0$ on the right hand side. But this means that the spatial dependence of the wave on the right-hand side near $z = \ell$, in terms of the amplitude A_o, at the center of the structure, is given by:

$$A \cong 2A_o e^{-\kappa\ell}\cosh\kappa(z-\ell); \text{ near } z = \ell, z < \ell. \tag{2.20}$$

The power escaping is:

$$P_{e2} = |A|^2 = 4|A_o|^2 e^{-2\kappa\ell} = P_{e1}. \tag{2.21}$$

The energy is:

$$W \cong 4\int_0^\infty \frac{|A_o|^2}{v_g} e^{-2\kappa z} dz = 2\frac{|A_o|^2}{\kappa v_g}, \tag{2.22}$$

where one factor of 2 accounts for the energy in both A and B, and the other factor of 2 accounts for the two sides of the structure. Thus the external Q is

$$\frac{1}{Q_{ei}} = 2\kappa v_g e^{-2\kappa\ell}. \tag{2.23}$$

The external Q's of the equivalent circuit are given by

$$\frac{1}{Q_{ei}} = \frac{Y_o}{\omega_o C}, \tag{2.24}$$

where Y_o is the characteristic admittance of the transmission line. The transmission characteristic of the quarter wave shifted resonator is as shown in Figure 2.9 for varying values of $\kappa\ell$. The center resonance is Lorentzian, and its width is accurately predicted by the external Q computed above for $\kappa\ell > 1$. The structure is reflecting over the remainder of the stopband, gradually providing less and less reflection with increasing deviation from the stopband. There are other low Q resonances outside the stopband.

A single grating pair gives a Lorentzian transmission characteristic near the center of the stopband. This is analogous to the equivalent circuit of Figure 2.8 which represents the quarter wave shifted resonator at and near the center resonance. It is well known that cascades of such resonators, separated by quarter wave segments of transmission line, produce very desirable filter characteristics. Since a quarter wave line inverts the impedance, a three segment structure as shown in Figure 2.10 has the equivalent lumped circuit as shown. If one varies the circuits in the cascade, various filter characteristics can be achieved. One can take advantage of handbook information on filter design and end up with Chebysheff, Butterworth, or Gaussian filters as shown in Figure 2.11. These are formed of resonant circuits with the same resonance frequency but with different L/C ratios. The relation for the external Q's relates the grating lengths to the normalized capacitance C/Y_o of the equivalent circuit.

In 1976 we realized some of these circuits with SAW filters (Haus and Schmidt, 1977a). Figure 2.12 shows some experimental results with SAW resonators, the three figures corresponding to one, two, and three resonators in cascade. One can see the excellent agreement of the experimental transmission characteristic with the theoretically predicted one in Figure 2.11a down to a −60 dB level.

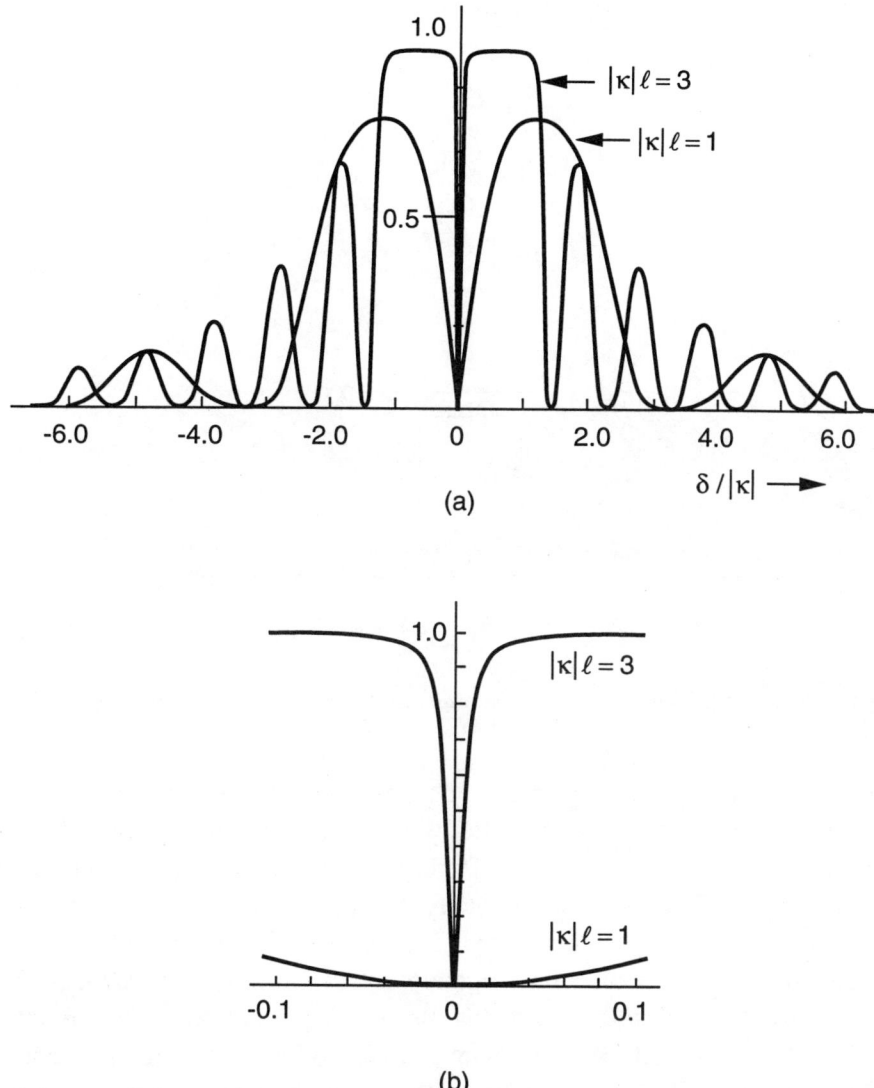

Figure 2.9 The transmission characteristic of quarter wave shifted DFB resonator.

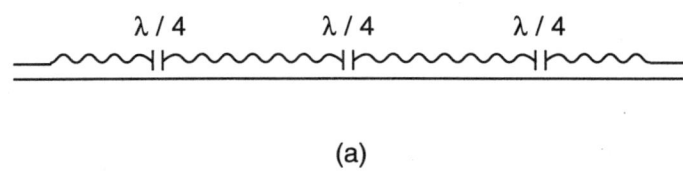

(a)

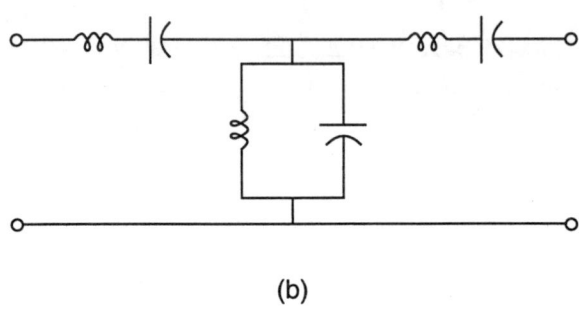

(b)

Figure 2.10 Three quarter wave shifted DFB resonators (a) in cascade and (b) their equivalent circuit.

Recently, in the Nanostructure Laboratory, in cooperation with Professor H. I. Smith and his students, we started fabricating quarter wave shifted DFB structures for optical filters centered at a wavelength of 1.54 microns, the wavelength of erbium amplifiers and of long distance optical communications. Figure 2.13 shows the experimentally observed response and the theoretical fit to it. The transmission at resonance is lower than predicted. This may be due to experimental resolution or due to radiation losses. This point needs further investigation. Otherwise, the agreement is excellent except for the deviation on the short wavelength side of the response. It is here that the structure experiences radiative loss by virtue of the fact that, for a constant grating vector k_g and for increasing propagation vectors β_o, the spatial spectrum starts to extend into the radiation sphere of the substrate (see Figure 2.4).

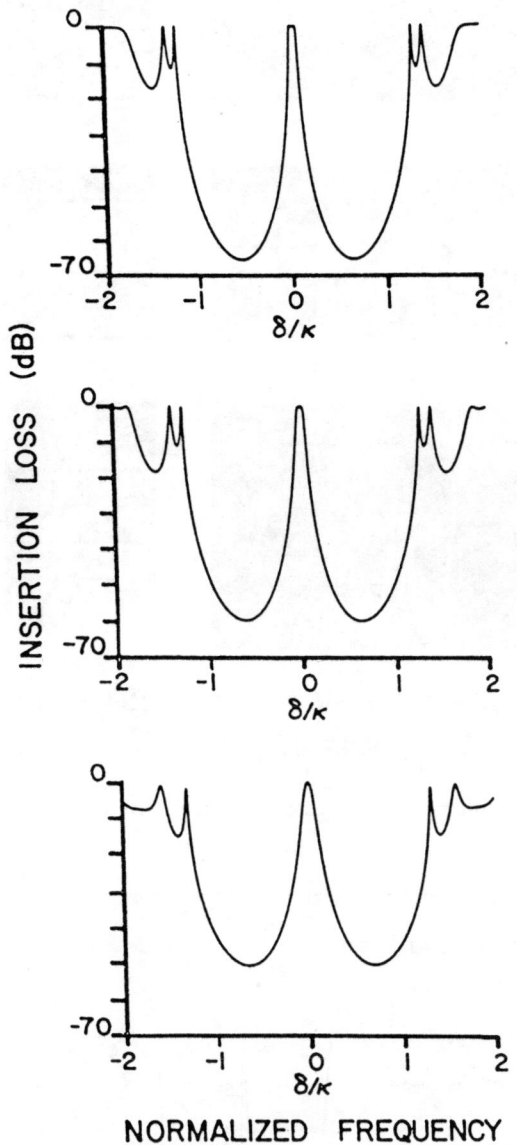

Figure 2.11 The transmission characteristic for (a) Chebysheff, (b) Butterworth, and (c) Gaussian filter designs.

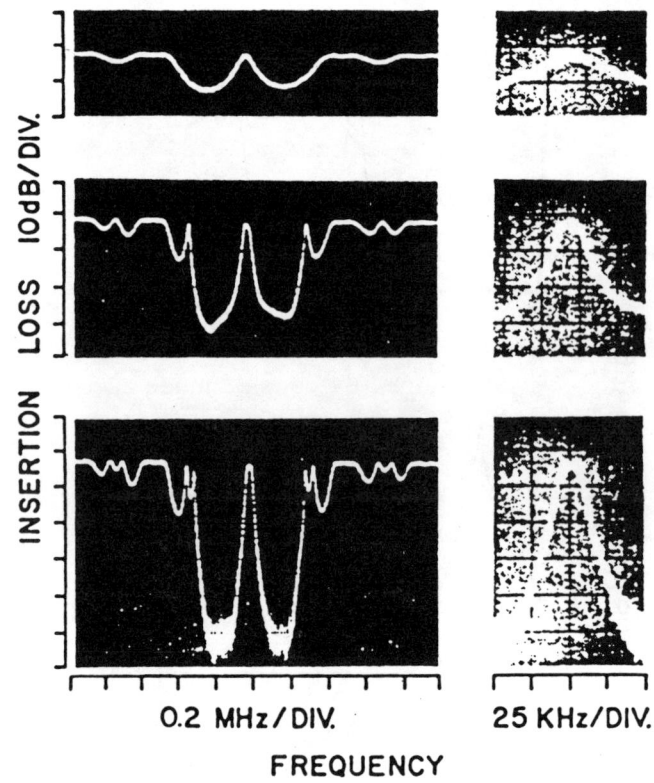

Figure 2.12 Experimental results with SAW filters: top, one resonator; middle, two resonators; bottom, three resonators.

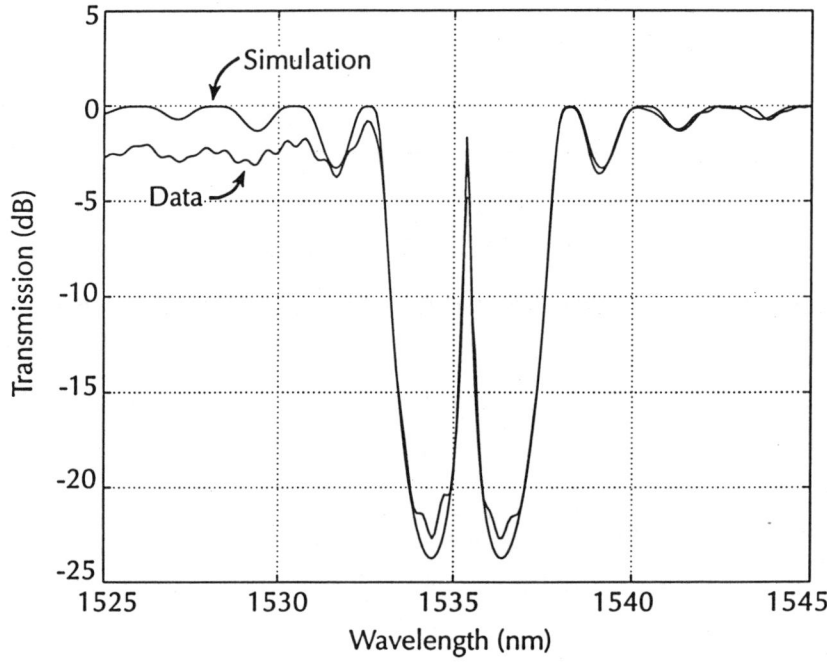

Figure 2.13 Experimentally observed transmission of quarter wave shifted optical grating and the theoretical prediction.

SIDE-COUPLED RESONATORS

When a microwave cavity is coupled to a waveguide via a small hole, the structure is known as a *Q-meter*. At the resonance of the cavity, transmission in the waveguide is interrupted. This is a means for frequency calibration.

The quarter wave shifted DFB resonators side-coupled to a signal waveguide also tap off the optical power at the resonance frequency of the resonator. If the power in the resonator is detected, a channel dropping filter is realized, leaving other signals at frequencies in the stopband of the grating essentially unaffected (Haus and Lai, 1991, 1992). Figure 2.14a shows a side-coupled quarterwave shifted DFB resonator and Figure 2.14b shows its equivalent circuit at and near resonance. The equivalent circuit follows from the response of the system obtained by coupled mode analysis, expanded in frequency around resonance.

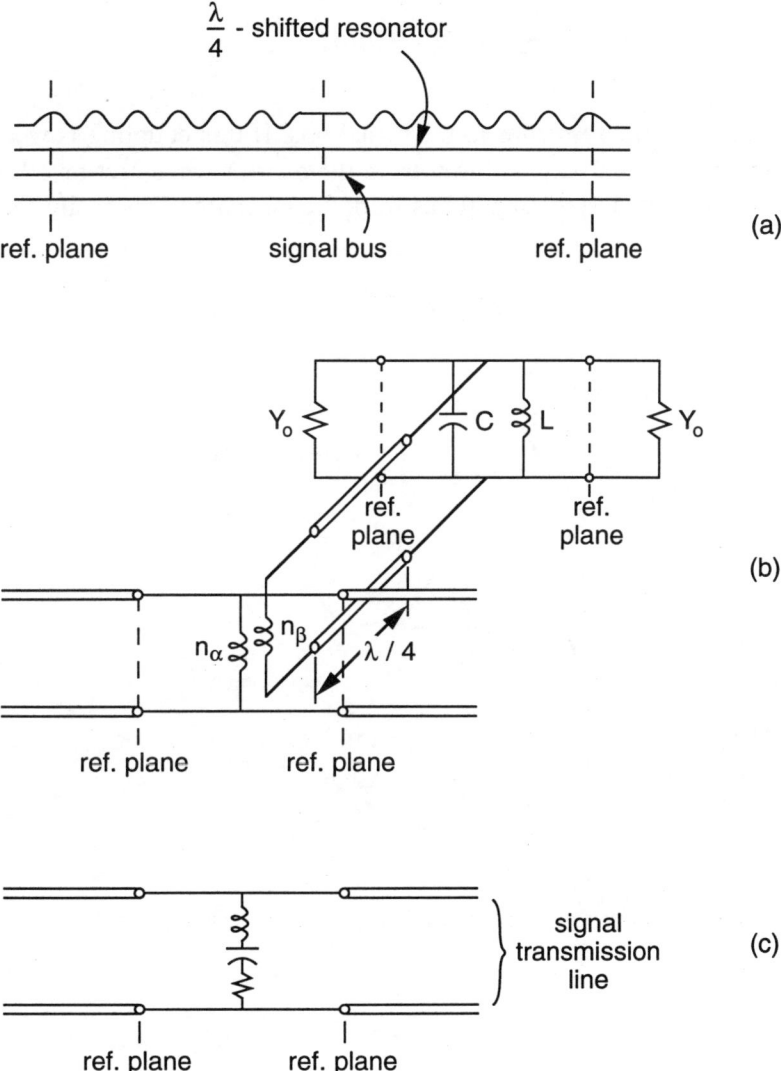

Figure 2.14 (a) Side-coupled quarter wave shifted DFB resonator, (b) its equivalent circuit, and (c) its equivalent circuit at resonance.

The parameters of the circuit follow from a simple analysis that determines the external Q of the resonator and that of the equivalent circuit. We suppose that all resonators are "long" enough so that the escape of power from their two ports is negligible compared with the coupling of power into the signal bus. Again we may resort to an analysis of Q's to get the equivalent circuit parameters. The coupled mode equations are

$$\frac{dA_\alpha}{dz} = -j\delta A_\alpha - j\kappa B_\alpha - j\mu A_\beta \tag{2.25}$$

$$\frac{dB_\alpha}{dz} = j\delta B_\alpha + j\kappa A_\alpha + j\mu B_\beta \tag{2.26}$$

$$\frac{dA_\beta}{dz} = -j\delta A_\beta - j\mu A_\alpha \tag{2.27}$$

$$\frac{dB_\beta}{dz} = j\delta B_\beta + j\mu B_\alpha, \tag{2.28}$$

where μ is the coupling from the resonator to the signal bus. If this coupling is weak, then the resonator at resonance still supports waves decaying away from the quarter wave sections. Equation (2.27) can be used to evaluate the wave excited in the signal bus from the resonator. It is easily found to be

$$A_\beta = -2j\frac{\mu}{\kappa}A_o. \tag{2.29}$$

Thus, the power escaping in the bus is

$$P_e = 8\frac{\mu^2}{\kappa^2}|A_o|^2. \tag{2.30}$$

The external Q's are

$$\frac{1}{Q_{ei}} = 2\frac{\mu^2}{\kappa^2}\frac{\kappa v_g}{\omega_o} = \frac{Z_o}{\omega_o L}. \tag{2.31}$$

We see that the equivalent L's and C's are

$$2L = \frac{\kappa^2}{\mu^2}\frac{1}{\kappa v_g}Z_o, \quad LC = \frac{1}{\omega_o^2}. \tag{2.32}$$

The power coupled to the bus is much larger than that out of the resonators if

$$e^{-2\kappa\ell} \ll \frac{\mu^2}{\kappa^2}.$$

The equivalent circuit at and near resonance is shown in Figure 2.14. At resonance, a single resonator couples out at most half of the incident power. If one wants to couple out all of the power at resonance, one must use two resonators in cascade, the second resonator acting as a blocking circuit. This is shown in Figure 2.15. Figure 2.16 shows the transmission characteristics for single- and two-resonator arrangements using the full coupled mode analysis of (2.5)-(2.8).

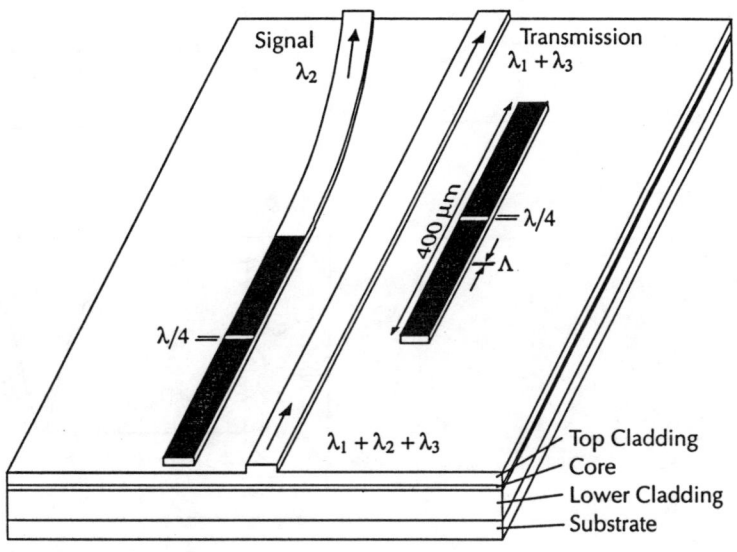

EQUIVALENT CIRCUIT

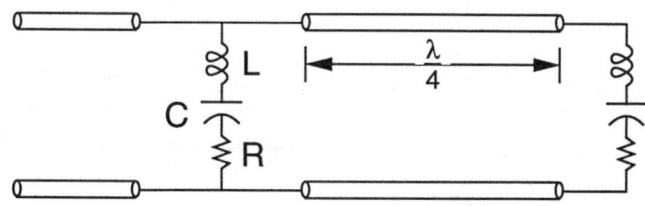

Figure 2.15 Two-resonator arrangement for full outcoupling at resonance.

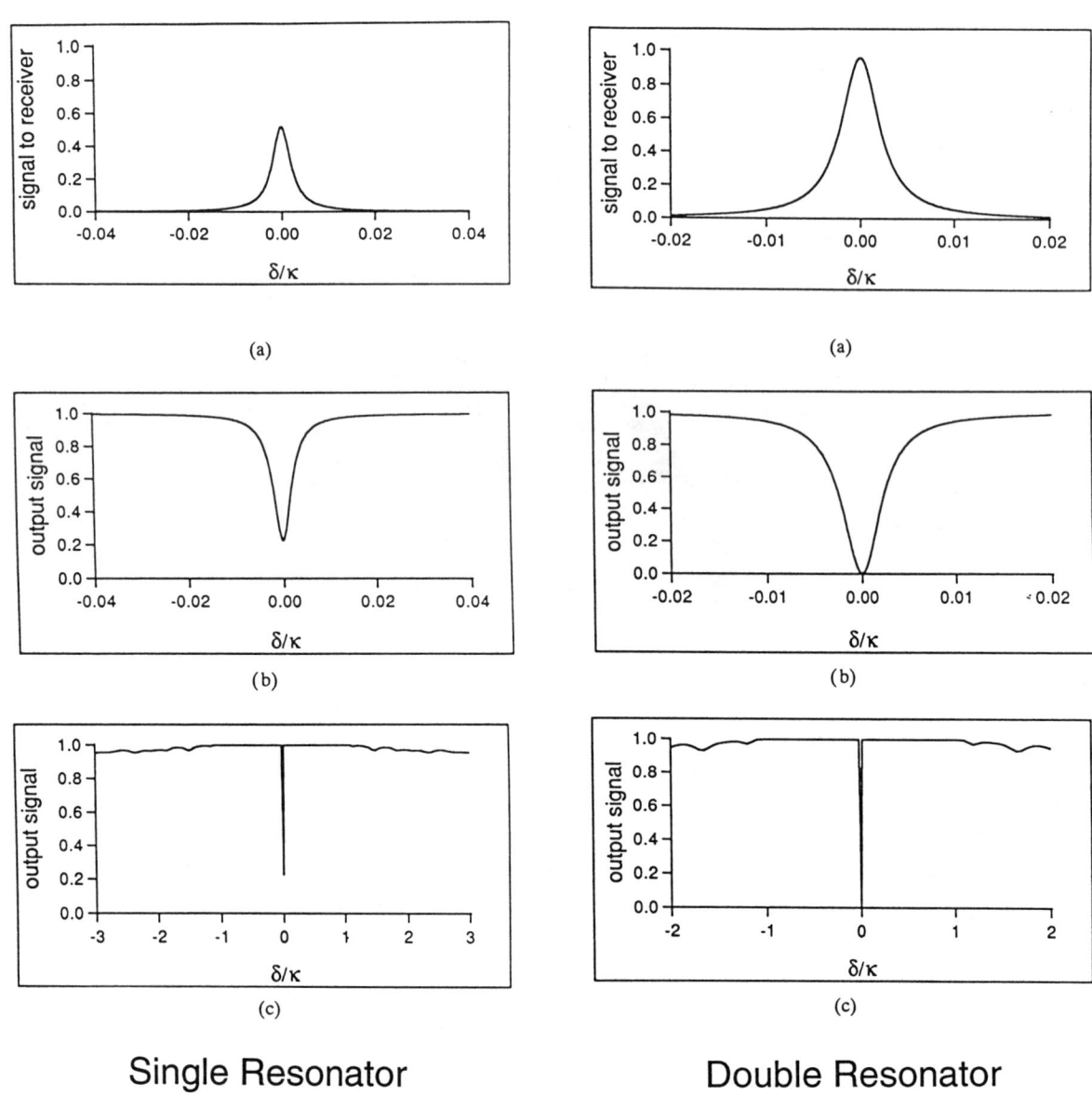

Figure 2.16 The responses of single- and double-resonator channel dropping filters.

PHOTONIC BANDGAP CRYSTALS

In recent years, photonic bandgap crystals have received a great deal of attention (Meade et al., 1991a,b, 1992, 1993a,b). One of the purposes of designing a structure which exhibits a gap for all electromagnetic waves is to construct a localized "state," i.e., a localized resonance. If gain is provided, the spontaneous emission couples only into this single resonance, providing lasing action with very special quantum properties.

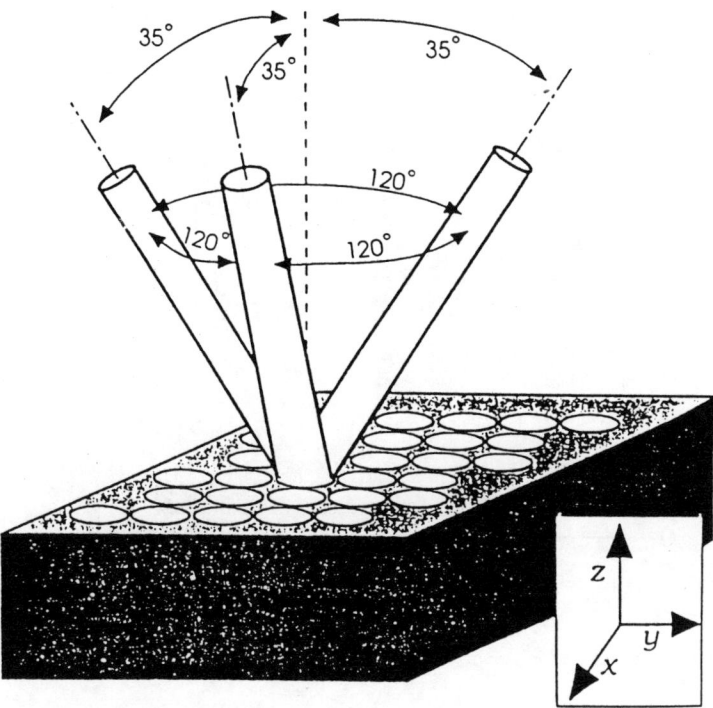

Figure 2.17 The construction of Yablonovite.

A structure proposed by E. Yablonovitch now goes under name of Yablonovite (Yablonovitch, 1987; Yablonovitch and Gmitter, 1989; Yablonovitch et al., 1991a,b). It can be constructed by drilling holes along the three axes of the diamond lattice as shown in Figure 2.17. The holes must be $\frac{1}{2}\lambda$ apart as measured in terms of the medium wavelength.

Photonic bandgap structures can only be realized with large reflection within each cell of the structure. This makes them interesting for broadband filter design along the lines of DFB filters described earlier. Modern optical communications utilizing the whole bandwidth offered by erbium doped fiber amplifiers require bandwidths, and thus stopbands, of the order of 5 THz.

For filter design it may not be necessary to use three-dimensional photonic bandgap structures. Figure 2.18 shows the kind of structure we are currently investigating for use as a quarterwave shifted DFB filter. Because the resonant mode is only a few wavelengths long, radiation becomes an important issue. Figure 2.4 shows the vector diagram for prevention of radiation-used single wave vectors. When the decay of the mode in the structure becomes large, the wave vector expands into a band of wave vectors, part of which may lie inside the radiation "sphere."

We calculate reflection, transmission, and radiation using a scalar finite difference time domain (FDTD) program (Chen, 1994) with first and second order absorbing boundary conditions (Engquist and Majda, 1977).

$$\frac{\partial^2 E(x,y,t)}{\partial x^2} + \frac{\partial^2 E(x,y,t)}{\partial y^2} - \frac{n(x,y)^2}{c^2}\frac{\partial^2 E(x,y,t)}{\partial t^2} = 0.$$

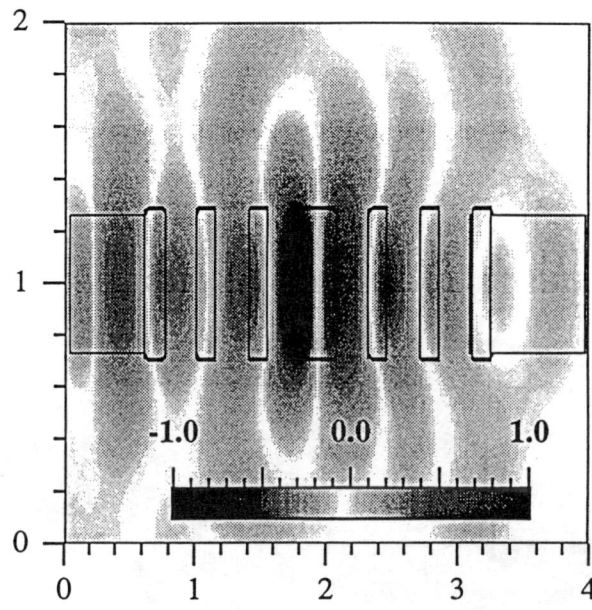

Figure 2.18 A filter structure with large reflection per cell and the field intensity distribution.

The transverse profile of the source is the fundamental mode of the waveguide; temporally, the source is a monochromatic sine wave. The field E is explicitly stepped forward in time until steady state is reached. From the field values at two different positions and times, we separate out the forward and backward propagating wave components. Taking the ratios of various field components before and after the filter gives the reflection and transmission coefficients. The radiation power is defined as the power not transmitted or reflected. This includes the power transferred to higher order transverse modes and to the substrate.

One promising filter is a more deeply etched version of the periodically segmented waveguide (Weissman and Hardy, 1993). Figure 2.18 shows a color contour plot of electric field overlaid on a contour plot of the index. The field comes in from the left side and becomes more localized about the center post. The extra thickness of this post acts as the quarter wave phase shift that produces the transmission resonance. Specifically, the posts are 0.58 by 0.14 wavelengths and have indices of $n = 1.88$. The air gap between adjacent posts is 0.2544 wavelengths and the center post is 0.2776 wavelengths wide. The entire device is less than 3 wavelengths long.

Figure 2.19 plots the reflection, transmission, and radiation powers as a function of normalized frequency. A large stopband, where transmitted power is zero, exists between 0.82 and 1.24. Large transmission occurs near frequencies of 0.68. Currently, we are trying to enhance the size of the transmission resonance within the stopband. We need to work more on coupling the power to the filter; much is reflected or radiated. In addition, narrower transmission resonances are expected by using even higher index contrasts.

The analysis of such structures is in its early stages and so is their fabrication. In order to diminish the radiation, the structure must be surrounded by air. Filters would have to be fabricated in the form of bridges between the input and the output. At MIT, Professor L. Kolodziejski has successfully etched bridge structures of this type. Only the future will tell whether these filters will be designable and manufacturable for the specifications of broad-band optical communication networks.

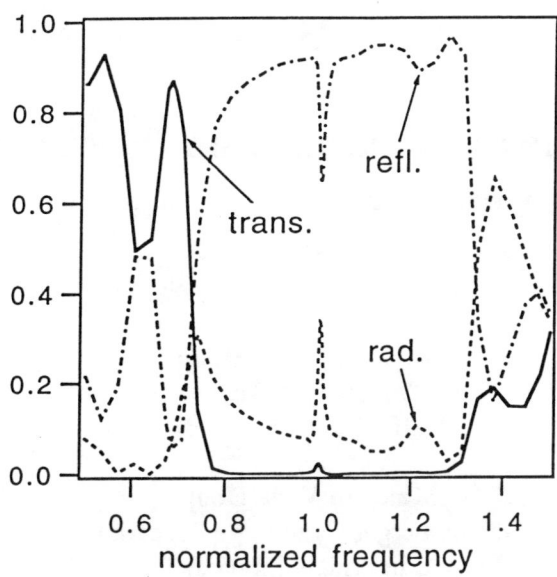

Figure 2.19 The transmission characteristic of the structure of Figure 2.18.

REFERENCES

Chen, J.C., 1994, in *Proceedings of the 1994 M.I.T. Student Workshop on Scalable Computing*, Cape Cod, Mass.

Engquist, B., and A. Majda, 1977, "Absorbing boundary conditions for the numerical simulation of waves," *Math. Comput.* **31**, 629-651.

Haus, H.A., 1975, "Grating filter transformation chart," *Electron. Lett.* **11**, 553-554.

Haus, H.A., 1984, *Waves and Fields in Optoelectronics*, Englewood Cliffs, N.J.: Prentice-Hall, Inc.

Haus, H.A., and C.V. Shank, 1976, "Antisymmetric taper of distributed feedback lasers," *IEEE J. Quant. Electron.* **QE-12**, 532-539.

Haus, H.A., and R.V. Schmidt, 1977a, "Transmission response of cascaded gratings," *IEEE Trans. Sonics Ultrason.* **SU-24**, 94-101.

Haus, H.A., and R.V. Schmidt, 1977b, "Cascaded SAW gratings as passband filters," *Electron. Lett.* **13**, 445-446.

Haus, H.A., and Y. Lai, 1991, "Narrow-band distributed feedback reflector design," *J. Lightwave Technol.* **9**, 754-760.

Haus, H.A., and Y. Lai, 1992, "Narrow-band channel-dropping filter," *J. Lightwave Technol.* **10**, 57-62.

Kogelnik, H., and C.V. Shank, 1972, "Coupled-wave theory of distributed feedback lasers," *J. Appl. Phys.* **43**, 2328-2335.

Meade, R.D., K.D. Brommer, A.M. Rappe, and J.D. Joannopoulos, 1991a, "Electromagnetic Bloch waves at the surface of a photonic crystal," *Phys. Rev. B* **44(19)**, 10961-10964.

Meade, R.D., K.D. Brommer, A.M. Rappe, and J.D. Joannopoulos, 1991b, "Photonic bound states in periodic dielectric materials," *Phys. Rev. B* **44(24)**, 13772-13774.

Meade, R.D., K.D. Brommer, A.M. Rappe, and J.D. Joannopoulos, 1992, "Existence of a photonic band gap in two dimensions," *Appl. Phys. Lett.* **61**, 495.

Meade, R.D., K.D. Brommer, A.M. Rappe, J.D. Joannopoulos, and O.L. Alerhand, 1993a, "Accurate theoretical analysis of photonic band gap materials," *Phys. Rev. B* **48**, 8434.

Meade, R.D., O.L. Alerhand, and J.D. Joannopoulos, 1993b, *Handbook of Photonic Band Gap Materials*, New York: JAMtex I.T.R.

Weissman, Z., and A. Hardy, 1993, "Modes of periodically segmented waveguides," *J. Lightwave Technol.* **11**, 1831-1838.

Yablonovitch, E., 1987, "Inhibited spontaneous emission in solid state physics and electronics," *Phys. Rev. Lett.* **58**, 2059.

Yablonovitch, E., and T.J. Gmitter, 1989, "Photonic band structures: the face-centered cubic case," *Phys. Rev. Lett.* **63**, 1950.

Yablonovitch, E., T.J. Gmitter, and K.M. Leung, 1991a, "Photonic band structures: the face-centered cubic case employing non-spherical atoms," *Phys. Rev. Lett.* **67**, 2295.

Yablonovitch, E., T.J. Gmitter, R.D. Meade, K.D. Brommer, A.M. Rappe, and J.D. Joannopoulos, 1991b, "Donor and acceptor modes in photonic band structure," *Phys. Rev. Lett.* **67**, 3380.

3

Acoustic, Elastodynamic, and Electromagnetic Wavefield Computation—A Structured Approach Based on Reciprocity

Adrianus T. de Hoop
Delft University of Technology
Delft, The Netherlands

Maarten V. de Hoop
Schlumberger Cambridge Research
Cambridge, England

> The reciprocity theorems for acoustic, elastodynamic, and electromagnetic wavefields in linear, time-invariant configurations show a common structure that can serve as a guideline for the development of computational methods for these wavefields. To this end, the wavefield reciprocity theorems are taken as points of departure. They are considered to describe the "interaction" between (a discretized version of) the actual wavefield in the configuration and a suitably chosen "computational state." The choice of the computational state determines which type of computational method results from the analysis. It is shown that finite-difference/finite-element methods and integral-equation/method-of-moment methods then do arise in a natural fashion. Time-domain methods are taken as a point of departure; the relationship with complex frequency-domain methods is indicated. In the total matrix of possibilities some schemes seem as yet to be underexplored.

INTRODUCTION

Acoustic, elastodynamic, and electromagnetic wavefields share a number of common features in their mathematical description. Their local, pointwise behavior in space-time is governed by a hyperbolic system of first-order partial differential equations that are representative for the physical phenomena involved on a local scale. When supplemented with initial conditions that relate the wave solution to its excitation mechanism and boundary conditions across interfaces where the coefficients in the system jump by finite amounts, the problem has a unique solution. The computational handling of the problem, however, often starts from a "weak" formulation, where the pointwise, or "strong," satisfaction of the equality signs in the equations is replaced by requirements on the equality of integrated or weighted versions of the differential equations. The resulting expressions have a counterpart in physics that is found in the pertaining reciprocity theorems of the Rayleigh (acoustic waves in fluids), Betti-Rayleigh (elastic waves in solids), or H.A. Lorentz (electromagnetic waves) types. This observation has led to the approach presented in the present paper, where the relevant reciprocity theorems are taken as points of departure. Through them, a computational scheme is conceptually taken to describe the interaction between the actual wavefield state to be computed and a suitably chosen "computational state" that is representative for the method at hand, just as in physics the reciprocity theorems describe the interaction between observing state and observed state, or quantify the reciprocity between transmitting and receiving properties of any device or system (transducer in acoustics and elastodynamics, antenna in electromagnetics, electromagnetic compatibility of interfering electromagnetic systems or devices). It is also believed that through this point of view one is guided to developing computational algorithms for each of the three types of wavefields in a manner that expresses the structures common to all of them. Background literature on reciprocity can be found in some papers by A.T. de Hoop (1987, 1988, 1989, 1990, 1991, 1992) and in a forthcoming book (de Hoop, 1995).

THE BASIC FIELD EQUATIONS

We consider linear acoustic, elastic, or electromagnetic wave motion in some subdomain D of three-dimensional Euclidean space $\Re^3$. The configuration in which the wave motion is considered to be present is assumed to be time invariant and linear in its physical behavior. The wave quantities involved are found to satisfy certain reciprocity properties which will be taken as the point of departure for our further considerations. Now, for the indicated type of configuration, there prove to be two kinds of reciprocity theorem: one of the time-convolution type, the other of the time-correlation type. Several operations on the wave quantities will occur throughout the paper. First, we shall introduce their notation.

Notation

Cartesian coordinates $x = \{x_1, x_2, x_3\}$ are used to specify position; t is the time coordinate. Differentiation with respect to x_p is denoted by ∂_p; ∂_t is a reserved symbol for differentiation with respect to t. The subscript notation for the vectorial and tensorial quantities occurring in the wave motion will be used whenever appropriate; the subscripts are to be assigned the values 1, 2, and 3.

The characteristic function of the domain D is denoted by χ_D and is given by

$$\chi_D(x) = \{1, 1/2, 0\} \quad \text{for} \quad x \in \{D, \partial D, D'\}, \tag{3.1}$$

where ∂D is the boundary of D, and D' is the complement of $D \cup \partial D$ in $\Re^3$.

Let $F = F(x, t)$ denote any space-time function. Then, the *time reversal* operator T is defined by

$$\mathsf{T}(F)(x, t) = F(x, -t). \tag{3.2}$$

It has the property

$$\partial_t \mathsf{T}(F) = -\mathsf{T}(\partial_t F). \tag{3.3}$$

Let $Q(x,t)$ denote another space-time function, then the *time convolution* $\mathsf{C}_t(F,Q)$ of F and Q is defined as

$$\mathsf{C}_t(F,Q)(x,t) = \int_{t'=-\infty}^{\infty} F(x,t')Q(x,t-t')dt'. \tag{3.4}$$

It has the properties
$$\mathsf{C}_t(F,Q) = \mathsf{C}_t(Q,F), \tag{3.5}$$
$$\mathsf{C}_t(\mathsf{T}(F),\mathsf{T}(Q)) = \mathsf{T}\mathsf{C}_t(F,Q), \tag{3.6}$$
$$\partial_t \mathsf{C}_t(F,Q) = \mathsf{C}_t(F,\partial_t Q) = \mathsf{C}_t(\partial_t F, Q). \tag{3.7}$$

The *time correlation* $\mathsf{R}_t(F,Q)$ of F and Q is defined as

$$\mathsf{R}_t(F,Q)(x,t) = \int_{t'=-\infty}^{\infty} F(x,t')Q(x,t'-t)dt'. \tag{3.8}$$

It has the properties
$$\mathsf{R}_t(F,Q) = C\big(F,\mathsf{T}(Q)\big), \tag{3.9}$$
$$\mathsf{R}_t(Q,F) = \mathsf{T}\mathsf{R}_t(F,Q), \tag{3.10}$$
$$\partial_t \mathsf{R}_t(F,Q) = -\mathsf{R}_t\big(F,\partial_t Q\big) = \mathsf{R}_t\big(\partial_t F, Q\big). \tag{3.11}$$

The First-order System of Wave Equations

Let the one-dimensional *field matrix* $F_P = F_P(x,t)$ of the wave motion be composed of the components of the two wavefield quantities whose inner product represents the area density of power flow (Poynting vector). Then, F_P satisfies a system of linear, first-order, partial differential equations of the general form

$$\left(D_{I,P} + M_{I,P}\partial_t\right)F_P = Q_I, \tag{3.12}$$

where uppercase Latin subscripts are used to denote the pertaining matrix elements and the summation convention for repeated subscripts applies. In (3.12), $D_{I,P}$ is a symmetrical, block off-diagonal *spatial differentiation operator matrix* that contains the operator ∂_p in a homogeneous linear fashion that is specific for each type of wave motion under consideration, $M_{I,P} = M_{I,P}(x)$ is the *medium matrix* that is representative for the physical properties of the (arbitrarily inhomogeneous, anisotropic) medium in which the waves propagate, and $Q_I = Q_I(x,t)$ is the *volume source density matrix* that is representative for the action of the volume sources that generate the wavefield.

The medium parameters are assumed to be piecewise continuous functions of position. Across a surface of discontinuity in medium properties, the parameters may jump by finite amounts. On the assumption that the interface is passive (i.e., free from surface sources) and that the wavefield quantities must remain bounded on either side of the interface, the wavefield must satisfy the boundary condition of the continuity type

$$N_{I,P} F_P \text{ is continuous across sourcefree interface,} \qquad (3.13)$$

where $N_{I,P}$ is the *unit normal operator* at the interface that arises from replacing ∂_p in $D_{I,P}$ by n_p where n_p is the unit vector along the normal to the interface.

For *acoustic waves in fluids*,

- $F_P = [p, v_1, v_2, v_3]^T$,

where p = acoustic pressure and v_r = particle velocity, and

- $Q_I = [q, f_1, f_2, f_3]^T$,

where q = volume source density of injection rate and f_k = volume source density of force. For *elastic waves in solids*,

- $F_P = [v_1, v_2, v_3, -\tau_{1,1}, -\tau_{1,2}, -\tau_{1,3}, -\tau_{2,1}, -\tau_{2,2}, -\tau_{2,3}, -\tau_{3,1}, -\tau_{3,2}, -\tau_{3,3}]^T$,

where v_r = particle velocity and $\tau_{p,q}$ = dynamic stress, and

- $Q_I = [f_1, f_2, f_3, h_{1,1}, h_{1,2}, h_{1,3}, h_{2,1}, h_{2,2}, h_{2,3}, h_{3,1}, h_{3,2}, h_{3,3}]^T$,

where f_k = volume source density of force and $h_{i,j}$ = volume source density of deformation rate. For *electromagnetic waves*,

- $F_p = [E_1, E_2, E_3, 0, -H_3, H_2, H_3, 0, -H_1, -H_2, H_1, 0]^T$,

where E_r = electric field strength and H_p = magnetic field strength, and

- $Q_I = [-J_1, -J_2, -J_3, 0, K_3/2, -K_2/2, -K_3/2, 0, K_1/2, K_2/2, -K_1/2, 0]^T$,

where J_k = volume source density of electric current and K_j = volume source density of magnetic current. The structures of $D_{I,P}$ and $M_{I,P}$ for the three types of wave fields are given in Appendix 3A.

The Reciprocity Concatenation Matrices

In the reciprocity theorems to be discussed below, two diagonal matrices $\delta^-_{Q,I}$ and $\delta^+_{Q,I}$ occur that concatenate, out of the wavefields pertaining to two admissible states, their interaction. For *acoustic waves in fluids* the diagonal matrix $\delta^-_{Q,I}$ is given by

- $\delta^-_{Q,I} = \text{diag}[1,-1,-1,-1],$

for *elastic waves in solids* by

- $\delta^-_{Q,I} = diag[1,1,1,-1,-1,-1,-1,-1,-1,-1,-1,-1],$

and for *electromagnetic waves* by

- $\delta^-_{Q,I} = diag[1,1,1,-1,-1,-1,-1,-1,-1,-1,-1,-1].$

The diagonal matrix $\delta^+_{Q,I}$ is just the unit matrix:

- $\delta^+_{Q,I} = 1$ for $Q = I$ and $\delta^+_{Q,I} = 0$ for $Q \neq I$.

For the reciprocity theorem of the time-convolution type to hold, a necessary and sufficient condition proves to be

$$\delta^-_{Q,I} D_{I,P} = -\delta^-_{P,J} D_{J,Q}. \tag{3.14}$$

This condition requires that the block-diagonal part of $D_{I,P}$ be anti-symmetric and that its block off-diagonal part be symmetric. For the reciprocity theorem of the time-correlation type to hold, a necessary and sufficient condition proves to be

$$\delta^+_{Q,I} D_{I,P} = \delta^+_{P,J} D_{J,Q}. \tag{3.15}$$

This condition requires that $D_{I,P}$ be symmetric. The two conditions are independent, but if they are satisfied simultaneously, $D_{I,P}$ is a symmetric, block off-diagonal matrix operator. For the three types of wave motion considered in this paper, this is indeed the case. It is therefore conjectured that the indicated structure of the spatial differential matrix operator could prove to be fundamental in order that a system of first-order partial differential equations be representative for a physical wave motion.

It is noted that the medium matrix $M_{I,P}$ is not subjected to any restriction of this kind.

Point-source Solutions; Green's Tensors

In view of the linearity of the wave motion, the principle of superposition ensures that the wavefield F_P that is generated by the volume source distribution Q_I can be written as the superposition of point-source contributions through the use of a Green's tensor. The latter is a solution of the system of differential equations

$$\left(D_{I,P} + M_{I,P}\partial_t\right)G_{P,I'} = \delta^+_{I,I'}\delta(x-x',t-t'), \tag{3.16}$$

where $\delta^+_{I,I'}$ is the unit matrix and $\delta(x-x',t-t')$ is four-dimensional Dirac delta distribution operative at $\{x,t\}=\{x',t'\}$. In view of the time invariance of the medium, the Green's tensor depends on t and t' only through the difference $t-t'$, i.e., $G_{P,I'} = G_{P,I'}(x,x',t,t') = G_{P,I'}(x,x',t-t')$. The Green's tensor plays an important role in the embedding formulations of the wavefield problem.

THE RECIPROCITY THEOREMS

In the wavefield reciprocity theorems certain *interaction quantities* are considered that are representative for the interaction between two admissible states of the pertaining wavefield in a given (proper or improper) subdomain D of $\mathfrak{R}^3$. Each of the two states applies to its own medium and has its own volume source distribution. Let the superscripts A and Z indicate the two states, then the wavefields in the two states are related to their respective sources via

$$\left(D_{I,P} + M^A_{I,P}\partial_t\right)F^A_P = Q^A_I, \tag{3.17}$$

$$\left(D_{J,Q} + M^Z_{J,Q}\partial_t\right)F^Z_Q = Q^Z_J. \tag{3.18}$$

Further, for each of the two states the boundary condition of the continuity type

$$N_{I,P}F^A_P \text{ is continuous across sourcefree interface,} \tag{3.19}$$

$$N_{J,Q}F^Z_Q \text{ is continuous across sourcefree interface} \tag{3.20}$$

holds.

The Reciprocity Theorem of the Time-convolution Type

The local interaction quantity to be considered in the reciprocity theorem of the time-convolution type is $\delta^-_{Q,I}R_t\left(D_{I,P}F^A_P, F^Z_Q\right) + \delta^+_{P,J}R_t\left(F^A_P, D_{J,Q}F^Z_Q\right) = \delta^+_{Q,I}D_{I,P}R_t\left(F^A_P, F^Z_Q\right)$, where the property of (3.14) has been used. With the aid of (3.17) and (3.18) this expression is rewritten as

$$\begin{aligned}&\delta^-_{Q,I}D_{I,P}C_t(F^A_P,F^Z_Q)+(\delta^-_{Q,I}M^A_{I,P}-\delta^-_{P,J}M^Z_{J,Q})\partial_t C_t(F^A_P,F^Z_Q)\\&=\delta^-_{Q,I}C_t(Q^A_I,F^Z_Q)-\delta^-_{P,J}C_t(F^A_P,Q^Z_J).\end{aligned} \quad (3.21)$$

Equation (3.21) is the local form of the reciprocity theorem of the time-convolution type. The global form, for the domain D, of this theorem follows upon integrating (3.21) over the domain D and applying Gauss' integral theorem to the first term on the left-handed side over each subdomain of D where the field quantities are continuously differentiable. Adding the contributions from these subdomains, the contributions from sourcefree interfaces of discontinuity in medium properties in the interior of D cancel in view of the boundary conditions given in (3.19) and (3.20) and only a surface integral over the boundary ∂D of D remains. The result is

$$\begin{aligned}&\int_{\partial D}\delta^-_{Q,I}N_{I,P}C_t(F^A_P,F^Z_Q)dA(\boldsymbol{x})\\&+\int_D(\delta^-_{Q,I}M^A_{I,P}-\delta^-_{P,J}M^Z_{J,Q})\partial_t C_t(F^A_P,F^Z_Q)dV(\boldsymbol{x})\\&=\int_D\left[\delta^-_{Q,I}C_t(Q^A_I,F^Z_Q)-\delta^-_{P,J}C_t(F^A_P,Q^Z_J)\right]dV(\boldsymbol{x}).\end{aligned} \quad (3.22)$$

Equation (3.22) is the global form, for the domain D, of the reciprocity theorem of the time-convolution type.

The terms in (3.21) and (3.22) containing the medium matrices define the contrast-in-medium contributions to the time-convolution interaction of the two states. They vanish at those positions where $\delta^-_{Q,I}M^A_{I,P}-\delta^-_{P,J}M^Z_{J,Q}=0$. If this condition holds, the media in the two states are denoted as each other's *adjoints*. If the condition holds for one and the same medium, such a medium is denoted as *self-adjoint*. An isotropic medium is always self-adjoint. The terms containing the volume source densities yield the contribution from the volume sources to the interaction of the two states. They vanish at sourcefree positions.

In a number of applications (3.22) will be applied to the entire $\mathfrak{R}^3$. Then, outside some sphere $S(O,\Delta_0)$ with radius Δ_0 and center at the origin O of the chosen reference frame, the media in the two states will be assumed to be the same and homogenous as well as isotropic. For such a medium, the tensor Green's function is known analytically and in particular the causal and anti-causal source-type integral representations are known analytically. For the application of (3.22) to the entire $\mathfrak{R}^3$, the theorem will be first applied to a sphere $S(O,\Delta)$ of radius Δ and center at the origin O of the chosen reference frame and the limit $\Delta\to\infty$ will be taken. If, now, in both states the wavefields are causally related to the action of their volume source distributions (assumed to have bounded supports), the integral over $S(O,\Delta)$ vanishes as $\Delta\to\infty$. However, if one of the two states is causally related to the action of its volume sources and the other anti-causally, the integral over $S(O,\Delta)$ does not vanish as $\Delta\to\infty$, but has a constant value for sufficiently large values of Δ.

The Reciprocity Theorem of the Time-correlation Type

The local interaction quantity to be considered in the reciprocity theorem of the time-correlation type is $\delta^+_{Q,I} \mathrm{R}_t \left(D_{I,P} F^A_P, F^Z_Q \right) + \delta^+_{P,J} \mathrm{R}_t \left(F^A_P D_{j,Q} F^Z_Q \right) = \delta^+_{Q,I} D_{I,P} \mathrm{R}_t \left(F^A_P, F^Z_Q \right)$, where the property of (3.15) has been used. With the aid of (3.11), (3.17), and (3.18) this expression is rewritten as

$$\delta^+_{Q,I} D_{I,P} \mathrm{R}_t \left(F^A_P, F^Z_Q \right) + \left(\delta^+_{Q,I} M^A_{I,P} - \delta^+_{P,J} M^Z_{J,Q} \right) \partial_t \mathrm{R}_t \left(F^A_P, F^Z_Q \right)$$
$$= \delta^+_{Q,I} \mathrm{R}_t \left(Q^A_I, F^Z_Q \right) + \delta^+_{P,J} \mathrm{R}_t \left(F^A_P, Q^Z_J \right). \tag{3.23}$$

Equation (3.23) is the local form of the reciprocity theorem of the time-correlation type. The global form, for the domain D, of this theorem follows upon integrating (3.23) over the domain D and applying Gauss' integral theorem to the first term on the left-hand side over each subdomain of D where the field quantities are continuously differentiable. Adding the contributions from these subdomains, the contributions from interfaces of discontinuity in medium properties in the interior of D cancel in view of the boundary conditions given in (3.19) and (3.20) and only a surface integral over the boundary ∂D of D remains. The result is

$$\int_{\partial D} \delta^+_{Q,I} N_{I,P} \mathrm{R}_t \left(F^A_P, F^Z_Q \right) dA(x)$$
$$+ \int_D \left(\delta^+_{Q,I} M^A_{I,P} - \delta^+_{P,J} M^Z_{J,Q} \right) \partial_t \mathrm{R}_t \left(F^A_P, F^Z_Q \right) dV(x)$$
$$= \int_D \left[\delta^+_{Q,I} \mathrm{R}_t \left(Q^A_I, F^Z_Q \right) + \delta^+_{P,J} \mathrm{R}_t \left(F^A_P, Q^Z_J \right) \right] dV(x). \tag{3.24}$$

Equation (3.24) is the global form, for the domain D, of the reciprocity theorem of the time-correlation type.

The terms in (3.23) and (3.24) containing the medium matrices define the contrast-in-medium contributions to the time-correlation interaction of the two states. They vanish at those positions where $\delta^+_{Q,I} M^A_{I,P} - \delta^+_{P,J} M^Z_{J,Q} = 0$. If this condition holds, the media in the two states are denoted as each other's *time reverse adjoints*. (The "time-reverse" is reminiscent of the fact that "adjoint" applies to the reciprocity theorem of the time-convolution type and that correlation can be considered as a compound operation consisting of convolution and time reversal.) If the condition holds for one and the same medium, such a medium is denoted as *time-reverse self-adjoint*. For an isotropic medium, the medium matrix is diagonal; an instantaneously reacting isotropic medium is therefore always time-reverse self-adjoint. The terms containing the volume source densities yield the contribution from the volume sources to the interaction of the two states. They vanish at source-free positions.

In a number of applications (3.24) will be applied to the entire $\Re^3$. Then, outside some sphere $S(O, \Delta_0)$ with radius Δ_0 and center at the origin O of the chosen reference frame, the media in the two states will be assumed to be the same and homogenous as well as isotropic. For such a medium, the tensor Green's function is known analytically and in particular the causal and anti-causal source-type integral representations are known analytically. For the application of (3.24) to the entire $\Re^3$, the theorem will be first applied to a sphere $S(O, \Delta)$ of radius Δ and center at the origin O of the chosen reference frame and the limit $\Delta \to \infty$ will be taken. If, now, in State A the wavefield is causally related to the action of its volume source distributions, and in State Z the wavefield is anti-causally related to the action of its volume source distributions (the volume source distributions being assumed to have bounded supports), the integral over $S(O,\Delta)$ vanishes as $\Delta \to \infty$. However, if both states are causally related to the action of their volume sources, the integral over $S(O,\Delta)$ does not vanish as $\Delta \to \infty$, but has a constant value for sufficiently large values of Δ.

For the choice State A = State Z and zero correlation time shift (i.e., $t = 0$), (3.23) reduces to the local energy balance for the wavefield and (3.24) to the global energy balance for the domain D, provided that $M_{Q,P} = M_{P,Q}$. This implies that for the energy considerations pertaining to a physical wavefield to hold, the medium matrix must be symmetric. In that case, also the quantity $(1/2)M_{P,Q}F_P F_Q$ (whose time derivative occurs in equations (3.23) and (3.24)) should represent the volume density of stored energy. For the latter, the symmetric medium matrix should, in addition, on physical grounds be positive definite.

Reciprocity Property of the Causal Green's Tensor

Equation (3.22) leads to a reciprocity property of the Green's tensor. Let $F_P^{A;G} = F_P^{A;G}(x,x',t)$ be the causal wavefield in Medium A generated by the point source $Q_I^A = a_I^A \delta(x - x',t)$ operative at $\{x,t\} = \{x',0\}$. Then (cf. (3.16)) $F_P^{A;G} = G_{P,I}^A(x,x',t)a_I^A$. Let, similarly, $F_Q^{Z;G} = F_Q^{Z;G}(x,x'',t)$ be the causal wavefield in Medium Z generated by the point source $Q_J^Z = a_J^Z \delta(x - x'',t)$ operative at $\{x,t\} = \{x'',0\}$. Then $F_Q^{Z;G} = G_{Q,J}^Z(x,x'',t)a_J^Z$. Take the media in the two states as each other's adjoints, i.e., $\delta_{Q,I}^- M_{I,P}^A = \delta_{P,I}^- M_{J,Q}^Z$, and apply (3.22) to the entire $\mathfrak{R}^3$. In this application, the contrast-in-media term and the contribution from the "sphere at infinity" vanish. The result is

$$\delta_{Q,I}^- G_{Q,J}^Z(x',x'',t)a_J^Z a_I^A = \delta_{P,J}^- G_{P,I}^A(x'',x',t)a_I^A a_J^Z \quad \text{for } x' \neq x''. \tag{3.25}$$

Since (3.25) has to hold for arbitrary values of a_I^A and a_J^Z, we end up with

$$\delta_{Q,I}^- G_{Q,J}^Z(x',x'',t) = \delta_{P,J}^- G_{P,I}^A(x'',x',t) \quad \text{for } x' \neq x''. \tag{3.26}$$

Equation (3.26) is the reciprocity relation for the causal Green's tensor.

EMBEDDING PROCEDURE, CONTRAST-PROPERTIES FORMULATION

On many occasions the wavefield computation in an entire configuration is beyond the capabilities because of the storage capacity required and the computation time involved. In that case, it is standard practice to select a target region D_{con} of bounded support in which a detailed computation is to be carried out, while the medium in the remaining part of the configuration (the *embedding*) is chosen to be so simple that the wave motion in it can be determined with the aid of analytical methods. In particular, this applies to scattering problems and to geophysical modeling, where the support of the model configuration is often taken to be the entire $\Re^3$. Examples of such simple embeddings are a homogeneous, isotropic medium (chosen for most scattering configurations) and a medium consisting of parallel layers of homogeneous, isotropic material (chosen in most geophysical applications). In these cases, time Laplace and spatial Fourier transform techniques provide the analytical tools to determine the wave motion or, in fact, the relevant Green's tensor. Once the embedding has been chosen, the problem of computing the wavefield in D_{con} can be formulated as a *contrast problem*. For this, we proceed as follows.

The State A is introduced consisting of the *A*ctual wavefield F_P^A, the actual sources Q_I^A that excite it, and the actual medium $M_{I,P}^A$ in which the propagation takes place. Next, we introduce a State B consisting of the wavefield F_P^B that the actual sources $Q_I^B = Q_I^A$ would generate in the medium $M_{I,P}^B$ of the em*B*edding. Denoting the Green's tensor of the embedding by $G_{P,I}^B = G_{P,I}^B(x, x', t)$, the latter wavefield is expressible as

$$F_P^B(x,t) = \int_{D_{scr}} C_t \left[G_{P,I}^B(x, x', \cdot), Q_I^B(x', \cdot) \right] dV(x') \quad \text{for } x \in \Re^3, \tag{3.27}$$

where D_{scr} is the support of the sources that generate the wavefield in the actual configuration. From the corresponding wavefield equations it then follows that

$$D_{I,P}(F_P^A - F_P^B) + M_{I,P}^A \partial_t F_P^A - M_{I,P}^B \partial_t F_P^B = 0. \tag{3.28}$$

This equation can be rewritten in two ways as a wavefield equation for the *C*ontrast state, to be denoted by the superscript C, in which the contrast wavefield is

$$F_P^C = F_P^A - F_P^B. \tag{3.29}$$

In one of them, the medium properties in the wave operator on the left-hand side are taken to be the ones of the actual medium; this is typically done in the combination of the embedding technique with finite-element or finite-difference modeling. In the other, the medium properties in the wave operator on the left-hand side are taken to be the ones of the embedding; this is typically done in the integral-equation or method-of-moments modeling. Both ways lead to a contrast-source formulation. The expressions for the two cases are given below.

Contrast Formulation for Finite-element/Finite-difference Modeling

For the contrast formulation for finite-element or finite-difference modeling, (3.28) is, in combination with (3.29), rewritten as

$$D_{I,P} F_P^C + M_{I,P}^C \partial_t F_P^C = Q_I^C, \tag{3.30}$$

with

$$M_{I,P}^C = M_{I,P}^A \qquad (3.31)$$

and

$$Q_I^C = -\left(M_{I,P}^A - M_{I,P}^B\right)\partial_t F_P^B. \qquad (3.32)$$

Note that in this contrast formulation, the contrast source density Q_I^C is known.

Contrast Formulation for Integral-equation/ Method-of-Moments Modeling

For the contrast formulation for integral-equation or methods-of-moments modeling, (3.28) is, in combination with (3.29), rewritten as

$$D_{I,P} F_P^C + M_{I,P}^C \partial_t F_P^C = Q_I^C, \qquad (3.33)$$

with

$$M_{I,P}^C = M_{I,P}^B \qquad (3.34)$$

and

$$Q_I^C = -\left(M_{I,P}^A - M_{I,P}^B\right)\partial_t F_P^A. \qquad (3.35)$$

Note that in this contrast formulation, the contrast source density Q_I^C is unknown, since F_P^A is unknown.

In the two sections following, it will be indicated how these different states are used in the reciprocity theorems of the previous section to lead to computational schemes for the evaluation of the wavefields.

FINITE-ELEMENT/FINITE-DIFFERENCE MODELING

In finite-element/finite-difference modeling over an entire configuration occupying the bounded domain D, the wavefield to be computed is the total wavefield. The latter is approximated by an expansion of the type

$$F_P^A(x,t) \simeq \sum_{n=1}^{N} \alpha_{[n]} \Phi_P^{[n]}(x,t) \quad \text{for } x \in D, \qquad (3.36)$$

where $\left\{\Phi_P^{[n]}; x \in D, t \in \Re, n = 1,\ldots,N\right\}$ is an appropriate sequence of known, linearly independent expansion functions with D as their supports, and $\left\{\alpha_{[n]}; n = 1,\ldots,N\right\}$ is the sequence of expansion coefficients to be computed. In typical finite-element/finite-difference modeling the support of each expansion function is an elementary subdomain of the (discretized) version of D (usually a simplex or a complex). Further, boundary conditions as needed for the uniqueness of the solution in D are prescribed on ∂D. Next, a sequence of "computational" states, denoted by the superscript Z, is selected, for which

$$F_Q^Z(x,t) \in \left\{ \Psi_Q^{[m]}(x,t); x \in D, t \in \Re, m = 1, \ldots, N \right\}, \tag{3.37}$$

where the right-hand side is a sequence of known, linearly independent weighting functions with D as their supports. Finally, we take

$$M_{J,Q}^Z = 0 \tag{3.38}$$

and hence

$$Q_J^Z = D_{J,Q} F_Q^Z. \tag{3.39}$$

Application of the earlier reciprocity theorems to the State A and the sequence of States Z leads to a system of linear algebraic equations in the expansion coefficients.

Embedding Procedure

In case an embedding procedure is applied, the wavefield to be computed is the contrast wavefield. The latter is approximated by an expansion of the type

$$F_P^C(x,t) \simeq \sum_{n=1}^{N} \gamma_{[n]} \Phi_P^{[n]}(x,t) \quad \text{for } x \in D_{con}, \tag{3.40}$$

where $\left\{ \Phi_P^{[n]}; x \in D_{con}, t \in \Re, n = 1, \ldots, N \right\}$ is an appropriate sequence of known, linearly independent expansion functions with D_{con} as their supports and $\left\{ \gamma_{[n]}; n = 1, \ldots, N \right\}$ is now the sequence of expansion coefficients to be computed. In typical finite-element/finite-difference modeling the support of each expansion function is an elementary subdomain of the (discretized) version of D_{con} (usually a simplex or a complex). Further, "absorbing boundary conditions" as needed for the uniqueness of the solution in D_{con} are prescribed on ∂D. These should model the radiation of the contrast wavefield into the passive embedding. Next, a sequence of "computational" states, denoted by the superscript Z, is selected, for which

$$F_Q^Z(x,t) \in \left\{ \Psi_Q^{[m]}(x,t); x \in D_{con}, t \in \Re, m = 1, \ldots, N \right\}, \tag{3.41}$$

where the right-hand side is a sequence of known, linearly independent weighting functions with D_{con} as their supports. Finally, we take

$$M_{J,Q}^Z = 0, \tag{3.42}$$

and hence

$$Q_J^Z = D_{J,Q} F_Q^Z. \tag{3.43}$$

Application of the reciprocity theorems to the State A and the sequence of States Z again leads to a system of linear algebraic equations in the expansion coefficients.

In finite-element/finite-difference modeling, the expansion and weighting functions are standardly taken to be polynomials in the time variable and the spatial variables. Their vector and tensor components in space can be organized such that the continuity conditions across an interface between two different media are taken into account automatically, while leaving those components that are not necessarily continuous free to jump by finite amounts. Such a procedure can be carried out consistently if the *simplex* is taken as the elementary subdomain of the discretized configuration and a consistent linear approximation within each simplex is used. Thus, the notions of "face element" and "edge element" for arbitrary vectors and tensors have been introduced. For literature on the subject, see Mur and de Hoop (1985) and Mur (1990, 1991, 1993) for the application to electromagnetic fields and Stam and de Hoop (1988, 1989, 1990) for the application to elastodynamic wavefields.

METHOD-OF-MOMENTS MODELING

The integral-equation or method-of-moments modeling is invariably based on an embedding procedure. As a consequence of this, the wavefield to be computed is the contrast wavefield. The latter is approximated by

$$F_P^C(x,t) \simeq \sum_{n=1}^N \gamma_{[n]} \Phi_P^{[n]}(x,t) \quad \text{for } x \in D_{con}, \tag{3.44}$$

where $\left\{\Phi_P^{[n]}(x,t); x \in D_{con}, t \in \Re, n = 1,\ldots,N\right\}$ is an appropriate sequence of known, linearly independent expansion functions with D_{con} as their supports, and $\left\{\gamma_{[n]}; n = 1,\ldots,N\right\}$ is the sequence of expansion coefficients to be computed. The contrast source density is written as

$$Q_I^C = -\left(M_{I,P}^A - M_{I,P}^B\right)\partial_t F_P^B - \left(M_{I,P}^A - M_{I,P}^B\right)\partial_t F_P^C, \tag{3.45}$$

in which the first term on the right-hand side is known and the second term on the right-hand side is unknown. Next, a "computational" state, denoted by the superscript Z, is selected, for which

$$Q_J^Z(x,t) \in \left\{\Psi_J^{[m]}(x,t); x \in D_{con}, t \in \Re, m = 1,\ldots,N\right\} \text{ for } x \in D \tag{3.46}$$

$$\delta_{Q,I}^- M_{I,P}^B = \delta_{P,J}^- M_{J,Q}^Z \tag{3.47}$$

and

$$F_Q^{Z[m]}(x,t) = \int_{D_{con}} C_t\left[G_{Q,J}^Z(x,x',\cdot)\Psi_J^{[m]}(x',\cdot)\right]dV(x') \quad \text{for } x \in \Re^3 \tag{3.48}$$

Substitution in the earlier reciprocity theorems then leads to a system of linear, algebraic equations in the expansions coefficients.

COMPLEX FREQUENCY-DOMAIN MODELING OF WAVE PROBLEMS

Although the real, physical wave phenomena take place in space-time, it can under certain circumstances be advantageous to parametrize the problem in the coordinates in which shift invariance in the configuration occurs. Since we have assumed that our configurations are, apart from linear, time-invariant in their physical behavior, such a procedure certainly applies to the time coordinate. Moreover, in this coordinate the principle of causality applies. In view of these two aspects, the time Laplace transformation performs the appropriate parametrization in the time coordinate. For any causal, bounded function $Q_I = Q_I(x,t)$ with temporal support $\{t \in \Re; t > t_0\}$ this transformation is

$$\hat{Q}_I(x,s) = \int_{t=t_0}^{\infty} \exp(-st) Q_I(x,t) dt \quad \text{for } s \in \mathbf{C}, \text{Re}(s) > 0. \tag{3.49}$$

Here, s is the time Laplace transform parameter or complex frequency. The time Laplace transformation has the following properties:

$$\hat{\partial}_t = s, \tag{3.50}$$

$$\hat{T}(Q_I) = \hat{Q}_I(x,-s), \tag{3.51}$$

$$\hat{C}(F_P Q_I) = \hat{F}_P(x,s)\hat{Q}_i(x,s), \tag{3.52}$$

$$\hat{R}_t(F_P Q_I) = \hat{F}_P(x,s)\hat{Q}_i(x,-s). \tag{3.53}$$

In view of Lerch's theorem (Widder, 1946), the correspondence between $\{\hat{Q}_I(x,s_n); s_n = s_0 + nh, s_0 \in \Re, s_0 > 0, h \in \Re, h > 0, n = 0, 1, 2, ...)\}$ and $Q_I(x,t)$ for $t > t_0$ is unique. Using these properties, the space-time wave motion can be recovered after having solved a sequence of space problems with appropriate values of the time Laplace transform parameter. For recent results in this direction, see Lee et al. (1994).

CONCLUDING REMARKS

In the preceding two sections it has been indicated how finite-difference/finite-element methods and integral-equation/method-of-moments methods for the computation of wavefields can be envisaged to arise from the time-convolution and time-correlation type reciprocity theorems pertaining to these wavefields. This does not mean that all possibilities in this respect have found application as yet. Apart from the different choices that can still be made in the selection of the sequences of expansion and weighting functions, it also happens that, for example, the application of the reciprocity theorem of the time-correlation type to the integral-equation modeling of *forward* wave scatttering problems has, as far as the present authors are aware, not been pursued yet. This is the more remarkable since this theorem finds prime application in the modeling of *inverse* scattering problems and has been extensively used in this realm. Whether or not the missing applications in a total matrix of possibilities might lead to better algorithms remains to be investigated.

ACKNOWLEDGMENT

The first author (A.T. de Hoop) acknowledges with gratitude the financial support from the Stichting Fund for Science, Technology, and Research (a companion organization to the Schlumberger Foundation in the United States) for carrying out the research presented in this paper.

REFERENCES

de Hoop, A.T., 1987, "Time-domain reciprocity theorems for electromagnetic fields in dispersive media," *Radio Science* **22**(7), 1171-1178.

de Hoop, A.T., 1988, "Time-domain reciprocity theorems for acoustic wave fields in fluids with relaxation," *The Journal of the Acoustical Society of America* **84**(5), 1877-1882.

de Hoop, A.T., 1989, "Reciprocity theorems for acoustic wave fields in fluid/solid configurations," *The Journal of the Acoustical Society of America* **87**(5), 1932-1937.

de Hoop, A.T., 1990, "Reciprocity, discretization, and the numerical solution of elastodynamic propagation and scattering problems." In: *Elastic Waves and Ultrasonic Nondestructive Evaluation*, S.K. Datta, J.D. Achenbach, and Y.S. Rajapakse (eds.), Amsterdam: Elsevier Science Publishers, 87-92.

de Hoop, A.T., 1991, "Reciprocity, discretization, and the numerical solution of direct and inverse electromagnetic radiation and scattering problems," *Proceedings of the IEEE* **79**(10), 1421-1430.

de Hoop, A.T., 1992, "Reciprocity, causality, and Huygens' principle in electromagnetic wave theory." In: *Huygens' Principle 1690-1990: Theory and Applications*, H. Blok, H.A. Ferwerda, and H.K. Kulken (eds.), Amsterdam: Elsevier Science Publishers, 171-192.

de Hoop, A.T., 1995, *Handbook of Radiation and Scattering of Waves*, London: Academic Press.

Lee, K.H., G.Q. Xie, T.M. Habashy, and C. Torres-Verdin, 1994, "Wavefield transform of electromagnetic fields," *Society of Exploration Geophysicists International Exposition and 64th Annual Meeting*, Los Angeles, October 23-28, 1994, Expanded Abstracts, pp. 633-635.

Mur, G., 1990, "A mixed finite-element method for computing three-dimensional time-domain electromagnetic fields in strongly inhomogeneous media," *IEEE Transactions on Magnetics* **MAG-26**(2), 674-677.

Mur, G., 1991, "Finite-element modeling of three-dimensional electromagnetic fields in inhomogeneous media," *Radio Science* **26**(1), 275-280.

Mur, G., 1993, "The finite-element modeling of three-dimensional electromagnetic fields using edge and nodal elements," *IEEE Transactions on Antennas and Propagation* **AP-41**(7), 948-953.

Mur, G., and A.T. de Hoop, 1985, "A finite-element method for computing three-dimensional electromagnetic fields in inhomogeneous media," *IEEE Transactions on Magnetics* **MAG-21**(6), 2188-2191.

Stam, H.J., and A.T. de Hoop, 1988, "Time-domain reciprocity theorems for elastodynamic wave fields in solids with relaxation and their application to inverse problems," *Wave Motion* **10**, 479-489.

Stam, H.J., and A.T. de Hoop, 1989, "A space-time finite-element method for the computation of three-dimensional elastodynamic wave fields (theory)." In: *Elastic Wave Propagation*, M.F. McCarthy and M.A. Hayes (eds.), Amsterdam: Elsevier Science Publishers, 483-488.

Stam, H.J., and A.T. de Hoop, 1990, "Theoretical considerations on a finite-element method for the computation of three-dimensional space-time elastodynamic wave fields," *Wave Motion* **12**, 67-80.

Widder, D.V., 1946, *The Laplace Transform*, Princeton, N.J.: Princeton University Press, 63-65.

APPENDIX 3A. STRUCTURE OF THE SPATIAL DIFFERENTIAL OPERATOR

In this appendix the structures of the spatial differential operator and the medium matrix in the system of (3.12) for acoustic waves in fluids, elastic waves in solids, and electromagnetic waves are given.

Acoustic Waves in Fluids

For acoustic waves in fluids, the spatial differential operator in the system of (3.12) has the following form:

$$[D] = \begin{bmatrix} 0 & \partial_1 & \partial_2 & \partial_3 \\ \partial_1 & 0 & 0 & 0 \\ \partial_2 & 0 & 0 & 0 \\ \partial_3 & 0 & 0 & 0 \end{bmatrix} \qquad (3.A1)$$

The medium matrix is given by

$$[M] = \begin{bmatrix} \kappa & 0 \\ 0 & \rho_{k,r} \end{bmatrix}, \qquad (3.A2)$$

where k is the compressibility and $\rho_{k,r}$ is the volume density of (inertial) mass.

Elastic Waves in Solids

For elastic waves in solids, the spatial differential operator in the system of (3.12) has the following form:

$$[D] = \frac{1}{2}\left([D^{row/col}] + [D^{diag}]\right), \qquad (3.A3)$$

in which

$$[D^{row/col}] = \begin{bmatrix} \begin{bmatrix} 0 & 0 & 0 \\ 0 & 0 & 0 \\ 0 & 0 & 0 \end{bmatrix} & \begin{bmatrix} \partial_1 & \partial_2 & \partial_3 \\ 0 & 0 & 0 \\ 0 & 0 & 0 \end{bmatrix} & \begin{bmatrix} 0 & 0 & 0 \\ \partial_1 & \partial_2 & \partial_3 \\ 0 & 0 & 0 \end{bmatrix} & \begin{bmatrix} 0 & 0 & 0 \\ 0 & 0 & 0 \\ \partial_1 & \partial_2 & \partial_3 \end{bmatrix} \\ \begin{bmatrix} \partial_1 & 0 & 0 \\ \partial_2 & 0 & 0 \\ \partial_3 & 0 & 0 \end{bmatrix} & \begin{bmatrix} 0 & 0 & 0 \\ 0 & 0 & 0 \\ 0 & 0 & 0 \end{bmatrix} & \begin{bmatrix} 0 & 0 & 0 \\ 0 & 0 & 0 \\ 0 & 0 & 0 \end{bmatrix} & \begin{bmatrix} 0 & 0 & 0 \\ 0 & 0 & 0 \\ 0 & 0 & 0 \end{bmatrix} \\ \begin{bmatrix} 0 & \partial_1 & 0 \\ 0 & \partial_2 & 0 \\ 0 & \partial_3 & 0 \end{bmatrix} & \begin{bmatrix} 0 & 0 & 0 \\ 0 & 0 & 0 \\ 0 & 0 & 0 \end{bmatrix} & \begin{bmatrix} 0 & 0 & 0 \\ 0 & 0 & 0 \\ 0 & 0 & 0 \end{bmatrix} & \begin{bmatrix} 0 & 0 & 0 \\ 0 & 0 & 0 \\ 0 & 0 & 0 \end{bmatrix} \\ \begin{bmatrix} 0 & 0 & \partial_1 \\ 0 & 0 & \partial_2 \\ 0 & 0 & \partial_3 \end{bmatrix} & \begin{bmatrix} 0 & 0 & 0 \\ 0 & 0 & 0 \\ 0 & 0 & 0 \end{bmatrix} & \begin{bmatrix} 0 & 0 & 0 \\ 0 & 0 & 0 \\ 0 & 0 & 0 \end{bmatrix} & \begin{bmatrix} 0 & 0 & 0 \\ 0 & 0 & 0 \\ 0 & 0 & 0 \end{bmatrix} \end{bmatrix} \qquad (3.A4)$$

and

$$[D^{diag}] = \begin{bmatrix} \begin{bmatrix} 0 & 0 & 0 \\ 0 & 0 & 0 \\ 0 & 0 & 0 \end{bmatrix} & \begin{bmatrix} \partial_1 & 0 & 0 \\ 0 & \partial_1 & 0 \\ 0 & 0 & \partial_1 \end{bmatrix} & \begin{bmatrix} \partial_2 & 0 & 0 \\ 0 & \partial_2 & 0 \\ 0 & 0 & \partial_2 \end{bmatrix} & \begin{bmatrix} \partial_3 & 0 & 0 \\ 0 & \partial_3 & 0 \\ 0 & 0 & \partial_3 \end{bmatrix} \\ \begin{bmatrix} \partial_1 & 0 & 0 \\ 0 & \partial_1 & 0 \\ 0 & 0 & \partial_1 \end{bmatrix} & \begin{bmatrix} 0 & 0 & 0 \\ 0 & 0 & 0 \\ 0 & 0 & 0 \end{bmatrix} & \begin{bmatrix} 0 & 0 & 0 \\ 0 & 0 & 0 \\ 0 & 0 & 0 \end{bmatrix} & \begin{bmatrix} 0 & 0 & 0 \\ 0 & 0 & 0 \\ 0 & 0 & 0 \end{bmatrix} \\ \begin{bmatrix} \partial_2 & 0 & 0 \\ 0 & \partial_2 & 0 \\ 0 & 0 & \partial_2 \end{bmatrix} & \begin{bmatrix} 0 & 0 & 0 \\ 0 & 0 & 0 \\ 0 & 0 & 0 \end{bmatrix} & \begin{bmatrix} 0 & 0 & 0 \\ 0 & 0 & 0 \\ 0 & 0 & 0 \end{bmatrix} & \begin{bmatrix} 0 & 0 & 0 \\ 0 & 0 & 0 \\ 0 & 0 & 0 \end{bmatrix} \\ \begin{bmatrix} \partial_3 & 0 & 0 \\ 0 & \partial_3 & 0 \\ 0 & 0 & \partial_3 \end{bmatrix} & \begin{bmatrix} 0 & 0 & 0 \\ 0 & 0 & 0 \\ 0 & 0 & 0 \end{bmatrix} & \begin{bmatrix} 0 & 0 & 0 \\ 0 & 0 & 0 \\ 0 & 0 & 0 \end{bmatrix} & \begin{bmatrix} 0 & 0 & 0 \\ 0 & 0 & 0 \\ 0 & 0 & 0 \end{bmatrix} \end{bmatrix}$$
(3.A5)

The medium matrix is given by

$$[M] = \begin{bmatrix} \rho_{k,r} & 0 \\ 0 & S_{i,j,p,q} \end{bmatrix},$$
(3.A6)

where $\rho_{k,r}$ is the volume density of (inertial) mass and $S_{i,j,p,q}$ is the compliance.

Electromagnetic Waves

For electromagnetic waves the spatial differential operator in the system of (3.12) is given by

$$[D] = \frac{1}{2}\left([D^{row/col}] - [D^{diag}]\right).$$
(3.A7)

The medium matrix is found to be

$$[M] = \begin{bmatrix} \varepsilon_{k,r} & 0 \\ 0 & \mu_{j,p}/2 \end{bmatrix},$$
(3.A8)

where $\varepsilon_{k,r}$ is the permittivity and $\mu_{j,p}$ is the permeability.

4

Numerical Modeling of the Interactions of Ultrafast Optical Pulses with Nonresonant and Resonant Materials and Structures

Richard W. Ziolkowski
Justin B. Judkins
University of Arizona

> We are developing full-wave, vector Maxwell equation solvers for use in studying the physics and engineering of linear and nonlinear integrated photonics devices and systems. These simulators and their applications are described. Particular emphasis is given to time-domain problems describing the interaction of ultrafast optical pulses with nonresonant and resonant optical materials and structures. These problems pose severe difficulties to numerical modeling because of the many time and length scales involved. The global structures that one deals with in integrated photonic systems are very large relative to their operating wavelengths, but their substructures are subwavelength in size. The corresponding computational procedures for studying these large-scale structures are computationally large and entail an extremely large number of degrees of freedom. We have developed hybrid simulators that incorporate the linear and nonlinear dynamics of the substructures and the overall response of the integrated system. We discuss one such simulator that couples Maxwell's equations and a Lorentz linear dispersion model with a near- to far-field transform capability to generate the global far-field patterns associated with the scattering of optical Gaussian beams from diffraction gratings coated with multiple, dielectric thin-film layers terminated with realistic metals. This simulator is being used to investigate a variety of wavelength-sized diffractive optic components. An extension of this model is given which includes both the Raman and the instantaneous Kerr nonlinear materials models to describe locally linear and nonlinear finite length corrugated optical waveguides for applications to grating-assisted couplers and beam steerers. We also describe another simulator under development which combines a multilevel atom materials model with our Maxwell's equations simulator to model self-induced transparency and laser gain-medium effects. This resonant systems model requires a careful marriage between a microscopic (quantum mechanical) materials model and the macroscopic Maxwell's equations solver. This simulator is being developed to model microcavity lasers and waveguide amplifiers. Examples from our efforts, which have civilian and military relevance, are given to illustrate the potential impact these classes of numerical simulators will have in the design and control of large-scale integrated photonics devices and systems.

INTRODUCTION

Laser pulses are continuing to be utilized in a variety of advanced commerical, civilian, and military systems. Their bandwidth and intensity have been increasing, to the point at which the materials they interact with no longer respond in a linear fashion. The material response is nonlinear and the properties of the materials depend on the shape of the pulse propagating in them. Moreover, the materials have memory effects so that trains of multiple pulses can produce effects similar to those occurring from one large pulse. Despite the increase in complexity of the associated physical properties, these nonlinear effects offer the potential for a variety of novel device and systems applications.

Nonlinear optical (NLO) devices are currently being explored for their applications in various systems associated with communications, remote sensing, optical computing, and so on. However, as the size of optical devices such as microcavity lasers is pushed to the size of an optical wavelength and less, the need for more exact materials and response models is paramount to the successful design and fabrication of those devices. Most current simulation models are based on known macroscopic, phenomenological models that avoid issues dealing with specific microscopic behavior of the materials in such NLO devices. Inaccuracies in the simulation results are then exacerbated as the device sizes shrink to subwavelength sizes and the response times of the excitation signals surpass the response times of the material. There are laser sources currently under development with submicron wavelengths that are pushing the boundaries of the subfemtosecond regime. Phenomenological nonresonant models lose their ability to describe the physics in this parameter regime; hence, they lose their accuracy there. Quantum mechanical effects begin to manifest themselves; the simulation models must incorporate this behavior to be relevant.

Until recently, the modeling of pulse propagation in and scattering from complex nonlinear media has generally been accomplished with one-dimensional, scalar models. These models have become quite sophisticated; they have predicted and explained many of the nonlinear as well as linear effects in present devices and systems. Unfortunately, they cannot be used to explain many observed phenomena; and expectations are that they are not adequately modeling multidimensional nonlinear phenomena. It is felt that vector and higher dimensional properties of Maxwell's equations that are not currently included in existing scalar models in addition to more detailed material and device structure models may significantly impact the scientific and engineering results. The associated propagation and scattering issues have a direct impact on a variety of applications, particularly on the design and engineering of integrated photonic components that have immediate utility to nonlinear soliton fiber optical communications systems currently under development. It is believed that the successful development of semi-classical simulators that combine numerical quantum mechanical models of materials and macroscopic Maxwell's equations solvers will significantly affect the concept and design stages associated with novel nonlinear optics phenomena.

The problem of accurate numerical modeling of the propagation of ultrafast pulses in nonlinear media and their use in NLO optical devices has been subject to increasing interest in recent years. Since the most interesting nonlinear phenomena are transient and superposition is not available, it is natural to try to carry out this modeling directly in the time-domain. For this reason the finite-difference time-domain (FDTD) method is receiving intensive study for modeling linear and nonlinear optical phenomena. In contrast to the case for frequency-domain linear analysis, a single value of permittivity ε is completely inadequate to describe nonlinear time-dependent phenomena, and it is essential to model the interaction of the electromagnetic field with the material medium.

Initial simulations of these ultrafast optical pulse interactions have been based upon several well-known phenomenological material models (Ziolkowski and Judkins, 1993a,b, 1994; Judkins and

Ziolkowski, 1995; Goorjian and Taflove, 1992; Goorjian et al., 1992; Joseph et al., 1993). They included the linear Lorentz dispersion model, the nonlinear Debye model, the nonlinear Raman model, and the instantaneous nonlinear Kerr model. This approach has allowed an investigation of the usually neglected longitudinal field component and polarization effects when optical beams self-focus in bulk materials, and of the physics underlying the design of optical beam steerers and output couplers constructed from corrugated linear and nonlinear waveguides. Nonetheless, while they have been adequate for the applications considered, these phenomenological material models do not handle well fully resonant interactions. To understand the physics underlying the small-distance scale and short-time scale interactions, particularly in the resonance regime of the materials and the associated device structures, a first principles approach is desirable. This in turn requires quantum-mechanical descriptions of the electronic states available in the medium. Accurate physical models must incorporate all propagation effects such as dispersion and nonlinearity, with the proper physical linkages between them.

MAXWELL EQUATIONS SOLVER FOR LINEAR DISPERSIVE MEDIA

A finite-difference time-domain (FDTD) technique-based simulation tool has been developed in two spatial dimensions to model the electromagnetic interaction of focused optical Gaussian beams incident on a corrugated surface coated with several layers of thin dielectric films and realistic metals. The technique is a hybrid approach that combines an intensive numerical method near the surface of the grating and that takes into account the optical properties of the dielectrics and metals, with a free space transform to obtain the radiated fields. It has been used to simulate a variety of configurations including the scattering behavior of an obliquely incident beam that is focused on a uniform grating and of a normally incident beam that is focused on a nonuniform grating. Both *TE* and *TM* polarization simulators have been developed and validated with solutions to canonical grating problems and with experimental results from actual grating structures. Using this hybrid FDTD technique results in a complete and accurate simulation of the total electromagnetic field in the near field as well as in the far field of the diffraction grating. Metal and dielectric gratings and corrugated surfaces are found in a wide variety of optics applications. Among them are optical storage disks, polarizers, and optical filters. Because lasers are often used in these systems, it is important to understand the interaction of finite optical beams with the grating surfaces associated with these devices. It has been shown (Judkins and Ziolkowski, 1995) that there are significant perturbations in the performance between the realistic metal and the perfect metal gratings, which necessitate the more detailed modeling offered by simulators such as ours.

Corrugated dielectric and metallic gratings have been modeled in the past by Fourier transform techniques and moment methods but these treatments have some disadvantages. They do not provide the mechanism for modeling beams in geometries containing both corrugated dielectrics and metal thin films. Furthermore, these methods do not work well for structures that are not strictly periodic, such as gratings with variations in the size or periodicity of the grating teeth. An FDTD modeling tool, which models the problem on a subsectional level, is ideally suited to these more complicated grating surfaces. This approach has been successfully applied in modeling near-field detection of subwavelength-sized rectangular wells in a perfect conductor surface in two dimensions (Kann et al., 1995). This model has been extended to simulate optical beams obliquely incident on a multi-layer thin film dielectric-metallic stack over a grooved dielectric backing. The hybrid FDTD approach gives a quick and accurate description of the near and radiated fields for almost any arbitrary grating design in two spatial dimensions. Extensions to three dimensions are straightforward, but computer intensive.

Devices in the optical regime which are constructed with nonideal materials do not always behave identically to the usual theoretical models, which treat the materials as ideal. An example of where a realistic metal model is important is when light impinges onto the surface of an optical disk. This geometry has a dielectric corrugated surface that is coated with a thin conducting film. A perfect conductor condition for the film layer does not allow energy to penetrate through it; and a conductivity σ model, for most good conductors at optical frequencies, gives the incorrect wave impedance for the corrugation. If this problem is to be examined from the perspective of an FDTD approach, then some physical considerations need to be taken into account in order to make the model complete. At optical frequencies the noble metals Au, Ag, and Cu and also the metal Al have dispersion relations similar to a free electron gas, which cannot be represented by a simple time independent ε, σ model. However, with a time-domain-based modeling approach, the time-dependent properties of materials can be taken into account allowing for a broader range of material possibilities. Thus, the metal regions can be modeled using an appropriate phenomenological relationship between the electric field and the polarization to give the correct refractive index.

In particular, we have numerically constructed solutions to the multidimensional, full-wave, vector form of Maxwell's equations describing the interaction of monochromatic beams with a dispersive material having a finite response time. These numerical solutions have been obtained in two space dimensions and time with a two-dimensional finite-difference time-domain (2D-FDTD) method, which combines a generalization of a standard, FDTD, full-wave, vector, linear Maxwell's equations solver with a Lorentz linear dispersion model. In particular, we are solving in a self-consistent manner the system of equations:

Maxwell Curl Equations

$$\frac{\partial}{\partial t}\vec{H} = -\frac{1}{\mu_0}\nabla \times \vec{E} \qquad (4.1a)$$

$$\frac{\partial}{\partial t}\vec{E} = \frac{1}{\varepsilon_\infty}\nabla \times \vec{H} - \frac{1}{\varepsilon_\infty}\frac{\partial}{\partial t}\vec{P}^L \qquad (4.1b)$$

Lorentz Model

$$\frac{\partial^2}{\partial t^2}\vec{P}^L + \Gamma\frac{\partial}{\partial t}\vec{P}^L + \omega_0^2\vec{P}^L = \varepsilon_0\chi_0\omega_0^2\vec{E}, \qquad (4.2)$$

where $\vec{P}^L$ is the polarization generated by the Lorentz model and does not contain the instantaneous permittivity, i.e., $\vec{P}_{total} = \varepsilon_\infty\vec{E} + \vec{P}^L = \varepsilon_o(1+\chi_\infty)\vec{E} + \vec{P}^L$.

In two dimensions, $\hat{x}$ and $\hat{z}$, Maxwell's equations decouple into two polarization sets, the *TM* set: $E_x\ H_y\ E_z\ (P_x^L\ P_z^L)$, and the *TE* set: $H_x\ E_y\ H_z\ (P_y^L)$. Each field set can be modeled independently of the other. Maxwell's equations are rendered in discrete form and solved on a square mesh in a stepwise manner. Nodes for the electric and magnetic fields are staggered in space and time. The second-order Lorentz model is solved as a sequence of first-order equations for the polarization (bound) current $\vec{J}$ in parallel with Maxwell's equations.

In addition, as with other hybrid approaches, our simulator uses a far-field transform to derive the far field pattern in free space from the fields generated in the vicinity of the geometry. This necessity arises because the FDTD mesh must be finite and all the structures relevant to the problem must be included in it, but the field patterns may be required at some great distance that the mesh cannot encompass due to computational size limitations. Only the complete field amplitude and phase distribution is generated within this near-field region. On the other hand, a detector placed many diffraction lengths away from this simulation region will see only the far-field pattern. Furthermore, the geometry, in this case a grating with a very thin metal surface layer, might demand higher than normal spatial discretization and thus, for the same amount of computer memory, will require the mesh to be confined to an even smaller spatial region. These issues demand either a large number of cells to be used in the FDTD grid or an alternate hybrid approach be taken in producing the signal that a detector would see. With our approach, the computational grid can be restricted to a much smaller spatial region and a Fourier transform method can be used to generate the desired far-field patterns if the region between the actual FDTD simulation domain and the detector is a homogeneous medium.

Following the field/material interaction, the components of the field that do not represent stored energy propagate some distance from the scatterer and eventually evolve into the far field. The pattern in the far field of a grating, where the detector distance d is much larger than Λ/λ, where Λ is the grating period, represents the angular distribution of this energy. Simulating the propagation of energy over this distance with the FDTD algorithm, given the discretization requirements of a structure with large dielectric constants and/or detailed structures, requires an extremely large computational grid. For example, to model typical optical disk cases with adequate discretization (i.e., $\lambda/20$ in the material with the highest refractive index) all the way out to the far-field observation point would involve a $1000\lambda \times 20\lambda$ grid and require in the *TE* case more than 40MB of memory. Truncating the mesh closer to the scatterer does not allow an adequate simulation region for the scattered field to develop into its far-field distribution.

The free space transform is one way to circumvent this problem. This transform takes a transverse component of the scattered field along a planar aperture near one of the four truncation boundaries in the simulation region and projects this field component onto a circle (in two dimensions, a sphere in three) a distance R away from this sampling plane. The method is accurate, within the limits of the Kirchoff approximation, for cases where nearly all the scattered energy generated in the total field region of the mesh is propagated back across the transform boundary and very little energy is leaked through the remaining three truncation boundaries of the FDTD domain. For problems where a beam impinges on a planar structure, the mesh does not need to be much bigger than the scattering object, but it must be large enough in one dimension to encompass in their entirety both the incident and the scattered beams. A diffraction grating configuration for a beam blazing at an angle from the diffraction grating is well suited to this method because the mesh can be made very long (along the grating) in comparison to its height (distance along the normal to the grating).

As shown in Figure 4.1, the simulation region is separated into a scattered field region and a total (incident plus scattered) field region where $(\vec{E}, \vec{H})_{total} = (\vec{E}, \vec{H})_{inc} + (\vec{E}, \vec{H})_{scat}$. The juncture of these two regions is the source boundary. This source boundary is separated into two adjacent planes located a half cell apart. It runs parallel to the grating and is located inside the mesh, near to the outer boundary of the simulation region, which is also taken parallel to the grating. The transverse field distributions along this plane represent exact solutions for an incident Gaussian beam, $\vec{E}^{inc}$ and $\vec{H}^{inc}$, that focuses to a desired location in space. The incident beam may have an arbitrary cross section; but for the cases we have modeled to date, it has been restricted to a lowest-order Gaussian cross-section beam. The incident beam is assumed to be focused at the grating surface $(z = z_s)$. The beam focus is at the center of the mesh along the direction of the grating. The source is driven at the source boundary at the distance z_s above the grating surface. The desired incident beam is achieved by using the free space transform to back-propagate the complex field structure in the focal region to the source boundary. After the source is turned on, it is allowed to run for the duration of the simulation, which is long enough for several cycles to scatter off the grating and reach the truncation boundary.

A variety of *TE* and *TM* cases for the metal film diffraction grating geometry have been considered and are reported in Judkins and Ziolkowski (1995). The basic problem configuration is summarized in Figure 4.2. The specific geometries are characterized by the grating period, the height of the grating teeth, the film thickness, the tooth size ratio, and the complex refractive index of the metal.

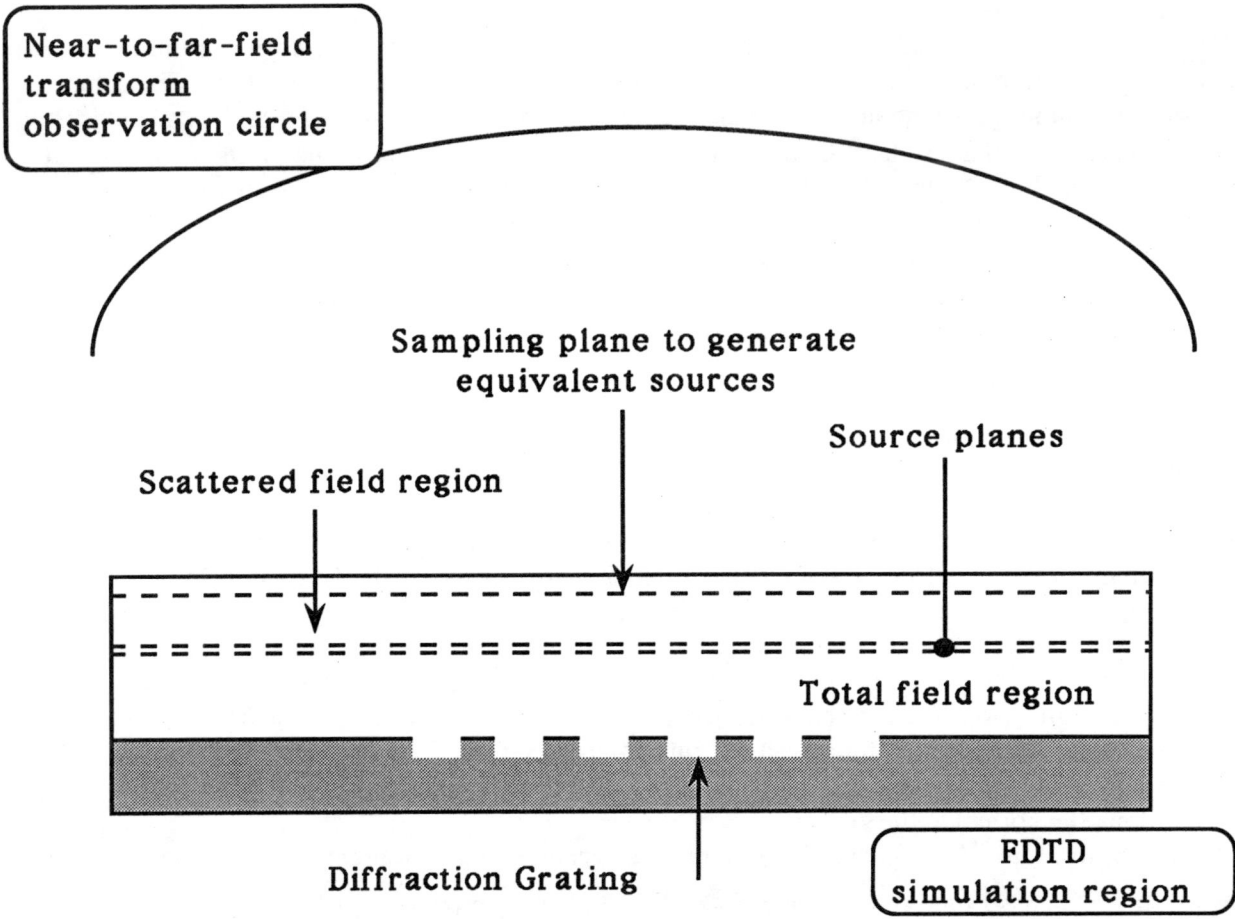

Figure 4.1 The FDTD-Lorentz Medium simulator allows one to model locally scatterers such as realistic multilayered thin film diffraction gratings and yet produce far-field patterns with a near- to far-field transform.

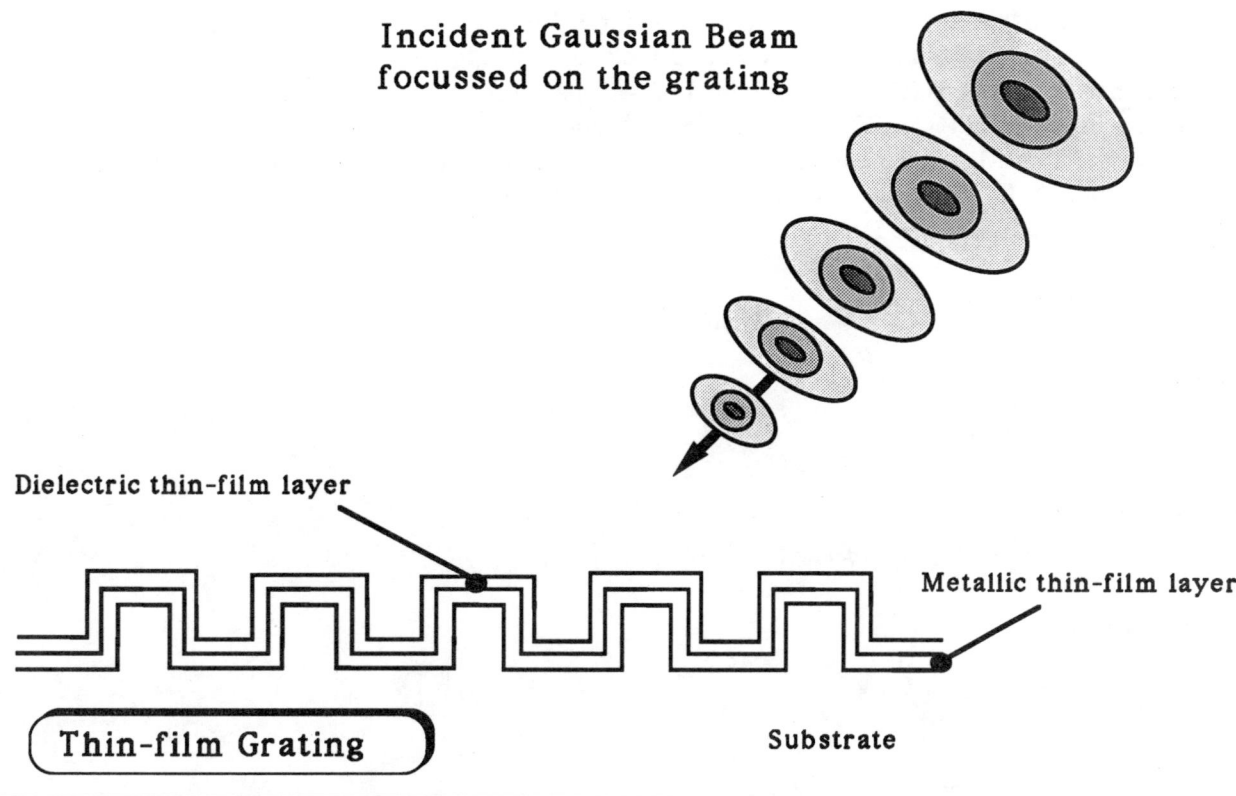

Figure 4.2 A typical optical grating problem involves a focused Gaussian beam and a multilayered thin film diffraction grating.

MAXWELL'S EQUATION SOLVER IN LINEAR AND NONLINEAR MEDIA

With the continuing and heightened interest in linear and nonlinear optically integrated devices, more accurate and realistic numerical simulations of these devices and systems are in demand. Such calculations provide an integrated optics/photonics testbed in which one can investigate new basic and engineering concepts, materials, and device configurations before they are fabricated. The time from device conceptualization to fabrication and testing should therefore be enormously improved with numerical simulations that incorporate more realistic models of the linear and nonlinear material responses and the actual device geometries. It is felt that vector and higher dimensional properties of Maxwell's equations that are not currently included in existing scalar models, in addition to more detailed materials models, may significantly impact the scientific and engineering results.

We have also been simulating a variety of linear and nonlinear corrugated waveguiding systems for their applications to integrated optics/photonics systems. For instance, corrugated waveguide structures have many potential uses as beam steerers and grating assisted couplers. We are developing a simulation toolbox that eventually will be used to design these and many other devices and systems. To meet self-imposed design goals that specify integrated optical/photonic devices that are only a few wavelengths or pulse lengths in size, we require a thorough understanding of the basic physics that we are modeling without the typical approximations generally used for this class of problems. This in turn has required our simulations to be based upon numerically solving the full-wave, vector Maxwell's equations. We have shown that this approach leads to a superior understanding of the underlying physics and to improved engineering designs (Ziolkowski and Judkins, 1993a,b, 1994; Judkins and Ziolkowski, 1995). These numerical solutions have been obtained in two space dimensions and time with a nonlinear finite-difference time-domain (NL-FDTD) method that combines a generalization of the Lorentz linear dispersive medium-Maxwell's equations solver with a nonlinear Raman model and an instantaneous Kerr nonlinear model. In particular, we are solving in a self-consistent manner the system of equations (Judkins and Ziolkowski, 1995):

$$\frac{\partial}{\partial t}\left[\mu_0 \vec{H}\right] = -\nabla \times \vec{E} \tag{4.3a}$$

$$\frac{\partial}{\partial t}\left[\varepsilon_L \vec{E}\right] = \nabla \times \vec{H} - \frac{\partial}{\partial t}\vec{P} \tag{4.3b}$$

Lorentz Model

$$\frac{\partial^2}{\partial t^2}\vec{P}^L + \Gamma_L \frac{\partial}{\partial t}\vec{P}^L + \omega_L^2 \vec{P}^L = \varepsilon_0 \chi_0 \omega_L^2 \vec{E} \tag{4.4}$$

Raman Model

$$\frac{\partial^2}{\partial t^2}\chi^{NL} + \Gamma_R \frac{\partial}{\partial t}\chi^{NL} + \omega_R^2 \chi^{NL} = \varepsilon_R \omega_R^2 \left|\vec{E}\right|^2, \tag{4.5}$$

where $\vec{P} = \vec{P}^L + \vec{P}^{NL}$ and

$$\vec{P}^{NL} = \varepsilon_0 \chi^{NL} \vec{E} + \varepsilon_0 \chi^{Kerr} \left|\vec{E}\right|^2 \vec{E}, \tag{4.6}$$

the last term representing the instantaneous Kerr nonlinearity, χ^{Kerr} being the instantaneous Kerr susceptibility. The resulting NL-FDTD simulator can model pulse propagation in complex environments under the influence of linear and nonlinear dispersive, linear and nonlinear diffractive, and time retardation effects of the materials in and surrounding the electromagnetic structures. By coupling the linear and nonlinear dispersion models together simultaneously with the natural boundary conditions arising from dielectric and metallic discontinuities, we are able to handle the gratings and corrugated interfaces readily. Moreover, as in the linear dispersion cases, both the *TE* and *TM* polarization cases can be simulated. Consequently, more complex, realistic integrated optics/photonics structures are straightforwardly modeled with the NL-FDTD approach.

The NL-FDTD approach can handle ultrafast single-cycle cases as readily as multiple-cycle cases having an intrinsic carrier wave. Since most current optical systems deal directly with a carrier-wave type signal, the NL-FDTD approach can simulate the propagation and scattering effects associated with those narrow bandwidth systems. However, it can also simulate the behaviors of the interactions of ultrafast pulses. Ultrafast pulses are single-cycle or multiple-cycle envelopes containing fewer than 15 cycles. Sources in the laboratory have produced pulses compressed to as fast as $4\,fs$ and the optics community is already investigating the *attosecond* regime. By using these ultrafast sources we illustrate two advantages of the time-domain approach: (1) the ability to carry phase information over a wide spectrum, and (2) the ability to model transient effects that occur either quickly or slowly relative to the time scale of the pulse. The evolution of the pulse in the medium can be dependent on both the material's resonances in the presence of the beam and the initial shape of the exciting pulse. Switching or steering of this type of pulsed beam requires one to take advantage of interference effects and the material's transient response.

The complex waveguiding structures we have considered (Judkins and Ziolkowski, 1995) are filled with either linear or nonlinear dispersive materials that have finite response times. The corrugations themselves can be modeled as dielectric teeth (an extension of the dielectric waveguide) or metallic teeth (deposited into or on top of the dielectric waveguide). A corrugated waveguide with these dielectric or metallic teeth can be viewed as a leaky-wave antenna. The corrugation section is a slow-wave structure whose impedance properties determine the properties of its radiated fields. The field radiated by an infinite linear or nonlinear corrugated structure can be modeled with a Floquet mode representation. The resulting fields have to satisfy a phase matching or Bragg condition resulting from the electromagnetic boundary conditions. Physically this means that because of the regular placement of the teeth in the corrugation section, the individual scattered fields will interfere constructively only along certain preferred directions and the "leaked" energy will appear in the form of pulsed beams that radiate at angles specified by the Bragg condition both into the air and into the substrate regions.

In particular, let θ_t be the angle that the radiated beam subtends with respect to the normal of the waveguide, n_0 be the index of refraction above the corrugations, and $n_G = n_B + n_2 I$ be the index of refraction in the waveguide, which includes the effective waveguide index n_B (which varies slightly from the TE and TM cases to achieve the desired TE_0 and TM_0 initial spatial amplitude distributions) and the intensity-induced index change $n_2 I$. This Bragg condition then takes the form

$$\frac{\omega}{c} n_0 \sin\theta_t = \frac{\omega}{c} n_B + m\frac{2\pi}{\Lambda}, \text{ where } m = 0, \pm 1, \pm 2, \ldots$$

or

$$\theta_t = \sin^{-1}\left[\frac{n_B}{n_0} + \frac{n_2}{n_0} I + m\frac{\lambda}{n_0 \Lambda}\right], \text{ where } m = 0, \pm 1, \pm 2, \ldots. \tag{4.7}$$

This immediately translates into a practical device: the output-beam from the corrugation section can be steered away from the normal by the strength of the intensity of the input waveguide pulsed-beam, the size of the unit cell or the strength of the nonlinearity.

A special case of this relationship suggests a useful output coupler design. If we specify that the corrugation spacing be $\Lambda = \lambda / n_B$, then the first order $(m = -1)$ output-beam from the corrugation section of the waveguide has the transmission angle:

$$\theta_t = \sin^{-1}\left(\frac{n_2}{n_0} I\right) \approx \frac{n_2}{n_0} I. \tag{4.8}$$

Thus, the output-beam from the corrugation section can be steered away from the normal simply by adjusting the strength of the intensity of the input waveguide pulsed-beam or the nonlinear index. Our simulations have confirmed this effect (Judkins and Ziolkowski, 1995).

The multidimensional NL-FDTD model has been applied to the modeling of the extraction of energy from a variety of linear and nonlinear waveguiding structures using corrugated waveguide sections.

Expected conversion efficiencies from the guided mode energy to the radiated field energy have been observed in the linear case. The nonlinear waveguiding structures are presenting interesting challenges in their analysis and interpretation. A variety of *TE* and *TM* cases with metallic corrugations have been considered (Judkins and Ziolkowski, 1995) to illustrate the desired linear and nonlinear output coupler and beam steering effects. Typical simulation geometries are shown in Figures 4.3 and 4.4. Since the electric field behavior near the edges of these metallic corrugations is significantly different between the two polarizations, the resulting radiated field structures reflect this difference. Output beam characteristics depending on the medium response time, the polarization, and the material parameters, have been studied and reported (Judkins and Ziolkowski, 1995). Near-field simulations obtained with the NL-FDTD approach are translated into Fresnel and Fraunhofer regime information with the same near- to far-field transforms introduced for the FDTD Lorentz media-Maxwell equations simulator. Particular emphasis has been given to ultrafast pulses whose time-record length is approximately the same size as the corrugation region. It has been found that even pulses that are short in comparison with the corrugation region can be effectively used to beam steer and couple energy through a grating-assisted coupler from one corrugated waveguide to another.

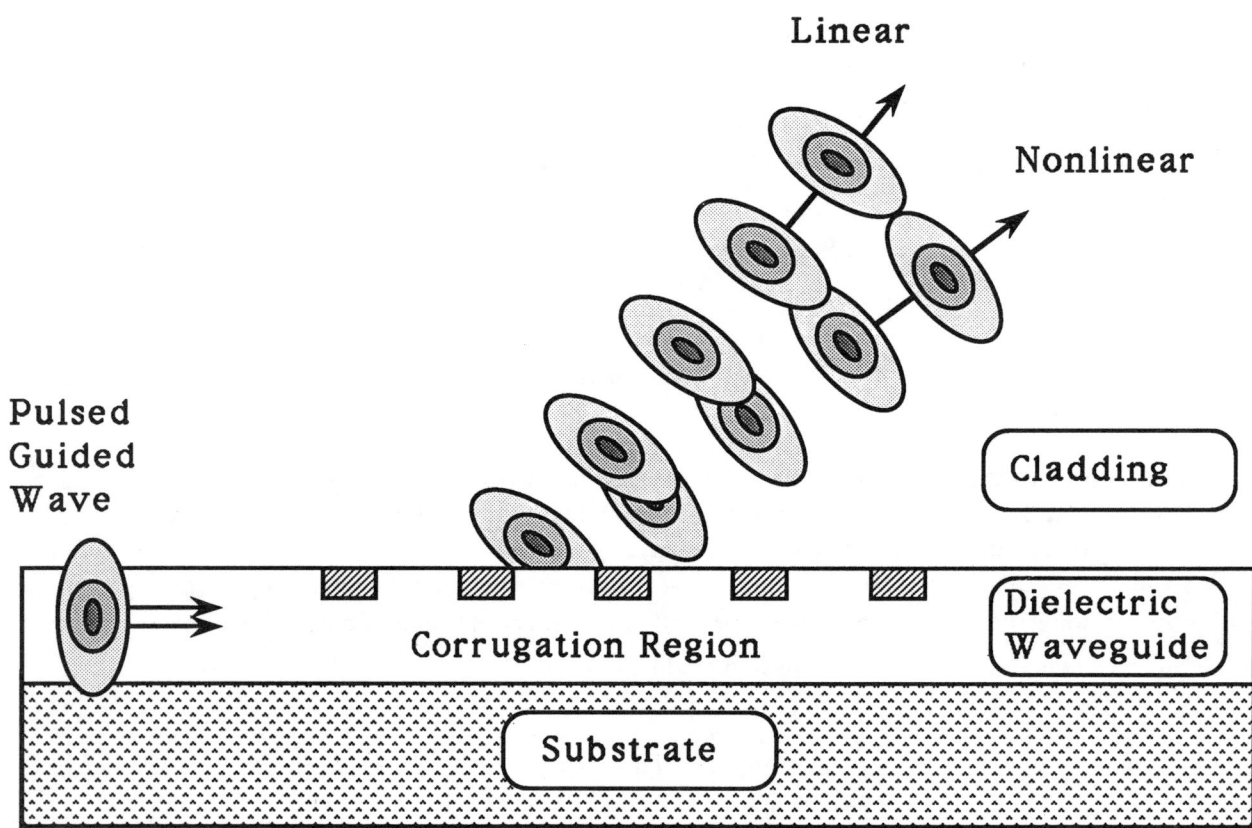

Figure 4.3 The output direction of nonlinear grating-assisted beam steering, integrated optics devices can be controlled by the intensity of the incident pulse.

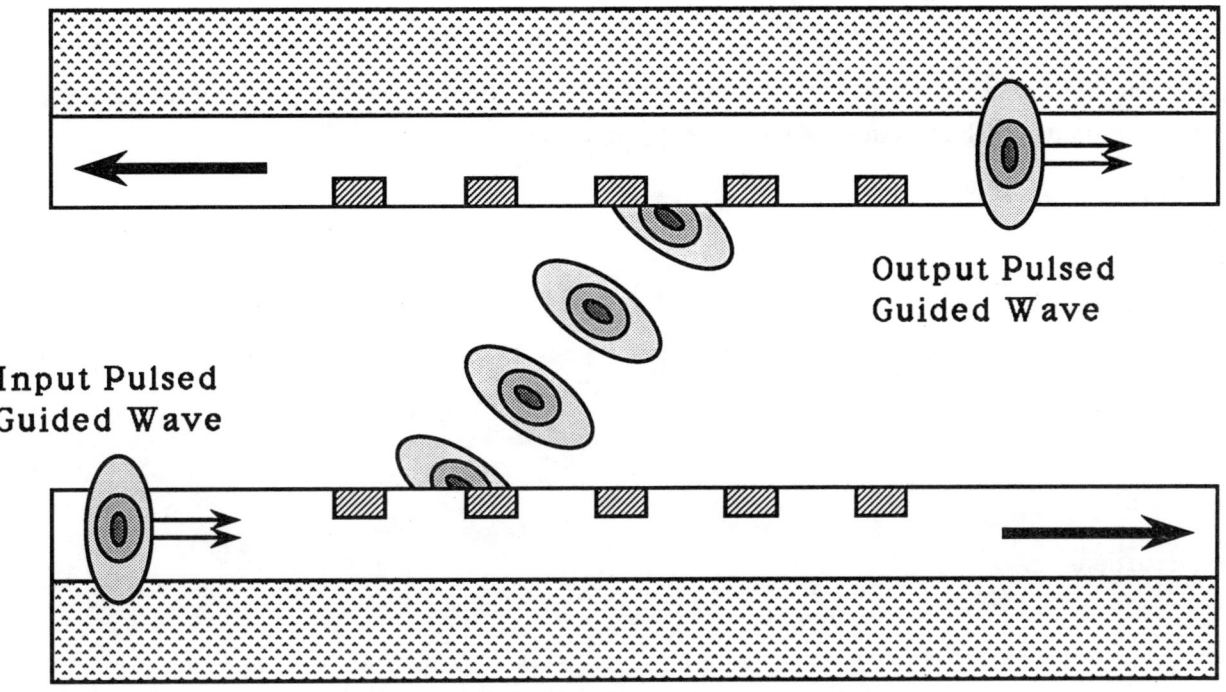

Figure 4.4 The S-parameters of nonlinear grating-assisted, waveguide output couplers can be controlled by the intensity of the incident pulse, the grating period, and the material parameters.

MAXWELL'S EQUATION SOLVER IN MULTILEVEL ATOM MEDIUM

We have also been developing a simulator that utilizes the Maxwell-Bloch system for multilevel atoms for our material models in Maxwell's equations. This effort is novel in that it combines a realistic material model that is quantum mechanically based with a full-wave, vector Maxwell's equations solver. The FDTD implementations of the Maxwell-Bloch modeling system in one space dimension and time have been accomplished (Ziolkowski et al., 1995) and have yielded some interesting physical consequences. In particular, in a density matrix approach to arrive at the Bloch equations describing a two-level atom medium we introduce the terms ρ_1, ρ_2, and ρ_3, which satisfy the relationship $\rho_1^2 + \rho_2^2 + \rho_3^2 = 1$, and represent, respectively, the dispersive or in-phase component of the polarization, the absorptive or in-quadrature component of the polarization, and the fractional difference in the populations for the two energy levels. The near-resonant behavior of nonlinear systems cannot be meaningfully discussed unless dissipative effects are taken into account. The usual method of achieving this in simple systems is to include in the Liouville equations for these terms, phenomenologically obtained diagonal terms consisting of characteristic decay rates. If we take the incident electromagnetic field to be a uniform plane wave that is propagating along the z-axis and is polarized along the x-axis, i.e., $\vec{E}(\vec{r},t) = E_x(z,t)\hat{x}$ and $\vec{H}(\vec{r},t) = H_y(z,t)\hat{y}$, and set the spatial orientation of the dipole to be $\hat{x}$, the polarization takes the form $\vec{P} = P_x \hat{x}$, where

$$P_x(t) = -N_{atom}\gamma p_1(t), \tag{4.9}$$

N_{atom} being the number density of atoms and γ the dipole coupling coefficient. The following one-dimensional Maxwell-Bloch system results from this reduction:

Maxwell Equations

$$\partial_t H_y = -\frac{1}{\mu_0}\partial_z E_x \tag{4.10a}$$

$$\begin{aligned}\partial_t E_x &= -\frac{1}{e_0}\partial_z H_y - \frac{1}{e_0}\partial_t P_x \\ &= -\frac{1}{\varepsilon_0}\partial_z H_y - \frac{N_{atom}\gamma}{\varepsilon_0 T_2}\rho_1 + \frac{N_{atom}\gamma\omega_0}{\varepsilon_0}\rho_2\end{aligned} \tag{4.10b}$$

Bloch Equations

$$\partial_t \rho_1 = -\frac{1}{T_2}\rho_1 + \omega_0 \rho_2 \tag{4.11a}$$

$$\partial_t \rho_2 = -\omega_0 \rho_1 - \frac{1}{T_2}\rho_2 + 2\frac{\gamma}{\hbar}E_x \rho_3 \tag{4.11b}$$

$$\partial_t \rho_3 = -2\frac{\gamma}{\hbar}E_x \rho_2 - \frac{1}{T_1}(\rho_3 - \rho_{30}), \tag{4.11c}$$

where T_1 is the excited state lifetime, T_2 is the dephasing time, and ρ_{30} is the initial population difference in the system. Note that the specification that $\rho_{30} = -1(+1)$ represents all the atoms initially being in their ground (excited) states. This system of equations can be discretized using finite differences in several different ways. We have characterized the performance of several of these discrete approaches and feel that we know how to extend this semi-classical model to higher space dimensions and to more complex media such as a three-level atom medium.

Using this FDTD approach to solving the semiclassical Maxwell-Bloch system, we have studied in a more exact manner (without removing the carrier wave) self-induced transparency effects in a two-level atom medium (Ziolkowski et al., 1995). Standard self-induced transparency (SIT), the so-called $\pi, 2\pi, 4\pi, \ldots$ results have been reproduced with this model. A SIT solution represents the nonlinear wave propagation dynamics in which a particular pulse shape, a carrier at the transition frequency with a hyperbolic secant envelope, having the appropriate high intensity completely loses its energy to a two-level atom medium by stimulating it from its ground state into its excited state, and then is completely reconstructed in a coherent manner via stimulated emission by having the excited medium completely decay back into its ground state. This SIT pulse thus propagates through the highly nonlinear two-level atom medium with no change in its shape, i.e., as though the medium is transparent; it is a soliton solution of the semiclassical Maxwell-Bloch system. The SIT effect is normally described with a rotating wave approximation of the Maxwell-Bloch system. As will be demonstrated, we have recovered these SIT pulse dynamics with our FDTD Maxwell-Bloch simulator. However, we have also found (Ziolkowski et al., 1995) novel features that appear at points where the electric field is null and have been identified as being associated with the maximums of the time derivative of the electric field. These features are not present in standard approximate solutions (e.g., the rotating-wave approximation) to this problem.

These nonlinear time-derivative effects have been emphasized further (Ziolkowski et al., 1995) by considering a variety of ultrafast pulse cases. It has been demonstrated (Ziolkowski et al., 1995), during ultrafast pulse interactions with a two-level atom medium, that a single-cycle pulse can be designed that completely inverts the two-level atom medium. A multiple ultrafast pulse train has been given that can completely invert the medium from the ground to the excited state and then completely reverse the process. These results confirm that the time-derivative-driven nonlinear properties of the two-level atom medium have a significant impact on the time evolution of this system in the limit of ultrafast pulses.

We have also used (Ziolkowski et al., 1995) the FDTD Maxwell-Bloch simulator to recover expected small-signal gain results for sinusoidal input signals. The designed ultrafast inversion pulse has been combined with a sinusoid input signal to form a pump-probe signal set. It has been illustrated (Ziolkowski et al., 1995) that a two-level atom medium could be inverted by the leading ultrafast pulse to yield a gain medium for the trailing sinusoidal probe pulse.

Full device and system integration complexities are presently being introduced into our model by considering multidimensional (spatial) extensions. We are investigating several different amplifier and microcavity laser configurations with the resulting multidimensional FDTD Maxwell-Bloch simulator. Current progress in higher dimensions will be reported elsewhere.

CONCLUSIONS

We are developing full-wave, vector Maxwell equation solvers for use in studying the physics and engineering of linear and nonlinear integrated photonics devices and systems. These simulators and their applications have been described. Particular emphasis is given to time-domain problems describing the interaction of ultrafast optical pulses with nonresonant and resonant optical materials and structures. These problems pose severe difficulties to numerical modeling because of the many time and length scales involved. The global structures dealt with in integrated photonic systems are large relative to their operating wavelengths; but their substructures are subwavelength in size. The corresponding computational procedures for studying these large-scale structures are computationally large and entail an extremely large number of degrees of freedom. We have developed hybrid simulators that incorporate the linear and nonlinear dynamics of the substructures and the overall response of the integrated system. We discussed one such simulator which couples Maxwell's equations and a Lorentz linear dispersion model with a near- to far-field transform capability to generate the global far-field patterns associated with the scattering of optical Gaussian beams from diffraction gratings coated with multiple layers of realistic thin film dielectrics and metals. This simulator is being applied to a variety of wavelength-sized diffractive optic components. An extension of this simulator was also described that includes both the Raman and the instantaneous Kerr nonlinear material models. This simulator has been used to describe locally the behaviors of linear and nonlinear finite length corrugated optical waveguides for applications to grating-assisted couplers and beam steerers. We also described another simulator under development that combines a multilevel atom materials model with our Maxwell's equations simulator to model self-induced transparency and laser gain-medium effects. This resonant systems model requires a careful marriage between a microscopic (quantum mechanical) materials model and the macroscopic Maxwell's equations solver. This simulator is being developed to model microcavity lasers and waveguide amplifiers.

Integrated optics/photonics devices and systems are truly representative of large-scale structures. The associated simulations require a corresponding large amount of high-performance computing (HPC) efforts. The need for more accurate simulators will increase as the integrated optics/photonics devices and systems complexity increases. Examples from our efforts in this area, which have civilian and military relevance, have been given to illustrate what classes of simulators have been developed to model these structures. Future efforts in this area will require even more complete materials models appropriately integrated into the Maxwell equations solvers and the associated advanced HPC methods.

REFERENCES

Goorjian, P.M., and A. Taflove, 1992, "Direct time integration of Maxwell's equations in nonlinear dispersive media for propagation and scattering of femtosecond electromagnetic solitons," *Opt. Lett.* **17**(3), 180-182.

Goorjian, P.M., A. Taflove, R.M. Joseph, and S.C. Hagness, 1992, "Computational modeling of femtosecond optical solitons from Maxwell's equations," *IEEE J. Quantum Electron.* **QE-28**(10), 2416-2422.

Joseph, R.M., P.M. Goorjian, and A. Taflove, 1993, "Direct time integration of Maxwell's equations in two-dimensional dielectric waveguides for propagation and scattering of femtosecond electromagnetic solitons," *Opt. Lett.* **18**(7), 491-493.

Judkins, J.B., and R.W. Ziolkowski, 1995, "FDTD modeling of nonperfect metallic thin film gratings," submitted to *J. Opt. Soc. Am. A*.

Kann, J., T. Milster, F. Froehlich, R.W. Ziolkowski, and J.B. Judkins, 1995, "Near-field optical detection of asperities in dielectric surfaces," *J. Opt. Soc. Am. A.* **12**(3), 501-512.

Ziolkowski, R.W., and J.B. Judkins, 1993a, "Full-wave vector Maxwell equation modeling of the self-focusing of ultrashort optical pulses in a nonlinear Kerr medium exhibiting a finite response time," *J. Opt. Soc. Am. B* **10**(2), 186-198.

Ziolkowski, R.W., and J.B. Judkins, 1993b, "Applications of discrete methods to pulse propagation in nonlinear media: self-focusing and linear-nonlinear interfaces," invited paper, special issue of *Radio Science* for the 1992 URSI EM Theory Symposium, *Radio Sci.* **28**(5), 901-911.

Ziolkowski, R.W., and J.B. Judkins, 1994, "NL-FDTD modeling of linear and nonlinear corrugated waveguides," *J. Opt. Soc. Am. B* **11**(9), 1565-1575.

Ziolkowski, R.W., J.M. Arnold, and D.M. Gogny, 1995, "Ultrafast pulse interactions with two-level atoms," submitted to *Phys. Rev. A*.

5

Algorithmic Aspects and Supercomputing Trends in Computational Electromagnetics

Vijaya Shankar
William F. Hall
Alireza Mohammadian
Chris Rowell
Rockwell International Science Center
Thousand Oaks, California

> Accurate and rapid evaluation of radar signatures for alternative aircraft/store configurations would be of substantial benefit in the evolution of integrated designs that meet radar cross-section (RCS) requirements across the threat spectrum. Finite-volume time-domain methods offer the possibility of modeling the whole aircraft, including penetrable regions and stores, at longer wavelengths on today's supercomputers and at typical airborne radar wavelengths on the teraflop computers of tomorrow. To realize this potential, practical means must be developed for the rapid generation of grids on and around the aircraft, and numerical algorithms that maintain high-order accuracy on such grids must be constructed. A structured grid and an unstructured grid-based finite-volume, time-domain Maxwell's equation solver have been developed incorporating modeling techniques for general radar absorbing materials. Using this work as a base, the goal of the computational electromagnetics (CEM) effort is to define, implement, and evaluate various issues suitable for rapid prototype signature prediction addressing many issues related to (1) the physics of electromagnetics, (2) efficient and higher-order accurate algorithms, (3) boundary condition procedures, (4) geometry and gridding (structured and unstructured), (5) computer architecture (SIMD and MIMD), and (6) validation.

INTRODUCTION

Computational Electromagnetics

The ability to predict radar return from complex structures with layered material media over a wide frequency range (100 MHz to 20 GHz) is a critical technology need for the development of stealth aerospace configurations. Traditionally, radar cross section (RCS) calculations have employed one of two methods: high-frequency asymptotics, which treats scattering and diffraction as local phenomena; or solution of an integral equation (in the frequency domain) for radiating sources on (or inside) the scattering body, which couples all parts of the body through a multiple scattering process. A third approach is the direct integration of the differential or integral form of Maxwell's equations in the time domain.

The time-domain Maxwell's equations represent a more general form than the frequency-domain vector Helmholtz equations, which are usually employed in solving scattering problems. A time-domain approach can, for instance, handle continuous wave (single frequency) as well as a single pulse (broadband frequency) transient response. Frequency-domain-based methods usually provide the RCS response for all angles of incidence at a single frequency, while time-domain-based methods provide solutions for many frequencies from a single transient calculation. Also, in a time-domain approach, one can consider time-varying material properties for treatment of active surfaces. By using Fourier transforms, the time-domain transient solutions can be processed to provide the frequency-domain response. Frequency-dependent (dispersive) and anisotropic material properties can also be included within the time-domain formulation.

CEM is a critical technology in the advancement of future aerospace development through supercomputing. As we make the transition from the present gigaflops to the next-generation teraflops computing, CEM will become integral to aerospace design not only as a standalone technology but also as part of the multidisciplinary coupling that leads to well-optimized designs.

Objectives

Toward establishing a computational environment for performing multidisciplinary studies, the initial goal is to advance the state of the art in CEM with the following specific objectives:

1) Apply algorithmic advances in Computational Fluid Dynamics (CFD) to solve Maxwell's equations in general form to study scattering (radar cross section), radiation (antenna), and a variety of eletromagnetic environmental (electromagnetic compatibility, shielding, and interference) problems of interest to both the defense and commercial communities. (Mohammadian et al., 1991)
2) Establish the viability of MIMD massively parallel architectures for tackling large-scale problems not amenable to present-day supercomputers.
3) Develop the CEM technology to the point of being able to perform coupled CFD/CEM optimization design studies.

CEM Issues

Proper development of a CEM capability appropriate for all aspects of aerospace design must consider various issues associated with electromagnetics. Some of them are addressed in the following seven subsections.

Maxwell's Equations

In order to apply conservation principles (for example, in fluid dynamics, mass, momentum, and energy are conserved), many of the governing equations representing appropriate physical processes are written in conservation form. The general form of a differential conservation equation can be written as

$$Q_t + E_x + F_y + G_z = \text{Source}, \tag{5.1}$$

where Q is the solution vector and E, F, and G are the fluxes in x, y, and z coordinate directions, respectively. The conservation form readily admits weak solutions such as shock waves.

The integral form of the conservation laws that can easily be derived from the differential form by integrating (5.1) with respect to x, y, z over any conservation cell whose volume is V is

$$\iiint_v \left(\frac{\partial Q}{\partial t} + \frac{\partial E}{\partial x} + \frac{\partial F}{\partial y} + \frac{\partial G}{\partial z} \right) dx\, dy\, dz$$
$$= \iiint_v S\, dx\, dy\, dz = \tilde{S}. \tag{5.2}$$

This can be rewritten in vector notation as

$$\frac{\partial}{\partial t} \iiint_v Q\, dx\, dy\, dz + dx \iiint_v \left(\vec{\nabla} \cdot \vec{\mathcal{F}} \right) dy\, dz = \tilde{S}. \tag{5.3}$$

In the above,

$$\vec{\mathcal{F}} = E\hat{j} + F\hat{k} + G\hat{\imath}. \tag{5.4}$$

Applying the Gauss divergence theorem, we can convert the volume integral into a surface integral

$$\frac{\partial}{\partial t}\left(\tilde{Q} V \right) + \iint_s \left(\vec{\mathcal{F}} \cdot \hat{n} \right) ds = \tilde{S}. \tag{5.5}$$

In the above equation, the cell average of the dependent variables are denoted by $\tilde{Q}$. The outward unit normal at any point of the boundary surface of a cell has been denoted by $\hat{n} = \hat{n}_x \hat{j} + \hat{n}_y \hat{k} + \hat{n}_z \hat{\imath}$,

$$\tilde{Q} = \frac{\iiint_v Q\, dV}{\iiint_v dV}. \tag{5.6}$$

The integral form of the conservation laws given by (5.5) defines a system of equations for the cell average values of the dependent variables.

Maxwell's equations in their vector form are

$$\frac{\partial B}{\partial t} = -\nabla \times \mathcal{E} \tag{5.7}$$

and

$$\frac{\partial D}{\partial t} = \nabla \times \mathcal{H} - \mathcal{J}. \tag{5.8}$$

The divergence conditions $\nabla \cdot D = \rho$ and $\nabla \cdot B = 0$ are derived directly from Maxwell's equations, where $\nabla \cdot \mathcal{J} = -\frac{\partial \rho}{\partial t}$. The vector quantities $\mathcal{E} = (\mathcal{E}_x, \mathcal{E}_y, \mathcal{E}_z)$ and $\mathcal{H} = (\mathcal{H}_x, \mathcal{H}_y, \mathcal{H}_z)$ are the electric and magnetic field intensities, $D = (D_x, D_y, D_z)$ is the electric displacement, $B = (B_x, B_y, B_z)$ is the magnetic induction, and $\mathcal{J} = (J_x, J_y, J_z)$ is the current density and ρ is the charge density. The subscripts x, y, z in the vector representation of $\mathcal{E}, \mathcal{H}, B, D$ refer to components in respective directions.

In order to apply conservation-law form finite-volume methods, (5.7) and (5.8) are rewritten in the form of (5.1),

$$Q = \begin{Bmatrix} B_x \\ B_y \\ B_z \\ D_x \\ D_y \\ D_z \end{Bmatrix}; \; E = \begin{Bmatrix} 0 \\ -D_z/\varepsilon \\ D_y/\varepsilon \\ 0 \\ B_z/\mu \\ -B_y/\mu \end{Bmatrix}; \; F = \begin{Bmatrix} D_z/\varepsilon \\ 0 \\ -D_x/\varepsilon \\ -B_z/\mu \\ 0 \\ B_x/\mu \end{Bmatrix};$$

$$G = \begin{Bmatrix} -D_y/\varepsilon \\ D_x/\varepsilon \\ 0 \\ B_y/\mu \\ -B_x/\mu \\ 0 \end{Bmatrix}; \; S = \begin{Bmatrix} 0 \\ 0 \\ 0 \\ -J_x \\ -J_y \\ -J_z \end{Bmatrix}.$$

(5.9)

In what follows, the permittivity coefficient ε and the permeability coefficient μ are taken to be isotropic, scalar material properties and satisfy the following relationship: $D = \varepsilon \mathcal{E}$, $B = \mu \mathcal{H}$. Generalization to tensor ε and μ is rather cumbersome but straightforward. The current density J is usually represented by $\sigma \mathcal{E}$, where σ is the material electrical conductivity.

For treatment of complex geometries, a body-fitted coordinate transformation is introduced to aid in the application of boundary conditions.

Under the transformation of coordinates implied by

$$\tau = t, \; \xi = \xi(t, x, y, z),$$
$$\eta = \eta(t, x, y, z), \; \zeta = \zeta(t, x, y, z),$$

equation (5.9) can be rewritten as

$$\overline{Q}_\tau + \overline{E}_\xi + \overline{F}_\eta + \overline{G}_\zeta = \overline{S} \tag{5.10}$$

where

$$\overline{Q} = \begin{pmatrix} \frac{\vec{D}}{J} \\ \frac{\vec{B}}{J} \end{pmatrix}, \; \overline{E} = \begin{pmatrix} \frac{-\vec{\xi} \times \vec{\mathcal{H}}}{J} \\ \frac{\vec{\xi} \times \vec{\mathcal{E}}}{J} \end{pmatrix}, \; \overline{F} = \begin{pmatrix} \frac{-\vec{\eta} \times \vec{\mathcal{H}}}{J} \\ \frac{\vec{\eta} \times \vec{\mathcal{E}}}{J} \end{pmatrix}, \tag{5.11}$$

$$\overline{G} = \begin{pmatrix} \dfrac{-\vec{\zeta}\times\vec{H}}{J} \\ \dfrac{\vec{\zeta}\times\vec{E}}{J} \end{pmatrix}, \text{ and } \overline{S} = \begin{pmatrix} \dfrac{-\vec{\mathcal{J}}}{J} \\ 0 \end{pmatrix},$$

where $J = |\partial(\xi,\eta,\zeta)/\partial(x,y,z)|$ is the Jacobian of the transformation, and, e.g., $\vec{\xi} = (\partial_x\xi, \partial_y\xi, \partial_z\xi)$. The quantities $\vec{\xi}\times\vec{H}$ and $\vec{\xi}\times\vec{E}$ in (5.11) represent tangential magnetic and electric fields at a constant ξ surface. Thus, the fluxes $\overline{E}, \overline{F},$ and $\overline{G}$ are nothing but the tangential fields.

Maxwell's equations can also be cast in integral form as

$$\frac{\partial}{\partial t}\iiint_v \begin{pmatrix} \vec{B} \\ \vec{D} \end{pmatrix} dV + \iint_{vs} \begin{pmatrix} \hat{n}\times\vec{E} \\ -\hat{n}\times\vec{H} \end{pmatrix} dS = 0, \qquad (5.12)$$

where the six components of $\vec{\mathcal{F}}\cdot\hat{n}$ in (5.5) are $(\hat{n}\times E, -\hat{n}\times H)$.

In general, the differential form, (5.11), will be employed for finite-volume schemes using a structured grid arrangement, and the integral form, (5.12), will be used for unstructured grid cell arrangements using finite-element-like finite-volume schemes.

Finite-Volume Treatment

Space/Time Discretization

The major feature of the present discretization approach that distinguishes it from other finite-volume and finite-difference procedures is that the electric and magnetic field unknowns are co-located in both space and time, rather than being assigned to two interpenetrating spatial grids and separated a half-step in time. These field unknowns are the volume averages of E and H within each cell in the space-filling grid.

Staggered-grid methods automatically achieve second-order accuracy in space, while co-located field algorithms require near-neighbor corrections. However, there is a fundamental equivalence between these methods in terms of the achievable accuracy and stability of the integration process.

Both approaches typically use explicit time integration, which means that the upper limit on the allowable size of the time step Δt is determined by the physical size and shape of the smallest cells, corresponding roughly to the time that light takes to cross one of these cells. Implicit integration schemes can choose larger time steps, but they require the inversion of a banded matrix the size of the whole grid, and their ability to preserve phase information is not known.

The unstructured algorithm developed here is applicable to any grid that fills the computational domain with polyhedral cells. In particular, necessary bookkeeping procedures are implemented to deal with hexahedra (such as cubes), tetrahedra, and prisms (translations of a triangle out of its plane of definition).

Each polyhedron in the computational domain is specified by the location (x,y,z) of its vertices in physical space. From these locations, all the necessary geometrical quantities, including areas, surface normals, and centroidal locations are computed. As stated earlier, each field unknown attributed to a given polyhedron is considered to be an average of the field over the volume of the polyhedron. The six components of E and H at one time level are thus stored according to an index α that runs over all polyhedra. Quantities related to the polyhedral faces, such as face normals, are stored according to another index that runs over all faces.

The interior faces of a given polyhedron are kept distinct from those of the neighboring polyhedra that share faces with it in a purely geometrical sense. This allows, for instance, for any type of impedance boundary condition to be applied at the boundary between cells. Thus, each polyhedral face has a co-face with a distinct face index, and each such co-face belongs to its own polyhedron.

Polynomial Representation and Least Squares

To go beyond representation of the fields as simple volume averages, we have chosen initially to implement linear polynomial functions for both E and H. Higher-order polynomial representations will follow the same general procedures. The essential question is how the higher-order terms in these polynomials are to be determined from near-neighbor data, so as to achieve the desired level of accuracy within each cell. In our unstructured approach, this evaluation is closely tied to the time integration procedure through the Riemann fluxes at each interface. Ultimately, this preserves time accuracy as well as accuracy in space.

In the first step of this method, first-order Riemann fluxes are constructed at each interface of a cell from the volume-averaged fields on either side of the interface. For Maxwell's equations, these fluxes are the tangential field components just inside the boundaries of the cell. To complete the specification at the cell surface, the normal components of E and H are taken to be the normal components of the volume-averaged fields. This maintains overall charge conservation within the cell.

These boundary data are sufficient to determine all the terms in a linear polynomial fit to either E or H by a procedure such as least-squares minimization of the fitting error integrated over the boundary. If we denote the vector polynomial to be fitted as $\vec{A}$ and its boundary values as $\vec{A}^*$, then the quantity to be minimized is

$$e = \int_{\text{(cell boundary)}} \left(\vec{A} - \vec{A}^*\right)^2 dS.$$

Taking derivatives of e with respect to each polynomial coefficient in $\vec{A}$ results in a nonsingular set of linear equations for these coefficients. For consistency, one constrains the constant term in $\vec{A}$ to be equal to the known volume average of $\vec{A}$ over the cell.

A separate set of equations is obtained for each Cartesian component of $\vec{A}$. If one writes, e.g.,

$$A_x(\vec{r}) = \langle A_x \rangle_\alpha + (x - x_\alpha)\frac{\partial A_x}{\partial x} + (y - y_\alpha)\frac{\partial A_x}{\partial y} + (z - z_\alpha)\frac{\partial A_x}{\partial z},$$

where the angular brackets denote volume averaging and, e.g., $x_\alpha = \langle x \rangle_\alpha$, then these equations become

$$\int_{\partial \alpha} (x - x_\alpha)\left[(x - x_\alpha)\frac{\partial A_x}{\partial x} + (y - y_\alpha)\frac{\partial A_x}{\partial y} + (z - z_\alpha)\frac{\partial A_x}{\partial z}\right] dS$$
$$= \int_{\partial \alpha} (x - x_\alpha)\left[A_x^* - \langle A_x \rangle_\alpha\right] dS$$

$$\int_{\partial\alpha} (y-y_\alpha)\left[(x-x_\alpha)\frac{\partial A_x}{\partial x}+(y-y_\alpha)\frac{\partial A_x}{\partial y}+(z-z_\alpha)\frac{\partial A_x}{\partial z}\right]dS$$
$$=\int_{\partial\alpha} (y-y_\alpha)\left[A_x^*-\langle A_x\rangle_\alpha\right]dS$$

$$\int_{\partial\alpha} (z-z_\alpha)\left[(x-x_\alpha)\frac{\partial A_x}{\partial x}+(y-y_\alpha)\frac{\partial A_x}{\partial y}+(z-z_\alpha)\frac{\partial A_x}{\partial z}\right]dS$$
$$=\int_{\partial\alpha} (z-z_\alpha)\left[A_x^*-\langle A_x\rangle_\alpha\right]dS,$$

where we have denoted the cell boundary as $\partial\alpha$. These equations can be solved by inverting the matrix M whose elements are the quadratic moments

$$M_{ij}=\int_{\partial\alpha} \hat{i}\cdot(\vec{r}-\vec{r}_\alpha)\hat{j}\cdot(\vec{r}-\vec{r}_\alpha)dS,$$

where $\hat{i}$ and $\hat{j}$ are unit vectors in the respective coordinate directions.

For a linear polynomial fit, there is a simpler alternative procedure that we have implemented to evaluate these linear terms. From the divergence theorem, the average value of any derivative over the cell volume can be rewritten as a surface integral:

$$\frac{1}{V_\alpha}\int_\alpha \frac{\partial\rho}{\partial x}dV=\frac{1}{V_\alpha}\int_{\partial\alpha}\hat{n}_x\rho dS,$$

where $\hat{n}$ is the unit outward normal on the boundary $\partial\alpha$ and V_α is the cell volume. In particular, if ρ is a linear function of $\vec{r}$, then $\partial\rho/\partial x$ is constant and equal to this volume average, which can be calculated just from the values of ρ on the boundary. For every component of A, we can replace its boundary values by the corresponding component of A^* to obtain the approximation

$$\vec{\nabla}\vec{A}_\alpha \approx \frac{1}{V_\alpha}\int_{\partial\alpha}\hat{n}\vec{A}^*dS \triangleq K_\alpha,$$

which is equivalent to using $\hat{n}$ as the weight in the method of weighted residuals applied to the difference $A-A^*$. The quantity K_α is a vector dyadic. Since we have chosen $\hat{n}\cdot\vec{A}^*=\hat{n}\cdot\langle\vec{A}\rangle_\alpha$, we can make use of the vector identity $\vec{a}=\hat{n}(\hat{n}\cdot\vec{a})-\hat{n}\times(\hat{n}\times\vec{a})$ to rewrite the integral as

$$K_\alpha=\frac{1}{V_\alpha}\int_{\partial\alpha}\hat{n}\,[\hat{n}\times\{\hat{n}\times(\langle\vec{A}\rangle_\alpha-\vec{A}^*)\}]dS,$$

which will be more convenient to compute in terms of the tangential components of $\vec{A}^*$. This particular weighting can be shown to result from a variational principal that assumes each Cartesian component of $\vec{A}^*$ is the boundary value of a solution of Laplace's equation inside the cell.

The Unstructured Second-order Algorithm

An algorithm that maintains second-order accuracy in both space and time can be constructed from the linear polynomial representation as follows:

$$\langle Q \rangle_\alpha^{m+1/2} = \langle Q \rangle_\alpha^m - \frac{\Delta t}{2V_\alpha} \int_{\partial \alpha} \hat{n} \cdot F(Q_\alpha^{*m}) dS$$

$$K_\alpha^m = \frac{1}{V_\alpha} \int_{\partial \alpha} \hat{n} Q_\alpha^{*m} dS = \frac{1}{V_\alpha} \int_{\partial \alpha} \hat{n}(\hat{n} \times \{\hat{n} \times [\langle Q \rangle_\alpha^m - Q_\alpha^{*m}]\}) dS$$

$$Q_\alpha^{m+1/2}(\vec{r}) = \langle Q \rangle_\alpha^{m+1/2} + (\vec{r} - \vec{r}_\alpha) \cdot K_\alpha^m \quad \text{for } \vec{r} \text{ in cell } \alpha$$

$$\langle Q \rangle_\alpha^{m+1} = \langle Q \rangle_\alpha^m - \frac{\Delta t}{V_\alpha} \int_{\partial \alpha} \hat{n} \cdot F(Q_\alpha^{*(m+1/2)}) dS.$$

Here we have written Maxwell's equations symbolically as

$$\frac{\partial Q}{\partial t} + \nabla \cdot F(Q) = 0, \quad Q = (\vec{D}, \vec{B}),$$

and the solution of the Riemann problem just inside a cell interface is denoted Q^*. This solution depends only on the values of $Q(\vec{r})$ immediately on either side of the interface. These are the cell-average values for Q^{*m} and the linear polynomial values for $Q^{*(m+1/2)}$.

Material Properties

The primary design variables affecting RCS are the shape of the conducting structure and the electric and magnetic polarizabilities of the materials that cover various parts of this structure. Because the illuminating fields are weak, the response of the materials can be taken as linear, so that an effective dielectric permittivity ε and magnetic permeability μ can be defined, both of which will in general be complex at the illuminating frequency. The treatment of various limiting material conditions relevant to radar absorbing structures within the framework of time-domain electromagnetics is described below.

Resistance Cards

A thin conducting sheet causes a jump in the tangential magnetic field proportional to the electric current in the sheet, which is given by $\sigma d (\hat{n} \times \vec{\mathcal{E}})$, where σ is the electric conductivity of the sheet, d is its thickness, $\hat{n}$ is the local normal to the surface of the sheet, and $\vec{\mathcal{E}}$ is the instantaneous local electric field.

Perfectly Conducting Surface

At typical radar frequencies, the electrical conductivity of metals and other aircraft structural composites is sufficiently high that they can be treated as perfect electrical conductors. In this limit, the electromagnetic fields do not penetrate the conducting surface, and the components of the electric field tangent to the surface vanish at every point. This relation, $\vec{n} \times \vec{\mathcal{E}} = 0$, thus appears as a boundary condition on the solution of Maxwell's equations.

Lossy Materials and Dispersive Media

The imaginary parts of ε and μ represent energy absorption. For a time harmonic radar wave at radian frequency ω, their effects are exactly equivalent to adding instantaneous electric and magnetic current conductivities equal to $\omega\,\mathrm{Im}(\varepsilon)$ and $\omega\,\mathrm{Im}(\mu)$, respectively, in the Maxwell curl equations:

$$\nabla \times \vec{\mathcal{E}} = -\mathrm{Re}(\mu)\partial \vec{\mathcal{H}} / \partial t - \omega\,\mathrm{Im}(\mu)\vec{\mathcal{H}}$$
$$\nabla \times \vec{\mathcal{H}} = \mathrm{Re}(\varepsilon)\partial \vec{\mathcal{E}} / \partial t + \omega\,\mathrm{Im}(\varepsilon)\vec{\mathcal{E}} + \sigma \vec{\mathcal{E}}.$$

For a transient pulse, the frequency dependence of ε and μ (dispersive media) must be properly modeled over the bandwidth of the pulse. This leads to an integration over the past history of the fields, but the value of this integral at every time step can be accurately updated using only the immediately previous values of the field and the integral.

Impedance Boundaries

When the product $\mu\varepsilon$ is large compared to $\mu_0\varepsilon_0$, the electrical wavelength in the material is correspondingly reduced from its free-space value. To achieve the same accuracy inside a layer of such a material as is needed on the outside, the number of grid points per unit area on the inside surface must be increased by the factor $(\mu\varepsilon / \mu_0\varepsilon_0)$ compared to the outside surface. In extreme cases, orders of magnitude employing more grid points would be used in the layer than in all the space surrounding the target. We avoid this problem by eliminating the points inside the layer through the use of an impedance boundary condition applied on its outer surface. The implementation of this condition in the time domain involves an integral over past history that is carried out by the same method that we use for transient-pulse integration with frequency-dependent ε or μ.

Anisotropic Media

In general, the polarization induced in a material by an applied field need not be parallel to the direction of the applied field. The electrical permittivity ε, relating $\vec{D}$ to $\vec{\mathcal{E}}$, and the magnetic permeability μ, relating $\vec{B}$ to $\vec{\mathcal{H}}$, are therefore tensors rather than scalar quantities, and there will typically be different characteristic wave speeds along the three principal axes associated with three tensors. Aside from this formal complication, Maxwell's equations as given in (5.7) and (5.8) still apply.

Chiral Media

Materials that are predominantly composed of elements of a single parity, such as right-handed helices, exhibit electrical polarization in response to an applied magnetic field and magnetic polarization in response to an applied electric field. At a given frequency, one can define a chiral admittance tensor ξ, in terms of which

$$D = \varepsilon \mathcal{E} + i\omega\xi B$$
$$\mathcal{H} = i\omega\xi \mathcal{E} + \mu^{-1} B,$$

and the propagation of electromagnetic fields through such a material will again be governed by (5.7) and (5.8). As in the case of dispersive media, an integration over the past history of the fields will be required to calculate the response to a transient pulse.

Cracks, Gaps, and Wires

Another limiting condition in which resolution requirements become intolerable is for a metal structure having one very small dimension. For instance, putting a grid cell inside a narrow crack can produce excessively large execution times. Fortunately, the local behavior of the fields near these singular geometries can be well approximated analytically. Formulas for the fluxes into reasonably sized grid cells neighboring the singularity can be derived from these asymptotic forms and used to update the local fields.

Boundary Conditions

Proper implementation of various boundary conditions associated with material properties such as perfectly conducting walls, resistive sheets, material interface, impedance boundaries, as well as computational boundary conditions such as nonreflecting outer boundaries are very crucial to accurately modeling problems in electromagnetics. In fact, higher-order accurate implementation of boundary conditions in any computational simulation in any discipline is the number one computational issue.

The physical boundary conditions on the electric and magnetic fields at a material interface follow directly from the requirement that Maxwell's equations be satisfied on the interface. In the limiting case of a perfect electrical conductor, these fields vanish inside the conductor, and a sheet of electric current and charge at the interface provides the necessary field discontinuities between the outer and inner surfaces of the conductor.

At the outer limits of the computational domain, the true behavior of the scattered wave is that it propagates outward without reflection. An approximate outgoing-wave condition that can be applied locally at points on the outer boundary will cause some reflection, but errors from this source can be minimized as discussed below.

Perfect Conductors

In a finite-volume scheme, unknown field values are normally computed only at the centroids of the grid cells. However, to capture potentially rapid field variations along a conducting surface, one will have to solve Maxwell's equations right on the conducting surface satisfying the boundary condition on the tangential electric field, $n \times E = 0$, to the same accuracy as any field points (preferably to at least second-order accuracy). A rigorous boundary condition implementation procedure based on characteristic theory can be applied to solving Maxwell's equations right on the body points. This step is crucial to capturing the right surface currents accounting for traveling waves. The final RCS results that are obtained from the surface currents are only as accurate as the values one computes for $n \times H$ on the conducting surface.

The boundary condition procedure for perfectly conducting walls can also be appropriately modified and applied for impedance walls, where the surface tangential electric field, instead of being 0, is proportional to the tangential magnetic field.

Outer Boundaries

Along a given direction in space, the local electromagnetic fields can be grouped into forward and backward propagating combinations. One can develop a hierarchy of nonreflecting boundary conditions using the characteristic theory of signal propagation. A simple first-order condition imposes the requirement that the incoming scattered field signal normal to the outer boundary be 0. Though this is sufficient for many scattering problems, research in numerical algorithms needs to address the development of higher-order nonreflecting boundary conditions. This will allow one to place the outer boundary very close to the scatter and minimize the number of grid cells in the computational domain.

Material Interface

Across a material interface where the material properties ε and μ can be different on either side, certain boundary conditions on the tangential fields are to be satisfied. For example, without the presence of any lossy medium at the interface, the tangential fields $n \times E$ and $n \times H$ are continuous even though the solution vectors D and B will be discontinuous. When a resistive medium is present at the interface, then appropriate jumps in $n \times E$ and $n \times H$ must be accounted for in the boundary condition implementation. In order to compute the right wave reflection and transmission at an interface, the boundary conditions will have to be satisfied to the same level of accuracy as the order of the scheme in the field points. For schemes with order of accuracy greater than two, developing corresponding higher-order material interface boundary condition procedures will be quite challenging.

Geometry/Gridding

Problems in CEM involve arbitrarily shaped three-dimensional geometries that need to be represented properly in the computer simulation. In addition to the external shape, CEM also requires modeling the interior of the penetrable structure. Depending on the formulation (differential or integral), one may choose either a structured grid or an unstructured grid setup.

Two gridding issues that need to be addressed in EM computations are

1) number of grid points per wavelength to properly represent the fields in and around a scatterer, and
2) how far the outer boundary should be placed from the scattering object to adequately simulate the nonreflecting boundary condition.

In general, the number of points/wavelength is not determined by wavelength alone, and involves the body dimensions (characteristic body size with respect to wavelength) also. The outer boundary location, theoretically, can be right on the body surface itself; however, the computational implementation of nonreflecting boundary conditions requires the outer boundary at a few (2 to 5) wavelengths away from the surface. Again, if one can construct higher-order accurate implementations of nonreflecting boundary conditions, the outer boundary can be brought very close to the scattering surface. In general, the necessary grid resolution is provided only around and near the body surface. Between the body and the outer boundary, the mesh is allowed to stretch resulting in very crude (3 to 5 points per wavelength) meshes near the outer boundary regions.

The free space wavelength is reduced to smaller values inside a material (as ε and μ become large, the speed of propagation, $c = \dfrac{1}{\sqrt{\varepsilon\mu}}$, goes down, causing the wavelength to scale accordingly). Thus, the grid resolution must take into account material properties to adequately resolve the fields inside material zones.

The number of grid points per wavelength required depends on the order of accuracy of the numerical scheme. A second-order accurate scheme usually requires at least 10 grid points per local wavelength. One may be able to use a higher-order scheme and minimize the number of grid points. However, as the order of accuracy goes up, the scheme will also require more computations per grid point, which may offset the execution savings with fewer grid points.

The requirement that the fields be resolved accurately with proper grid resolution makes CEM problems computationally intensive, requiring large-scale supercomputing. For example, to compute the radar cross section of a typical aircraft at 1 GHz, even if one uses 10 grid cells per wavelength, will require tens of millions of grid points.

Massively Parallel Computing

With the emergence of massively parallel computing architectures with potential for teraflops performance, any code development activity must effectively utilize the computer architecture in achieving the proper load balance with minimum internodal data communication.

Some of the massively parallel computing architecture issues addressed in the present study are

- Domain decomposition and load balancing
- Internodal message passing with minimum communication delays

- Synchronization for time-accurate computation
- Storage of data for FFT processing of large RCS pulse cases
- Measure of MFLOP rating
- Scalability measure
- Pre- and post-processing

A 512-node nCUBE and a 208-node Intel Paragon are currently used to study these issues in the development of the CEM code.

Validation

Once a CEM code is developed, the results must be validated against known exact solutions and carefully tailored experimental data. There are many computational issues such as grid resolution, location of the outer boundary, and accuracy of the boundary condition procedures that can only be addressed through a careful study of many validation cases. The Electromagnetic Code Consortium (EMCC) has a list of validation cases comprising many target shapes specifically designed for validating codes.

PRESENT CEM CAPABILITY

Both a structured grid version and an unstructured grid version of the CEM code are in development. The structured grid version is in a relatively advanced mature stage, while the unstructured grid version is the subject of state-of-the-art algorithm and code development. Figure 5.1 shows the progression of various CEM developments at Rockwell.

Some of the salient features of the current CEM capabilities are

- Time-domain Maxwell's equations
- Proven algorithms from CFD
- Single pulse (multiple frequency, transient) or continuous incident wave (single frequency, time harmonic steady state)
- Numerical grid generation-structured multizone grid or unstructured grid
- Lossy or lossless material properties
    —Frequency- and time-dependent properties
    —Thin structures (resistive card, lossy paint)
- Vector/parallel code architecture—2 GFLOPS demonstrated on the CrayYMP with 8 processors, and 10 GFLOPS on the Cray-C90 with 16 processors. Scalable performance demonstrated on both the NCUBE and the Intel Paragon.
    Received the 1990 CRAY Gigaflop Performance Award
    Received the 1993 Computerworld Smithsonian Award

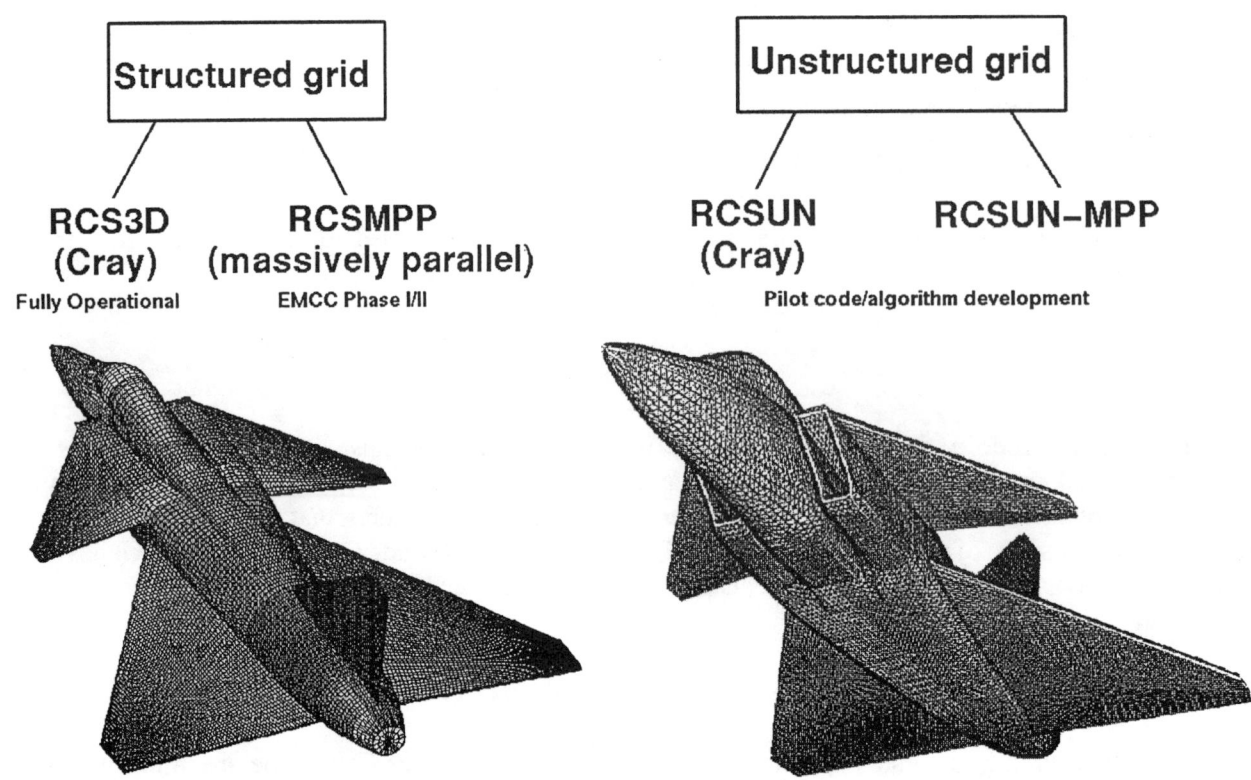

Figure 5.1 Structured grid and unstructured grid-based CEM development.

- Pre- and post-processor graphics/animation
- Application to scattering (RCS), radiation (antenna), EMP/EMI/EMC, and bioelectromagnetics problems
- Ideal for CFD/CEM optimization studies
- The CEM code has been extensively tested for the following geometries:
    1) Canonical obects such as spheres, cylinders, ogives, thin rods, cones, airfoils, and a circular disc
    2) Almond-shaped target
    3) Inlets of various shapes (square, circular, curved, . . .) including the presence of infinite ground plane
    4) Flat plates of various planforms
    5) Double sphere
    6) Complete wing geometries with layers
    7) Finned projectile and cone-cylinder combinations
    8) Scattering from ship-like targets
    9) Complete fighter targets

Three-Dimensional Sphere of Increasing Ka. 500 time steps

# of Grid Points	6 Nodes	24 Nodes	96 Nodes	384 Nodes
6354 $Ka=2.5$	**389** sec (1) 11×11×9 per node	**135** sec (4) 6×6×9 per node		
23818 $Ka=5$	**1300** sec (3) 21×21×9 per node	**389** sec (6) 11×11×9 per node	**135** sec (14) 6×6×9 per node	
90774 $Ka=10$		**1301** sec (17) 21×21×9 per node	**390** sec (25.5) 11×11×9 per node	**135** sec (68) 6×6×9 per node
354294 $Ka=20$			**1302** sec (40) 21×21×9 per node	**390** sec (75) 11×11×9 per node

- For $Ka = 20$, a one-CPU C–90 takes 425 seconds running at 550 MFLOPS
- (·) represents set up time taken by the host
- $Ka = 20$ case runs at 1.5 MFLOPS/node on the nCUBE (60% of peak performance)

Figure 5.2 Scalable performance on a massively parallel architecture.

Some sample results are shown here to illustrate the present capability. The scalable performance of the structured grid CEM code is shown in Figure 5.2 for the nCUBE architecture. As the number of nodes and the problem size increase, the turnaround clock time is maintained, demonstrating one of the scalability measures. Similar scalable performances are achieved on the Intel Paragon and the Cray T3D.

Figure 5.3 shows the results for a square inlet for both a CW case and a pulse case. The comparison of monostatic RCS with experimental data is very good. Figure 5.4 shows an application of the code for predicting the RCS of a complete fighter.

Some of the other applications of the CEM code are shown in Figure 5.5. Figure 5.6 shows an application of the CEM code for bioelectromagnetics to study the effects of microwave heating of cancer tissues (hyperthermia treatment).

Currently work is continuing to further develop the unstructured grid-based CEM code for massively parallel architectures with special emphasis on domain decomposition techniques and applications to problems involving hundreds of millions of grid points. Future efforts will include coupling of the CEM codes with CFD and Computational Structural Mechanics for multidisciplinary optimization studies.

ACKNOWLEDGMENT

This work is funded by Rockwell IR&D, AFOSR, Army Research Laboratory at Aberdeen Proving Ground, NASA Ames Research Center, and the Elecromagnetic Code Consortium.

REFERENCE

Mohammadian, A., V. Shankar, and W.F. Hall, 1991, "Computation of electromagnetic scattering and radiation using a time-domain finite-volume discretization procedure," *Comput. Phys. Commun.* **68**, 175-196.

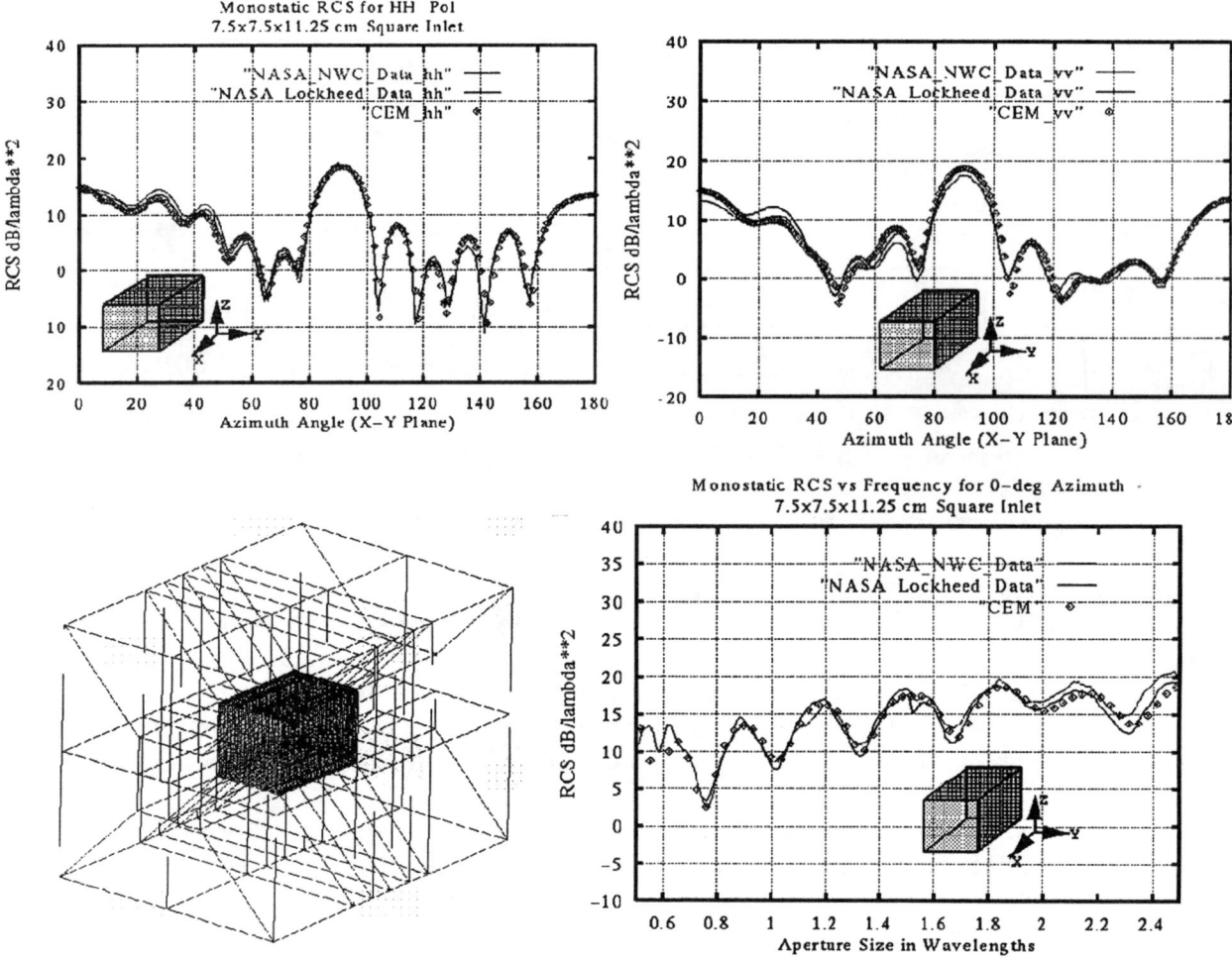

Figure 5.3 Monostatic RCS for a square inlet.

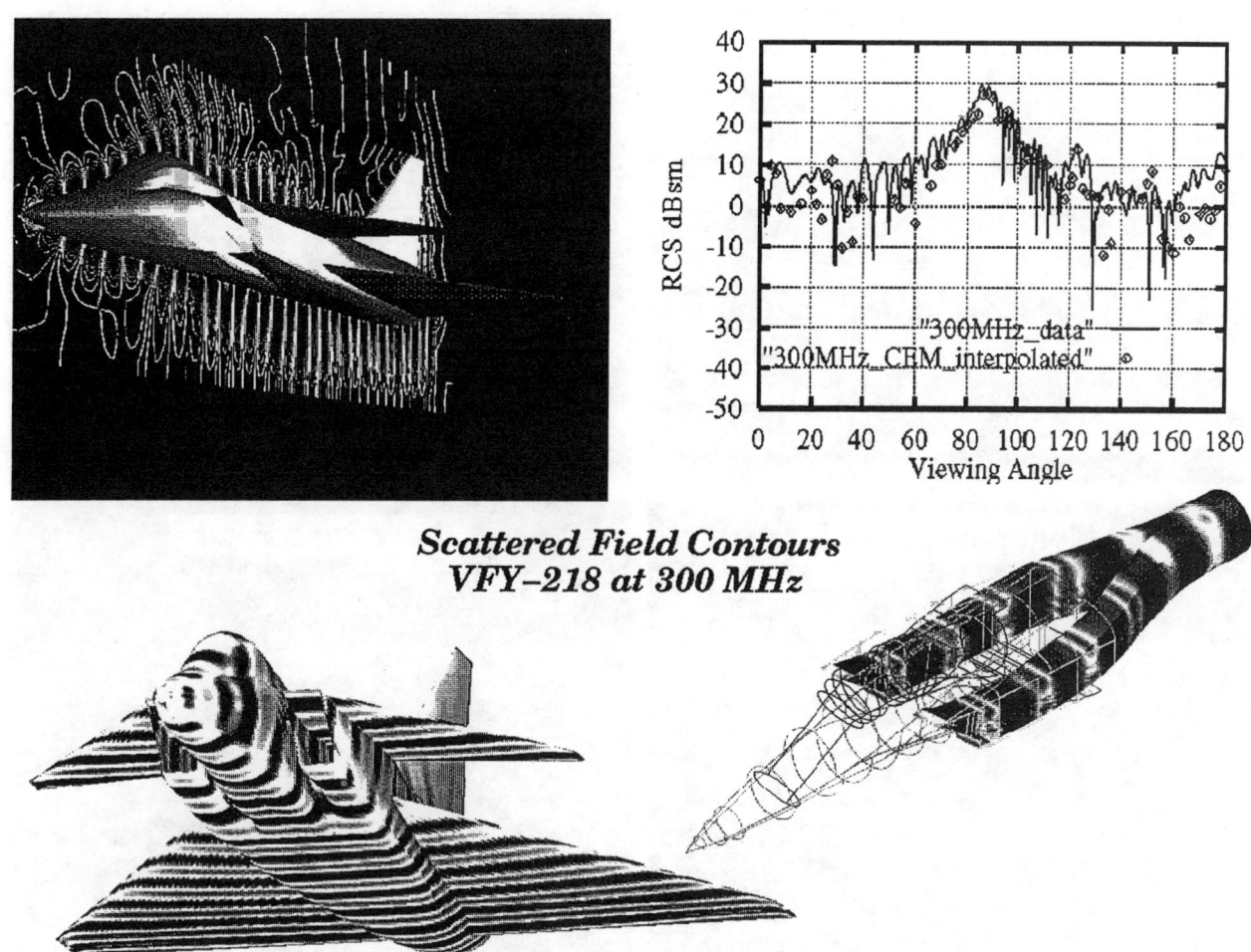

Figure 5.4 RCS for a complete fighter

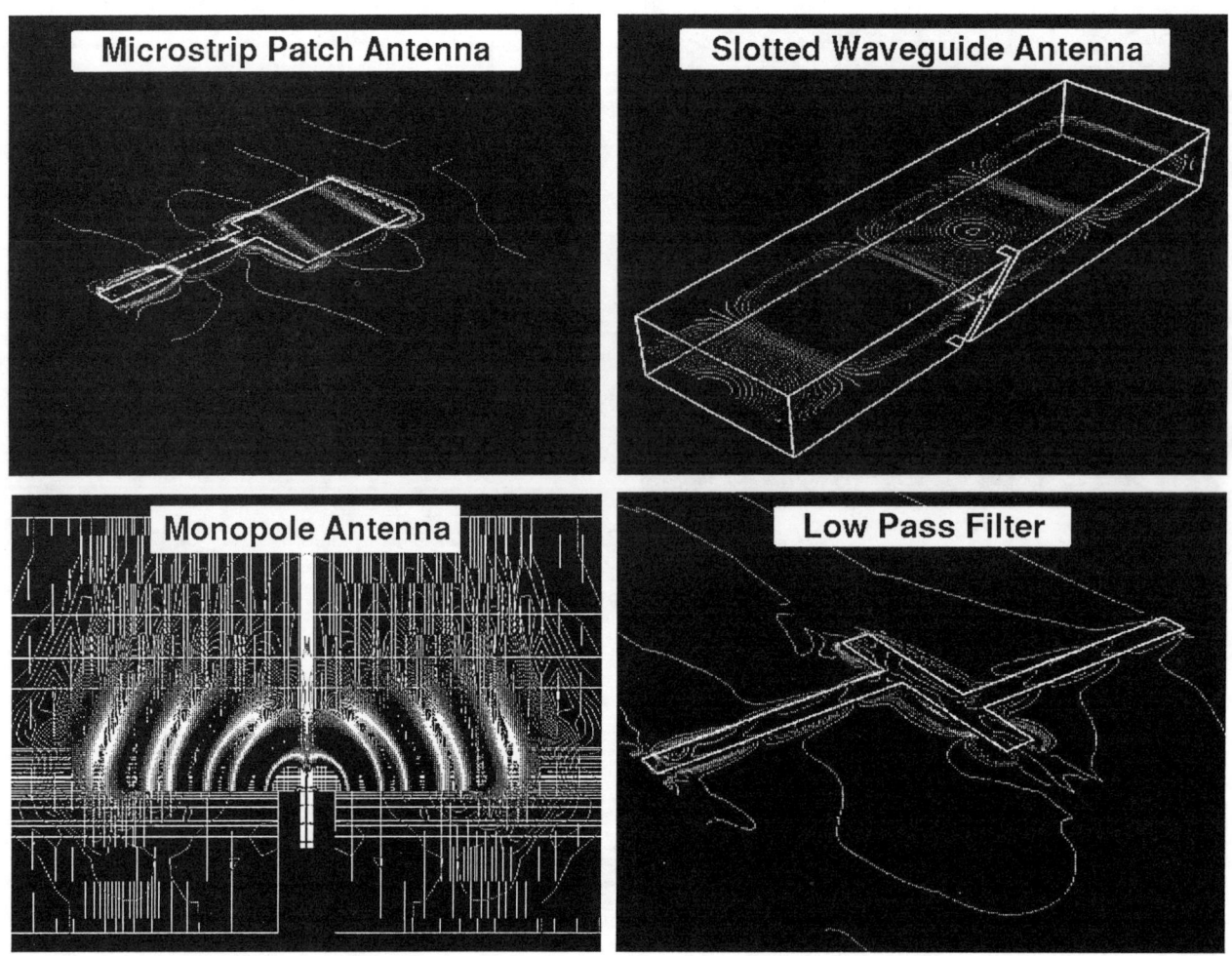

Figure 5.5 Different applications of the CEM code.

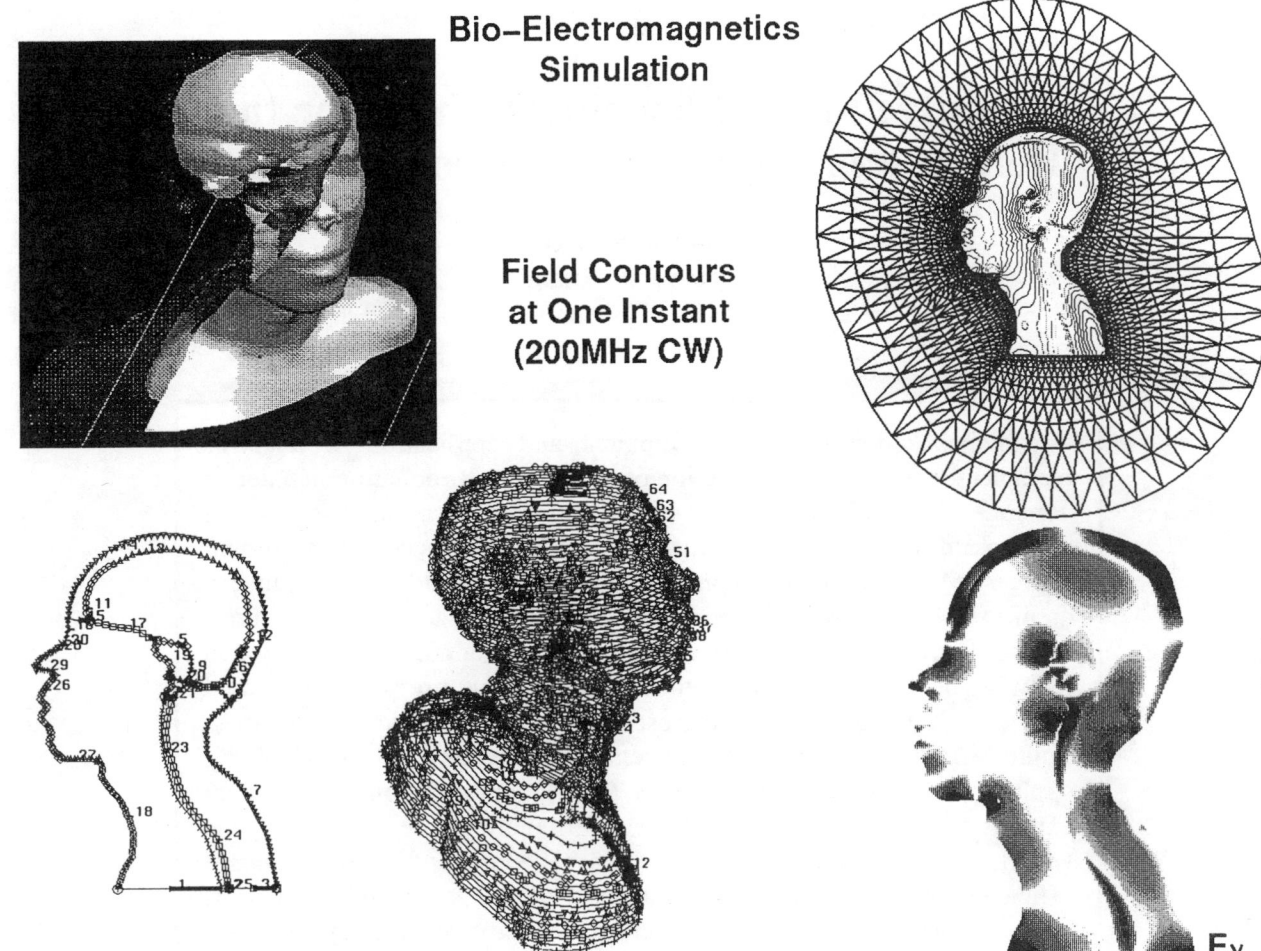

Figure 5.6 Bioelectromagnetic application of the CEM code.

6

Adaptive Finite Element Methods for the Helmholtz Equation in Exterior Domains

James R. Stewart
Thomas J.R. Hughes
Stanford University

> This paper addresses the development and application of adaptive methods for finite element solution of the Helmholtz equation in exterior domains. Adaptivity allows for efficient resolution of both large- and small-scale solution features by minimizing the necessary computational degrees of freedom. This provides a cost-effective mechanism for application of domain-based procedures to large-scale structures. Although we restrict attention to the frequency domain, adaptivity can likewise be beneficial in the time domain. The use of unstructured meshes allows for efficient representation of geometric complexities. Finite element computations in exterior domains are enabled by applying the Dirichlet-to-Neumann (DtN) condition on an artificial exterior boundary. Two finite element formulations are considered: the Galerkin and the Galerkin Least-Squares (GLS) methods. Computational efficiency is achieved by applying an explicit residual-based a posteriori error estimator coupled with a global h-adaptive strategy. An hp-adaptive strategy, in which the element size h and polynomial order p can be refined simultaneously, is also presented. Through two-dimensional simulation of nonuniform radiation from a rigid infinite circular cylinder, it is shown (using h-refinement only) that the adaptive process produces a final mesh with a greatly reduced number of elements for any given accuracy. Based on a solution error of 1 percent, the adaptive mesh contains 3.5 times fewer elements than a uniform mesh (for the Galerkin formulation). When adaptivity is coupled with the GLS formulation, the number of elements required for any given accuracy is reduced even further.

Note: This research was supported by the Office of Naval Research under Grant N00014-92-J-1774.

INTRODUCTION

Adaptive Methods

The purpose of this work is to apply previously developed adaptive finite element methodologies to the solution of the Helmholtz equation in exterior domains. In doing so, the viability of adaptivity in reducing the cost of computation will be demonstrated. We compute solutions to a model problem of acoustics in two dimensions, using Galerkin and Galerkin Least-Squares (GLS) finite element formulations with the fully coupled (truncated) DtN boundary condition (Givoli and Keller, 1989; Harari and Hughes, 1994). The development of the DtN boundary condition has allowed the exterior problem to be posed on a finite domain, which enables a finite element analysis. The finite element meshes consist of linear triangles. We present a general hp-adaptive strategy, which allows specification of simultaneous refinement in both the element size h and spectral order p. However, we show results for h-refinement only. In addition, we consider only the Helmholtz equation, which governs the frequency domain; adaptive solutions in the time domain are not addressed. Although adaptivity in the time domain is more complex due to the need to track solution features over time, much of the methodology contained herein could be adapted for that purpose.

Traditionally, the Helmholtz equation in exterior domains has been solved using boundary element formulations applied to the boundary integral form of the equation (see, e.g., Burton and Miller, 1971; van den Berg et al., 1991; Kleinman and Roach, 1974; Seybert and Rengarajan, 1987; Cunefare et al., 1989; Kirkup, 1989; Demkowicz et al., 1991; Demkowicz and Oden, 1994). Another approach is the application of infinite elements to the exterior problem (Bettess, 1977; Burnett, 1995). One of the main issues concerning the choice of solution methodologies is the cost of achieving a desired level of accuracy. Boundary element formulations engender considerable savings in that the problem size is much smaller, requiring meshing of only the domain boundary. However, these formulations require significant costs in equation formation, solution of the linear system (since the coefficient matrix is full and not symmetric), and so on. Harari and Hughes (1992a) have compared the costs of boundary element formulations to the costs of finite element formulations using the fully coupled DtN boundary condition, and have found that finite element methods are competitive. Burnett (1995) has shown a significant cost advantage of the infinite element approach compared to boundary element methods.

There are several ways in which the cost competitiveness of finite element techniques can be increased. The first is to change the finite element discretization. For example, Harari and Hughes (1990, 1991, 1992b) developed a GLS method that provides higher accuracy than Galerkin, and maintains stability even on coarse meshes. We will compare GLS and Galerkin solutions in this paper. A second means of decreasing the cost of finite element methods is to use a local DtN boundary condition (Harari and Hughes, 1991; Bayliss and Turkel, 1980; Givoli, 1991; Givoli and Keller, 1990), which leads to savings in both equation formation and solution time, since much smaller bandwidths are obtained in the coefficient matrix. The penalty for such an approach seems to be a reduced accuracy, or the need for C^1-continuous shape functions across element interfaces on the artificial boundary. Still other cost-saving measures are possible. Among these are development of iterative solution techniques and parallelization, as well as reducing the cost of mesh generation. The meshing needs of finite element methods can be formidable, particularly in three dimensions. Mesh generation is currently an area of intensive research, and much progress has been made in recent years (see, e.g., Shephard and Georges, 1991; Schroeder and Shephard, 1989, 1990; Peraire et al., 1988; Cavendish et al., 1985; Lohner and Parikh, 1988; Baker, 1989; Blacker and Stephenson, 1991).

Another cost-saving approach is adaptivity, and it is the focus of this paper. The goal of adaptivity is efficiency of the finite element mesh, in that a solution of a given accuracy is to be obtained with a minimum number of degrees of freedom. The role of adaptivity depends heavily on the problem—if highly localized solution features are present, then adaptivity will lead to a larger cost reduction. Adaptivity can be used in conjunction with most of the other cost-saving approaches discussed above. It involves the development and application of technologies outside the finite element solver. In addition to reliable mesh generation and mesh refinement techniques, the primary technologies are a posteriori *error estimation* and *adaptive strategies*.

Figure 6.1 provides a simplistic illustration of how adaptivity works; all adaptive computations fit more or less into this general framework. The diagram shows how the various components of adaptivity work together to drive the efficient placement of the mesh nodes. The a posteriori error estimator takes the finite element solution as input and computes an estimate of the solution error. The adaptive strategy then uses the estimated error distribution to compute a new, more efficient, distribution of element sizes (in the case of *h*-refinement), element spectral orders (in the case of *p*-refinement), or combination thereof (*hp*-refinement). The mesh generator then creates the adaptive mesh with the requested size distribution (in our case by regenerating the mesh globally), and the process repeats until a suitable stopping criterion is satisfied. The idea is to begin with a relatively coarse, uniform mesh, and drive the adaptive refinement through the selection of smaller and smaller error tolerances in each adaptive iteration. Usually three or four iterations of this loop are sufficient. Efficiency is attained by placing more degrees of freedom in areas where solution errors are large, and fewer degrees of freedom in areas of comparatively smaller errors. In doing so, the adaptive strategy attempts to compute a distribution of degrees of freedom such that the error is equidistributed among the elements of the adaptive mesh.

Each of the component technologies—mesh generation, finite element methods, a posteriori error estimation, and adaptive strategies—individually plays a critical role in the realization of an efficient mesh. The meshes in this paper were generated using an advancing front mesh generator written by Jaime Peraire (Peraire et al., 1987). The error estimator and adaptive strategy are derived in Stewart and Hughes (1995). The derivation of the error estimator is based on a general methodology developed by Johnson and coworkers in other contexts, such as model elliptic and advection-diffusion operators and nonlinear conservation laws (see Johnson and Hansbo, 1992; Eriksson and Johnson, 1988; Johnson, 1990, 1992, and references therein). The error estimator is an explicit function of residuals, and provides an upper bound on the L_2-norm of the error. In addition, it does not assume positive-definiteness of the operator. This is important for the propagating case $(k^2 > 0)$, in which the Helmholtz operator may lose positivity. Adaptive computation of propagating waves is thus made possible. The adaptive strategy computes simultaneous *hp*-refinement, and for the problem shown in this paper we specialize to the case of *h*-refinement only.

The outline of this paper is as follows. In the second section the problem statement is given. In addition, the mesh generator, finite element formulations, a posteriori error estimator and *hp*-adaptive strategy are described. The remainder of the paper is dedicated to the presentation of results (for *h*-refinement only). Through solution of nonuniform radiation from an infinite circular cylinder $(ka = 2\pi)$, we demonstrate the adaptive solution procedure. Galerkin results are presented in the first part of section 3 and GLS results are presented in the second part of that section. Comparisons are made throughout the third section to the case of uniform refinement, providing a context for the adaptive results. Finally, conclusions are given in the fourth section.

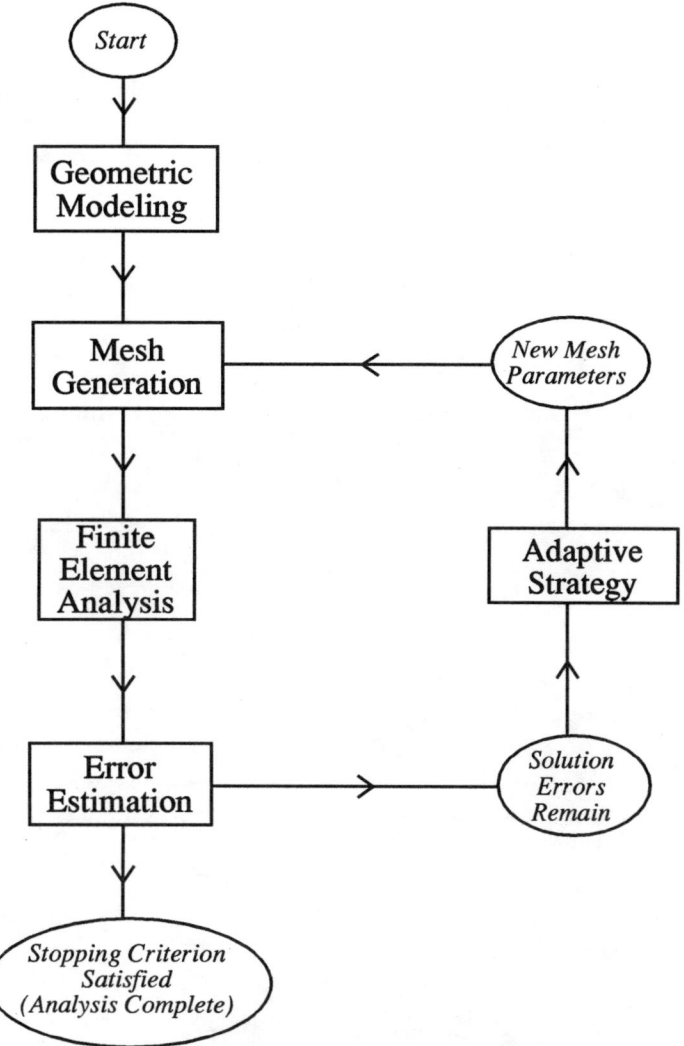

Figure 6.1 Schematic diagram of adaptive solution algorithm, showing the critical technologies involved.

PROBLEM STATEMENT AND ADAPTIVE SOLUTION METHODOLOGIES

In this section we present the problem statement (in the first part) and briefly describe the technologies used to perform the adaptive computations—mesh generation, finite element methodologies, a posteriori error estimation, and the *hp*-adaptive strategy (in the second part).

Problem Statement

The continuous problem is as follows: Find $\phi: \overline{\Omega} \to \mathbf{C}$, the spatial component of the acoustic pressure or velocity potential, such that

$$-\mathcal{L}\phi = f \quad \text{in } \Omega \tag{6.1}$$

$$\phi = g \quad \text{on } \Gamma_g \tag{6.2}$$

$$\phi_{,n} = ikh \quad \text{on } \Gamma_h \tag{6.3}$$

$$\phi_{,n} = -M\phi \quad \text{on } \partial B_R, \tag{6.4}$$

where $\mathcal{L}\phi := \nabla^2 \phi + k^2 \phi$ is the Helmholtz operator, $k \geq 0$ is the wave number, and a comma denotes partial differentiation. The interior boundary of Ω is

$$\Gamma = \overline{\Gamma_g \cup \Gamma_h}, \tag{6.5}$$

where we assume

$$\Gamma_g \cap \Gamma_h = \emptyset. \tag{6.6}$$

The outward normal to Ω is n, $i = \sqrt{-1}$ is the imaginary unit, and $f: \Omega \to \mathbf{C}$, $g: \Gamma_g \to \mathbf{C}$, and $h: \Gamma_h \to \mathbf{C}$ are the prescribed data. The DtN boundary condition is given by (6.4), where M is the DtN map and ∂B_R is the artificial exterior boundary of Ω. The geometry is depicted in Figure 6.2.

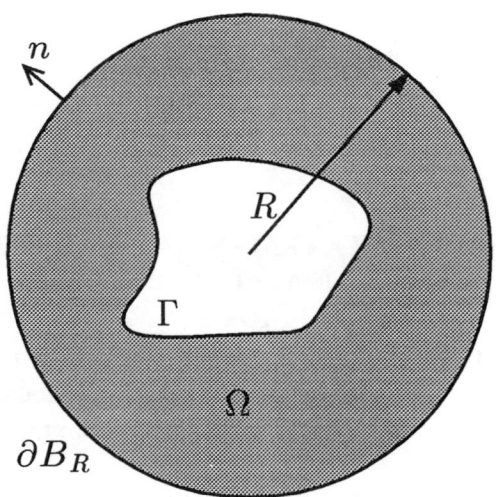

Figure 6.2 Problem geometry for the solution of the Helmholtz equation in an exterior domain.

The weak form of the continuous problem is now stated: Find $\phi \in S$ such that $\forall w \in V$

$$A_k(w,\phi) = L(w), \tag{6.7}$$

where

$$A_k(w,\phi) := a_k(w,\phi) + b_k(w,\phi) \tag{6.8}$$
$$a_k(w,\phi) := (\nabla w, \nabla \phi) - k^2(w,\phi) \tag{6.9}$$
$$b_k(w,\phi) := (w, M\phi)_{\partial B_R} \tag{6.10}$$
$$L(w) := \int_\Omega \overline{w} f d\Omega + \int_{\Gamma_h} \overline{w} ikh d\Gamma \tag{6.11}$$

and S and V denote the usual spaces of trial solutions and weighting functions, respectively, and an overbar denotes conjugation. In addition, $(\cdot,\cdot): V \times S \to \mathbf{C}$ is the $L_2(\Omega)$ inner product (with the first argument conjugated), and subscripts on inner products denote domains of integration other than Ω.

Adaptive Solution Methodologies

Mesh Generation

The adaptive meshes are produced via global regeneration; i.e., for each adaptive refinement, the old mesh is discarded and a completely new mesh is generated. This allows us to use the same mesh generator for the initial mesh and the adaptive meshes. All meshes in this paper were generated using a 2D advancing front mesh generator written by Jaime Peraire (Peraire et al., 1987). The method utilizes a background mesh to provide a discrete distribution of element sizes. This is particularly convenient for adaptive mesh generation, since the adaptive element size distribution can be specified using the old mesh. The h-adaptive strategy, presented below, is used to compute a new element size at each x_K, where K is an element of the old mesh. The element sizes are scattered to the nodes (where they are stored), and the old mesh then serves as the background mesh.

Finite Element Formulations

The Galerkin form of the continuous problem (6.7) is as follows: Find $\phi^h \in S^h$ such that $\forall w^h \in V^h$

$$A_k(w^h, \phi^h) = L(w^h), \tag{6.12}$$

where $S^h \subset S$ and $V^h \subset V$ are finite dimensional spaces containing continuous piecewise linear polynomials, and h refers to a suitable measure of element length.

The Galerkin Least-Squares (GLS) form of (6.7) is given by

$$A_{GLS}(w^h, \phi^h) = L(w^h) - \tau(\mathcal{L}w^h, f)_{\tilde{\Omega}}, \tag{6.13}$$

where

$$A_{GLS}(w^h, \phi^h) := A_k(w^h, \phi^h) - \tau(\mathcal{L}w^h, \mathcal{L}\phi^H)_{\tilde{\Omega}} \tag{6.14}$$

and the subscript $\tilde{\Omega}$ denotes integration over the union of element interiors. The parameter τ has dimensions of length-squared and is given by

$$\tau = \left\{ \frac{1}{(kh)^2} - \frac{6}{(kh)^4} \left[\frac{1-\cos(kh)}{2+\cos(kh)} \right] \right\} h^2. \tag{6.15}$$

A detailed derivation is given in Harari (1991). The GLS formulation was designed to provide phase accuracy for propagating solutions with fewer mesh points than required for comparable Galerkin solutions. In addition, consistency and stability are maintained. Later we compare GLS and Galerkin results.

A Posteriori Error Estimation

The a posteriori error estimator was derived by Stewart and Hughes (1995) using the method developed by Johnson and coworkers (see, e.g., Johnson and Hansbo, 1992; Johnson, 1992; Eriksson and Johnson, 1988; Johnson, 1990). It is in the form of an upper bound on the global L_2-norm of the error, and is an *explicit* function of residuals (no local problems need to be solved). Important attributes of the error estimator, in the context of the Helmholtz equation in exterior domains, are that nonpositive operators are easily handled (due to the use of the L_2-norm) and the DtN boundary condition is incorporated in a natural way. The a posteriori error bound for the Galerkin formulation is given by

$$\|e\| \leq C_i C_s \left[\|h^2 r^h\|_{L_2(\tilde{\Omega})} + \|h^2 R^h\|_{L_2(\tilde{\Omega})} \right] \tag{6.16}$$

and for the GLS formulation,

$$\|e\| \leq C_i C_s \left[\|h^2 r^h\|_{L_2(\tilde{\Omega})} + \|h^2 R^h\|_{L_2(\tilde{\Omega})} \right] + C_{GLS} \|h^2 r^h\|_{L_2(\tilde{\Omega})} \tag{6.17}$$

The first term on the right-hand side of (6.16) is the error contribution from element interiors, and the second term is the error contribution from element boundaries (which is assigned a constant value over each element interior for the purpose of computing the norm). C_i, C_s, and C_{GLS} are constants.[1] Thus, the GLS bound (6.17) is equal to the Galerkin bound (6.16) plus an additional contribution from the error on element interiors.

The residual on element interiors, r^h, is given by

$$r^h = f + \nabla^2 \phi^h + k^2 \phi^h. \tag{6.18}$$

---

[1] C_i emanates from a standard interpolation estimate, C_s from the statement of strong stability (i.e., regularity) of a continuous dual problem posed in Ω, and $C_{GLS} = 1.09 C_i C_s + 0.09$. Details are provided in Stewart and Hughes (1995).

The element boundary residual, R^h, is defined as

$$R^h\Big|_K = \max_{S \subset \partial K} \max_S (|r_2^h|) \qquad (6.19)$$

where

$$h_K r_2^h(\phi^h) = \begin{cases} \frac{1}{2}[\![\phi^h_{,n_s}]\!] & \text{on } S \subset \partial K_{int} \\ \phi^h_{,n} - ikh & \text{on } S \subset \partial K \cap \Gamma_h \\ \phi^h_{,n} - (-M\phi^h) & \text{on } S \subset \partial K \cap \partial B_R \\ 0 & \text{on } S \subset \partial K \cap \Gamma_g \end{cases} \qquad (6.20)$$

In two dimensions, S denotes an element edge and ∂K_{int} denotes an interior mesh edge belonging to element K. In words, (6.19) states that R^h on element K is the maximum value of the boundary residual, r_2^h, over all the edges (i.e., the boundary) of the element. The boundary residual involves a jump in the normal derivative of ϕ^h across interior edges, which emanates from the use of C^0-continuous shape functions. It is assumed that on Γ_g the given data are exactly representable by the finite element functions, i.e., $g \in V^h$.

hp-adaptive Strategy

The hp-adaptive strategy is derived in detail in Stewart and Hughes (1995). We summarize the main results here. For h-refinement only, p is unchanged from the old mesh and the new element size h is given by

$$h_{new}(x_K) = \left(\frac{\bar{e}_{tol}}{\bar{e}_K}\right)^{\frac{1}{p_K + \beta}} h_K, \qquad (6.21)$$

where the subscript K refers to quantities in the old mesh, $\bar{e}_{tol}$ is a user-defined element error tolerance for the subsequent adaptive mesh solution, and $\beta = 2$ for $2D$ and $\beta = 2.5$ for $3D$. Thus, (6.21) computes a new h for the location x_K of element K in the old mesh. The quantity $\bar{e}_K$ is the (scaled) estimated error in element K, given by

$$\bar{e}_K \stackrel{def}{=} \left[\|h_K^2 r_K^h\|_{\tilde{\Omega}_K}^2 + \|h_K^2 R_K^h\|_{\tilde{\Omega}_K}^2\right]^{1/2}. \qquad (6.22)$$

Note that (6.22) does not require the constants C_i, C_S or C_{GLS} (see (6.16) and (6.17)) to be known. This engenders considerable computational savings. The computation of these constants is not discussed herein, but will be the focus of a future paper. The element error $\bar{e}_K$, therefore, represents the approximate element contribution to the scaled global error, where the scaling is with respect to C_i, C_S (for GLS, C_{GLS} is essentially ignored in order for the scaling to be carried out). The element error tolerance $\bar{e}_{tol}$ must reflect this scaling. We compute $\bar{e}_{tol}$ using a formula given in Stewart and Hughes (1995),

$$\bar{e}_{tol} = \frac{\alpha}{n_{el}} \sum_{K=1}^{n_{el}} \bar{e}_K, \qquad (6.23)$$

where n_{el} is the number of elements in the old mesh, and $\alpha \in \,]0,1[$ is a user-defined parameter.

If p is variable, then the dependence of C_i on p can be accounted for by incorporating C_i into the definition of element error:

$$\bar{e}_{i,K} \stackrel{def}{=} C_{i,K} \left[\left\| h_K^2 r_K^h \right\|_{\tilde{\Omega}_K}^2 + \left\| h_K^2 R_K^h \right\|_{\tilde{\Omega}_K}^2 \right]^{1/2}, \qquad (6.24)$$

where $C_{i,K}$ is the interpolation constant on element K, and $\bar{e}_{i,K}$ represents an error scaled only by C_S instead of $C_i C_S$ (compare with (6.22)). This change must be suitably accounted for in the specification of $\bar{e}_{tol}$. For p-refinement only, h is unchanged from the old mesh and the new element shape function polynomial order for location x_K is given by

$$p_{new}(x_K) = \frac{\log\left(\frac{\bar{e}_{tol}}{\bar{e}_{i,K}}\right)}{\log h_K} + p_K, \qquad (6.25)$$

where $\bar{e}_{i,K}$ has been used instead of $\bar{e}_K$.

For a general hp-refinement, where h and p can be changed simultaneously, p_{new} is computed using (6.25) and this result is incorporated into the computation of h_{new}:

$$h_{new}(x_K) = \left(\frac{\bar{e}_{tol}}{\bar{e}_{i,K}}\right)^{\frac{1}{p_K+\beta}} (h_K)^{\frac{p_K+\beta}{p_{new}+\beta}}. \qquad (6.26)$$

The hp-adaptive strategy is given in Box 6.1.

Remarks:

1. Computation of the dual problem stability constant, C_S, is not necessary.

2. The hp-adaptive strategy allows for either increasing or decreasing h, but p can only be increased (or held constant). In addition, p_{new} must be an integer, but h_{new} is generally a noninteger. Directional stretching of the mesh is not accounted for.

3. It may be necessary to smooth the distribution of p_{new}, and/or enforce a limit on the maximum value.

> Box 6.1 - *hp*-Adaptive Strategy
>
> 1. Define an adjustable parameter ε_p, where $\varepsilon_p \geq 0$.
> 2. Compute $\bar{e}_{tol}$ using (6.23).
> 3. For element K of the old mesh, compute $p_{new}(x_K)$ using (6.25).
> 4. Set $p_{temp} = p_K$.
> 5. **If** $p_{new}(x_K) \geq p_{temp} + \varepsilon_p$,
>   **then** $p_{new} \leftarrow p_{temp} + 1$; go to beginning of Step 5,
>   **else** $p_{new}(x_K) \leftarrow p_{temp}$.
> 6. Compute $h_{new}(x_K)$ using (6.26).
> 7. Repeat Steps 3-6 for each K.

4. The parameter ε_p defines the relative amounts of *p*-refinement and *h*-refinement. For example, $\varepsilon_p > 1$ favors *h*-refinement, and $\varepsilon_p \gg 1$ essentially prevents any *p*-refinement. On the other hand, $\varepsilon_p < 1$ favors *p*-refinement.

5. The problem of generating the mesh with the requested $\{h_{new}, p_{new}\}$ distribution has not been addressed. The method used to refine the mesh dictates what form this distribution must take, and significantly influences the *achievability* of the requested adaptive mesh. As given in Box 6.1, the adaptive strategy defines $\{h_{new}, p_{new}\}$ at discrete locations x_K, which may be taken as the centroids of elements K in the old mesh. This is a convenient form for an advancing front mesh generator, for example, in which the $\{h,p\}$ distribution is interpolated from a background mesh. If a finite quadtree/octree mesh generator is used, the h_{new} distribution may have to be converted to desired quadrant/octant levels. If, instead, a local mesh enrichment procedure is used, in which existing elements are subdivided, $h_{new}(x_K)$ may have to be converted to an integer value, indicating the number of new elements to form from element K. Other techniques can also be used for mesh refinement.

6. As mentioned previously, a sequence of several meshes is used to obtain a complete adaptive solution, where in each step the solution becomes gradually more resolved. The parameter α in (6.23) controls how fast this resolution takes place, or, in other words, the amount of error reduction in each subsequent mesh. This parameter plays an important role in the overall efficiency of the adaptive algorithm. The goal is to drive down the error as quickly as possible without sacrificing efficiency of the adaptive mesh. If α is too low, then the adaptive meshes will contain large areas of overrefinement, decreasing their efficiency. On the other hand, if α is too high, then error reduction will be slow and too many meshes will be necessary. One can also change α from mesh to mesh, or derive an alternative formula for computing $\bar{e}_{tol}$.

For the computations herein, p is fixed (for linear elements, $p = 1$), $\beta = 2$ (for 2D), and a distribution of the new mesh size, h_{new}, is all that is required. From (6.21), the computation of h_{new} reduces to

$$h_{new}(x_K) = \left(\frac{\bar{e}_{tol}}{\bar{e}_K}\right)^{\frac{1}{3}} h_K . \quad (6.27)$$

The same adaptive strategy is used for both the Galerkin and the GLS formulations. This does not imply, however, that the element *errors* are equivalent. Note that $\bar{e}_K$ depends on residuals, which are not necessarily the same for Galerkin and GLS. This will be further explored in the GLS results section below.

As discussed previously, the h-adaptive strategy computes a pointwise distribution of h_{new}. The values are stored at the nodes of the old mesh, which in turn provides the necessary input for the mesh generator to build the new, adaptive mesh.

NONUNIFORM RADIATION FROM A RIGID INFINITE CIRCULAR CYLINDER

In this section we compute nonuniform radiation from a rigid infinite circular cylinder for a wave number of $ka = 2\pi$, where a is the radius of the cylinder. This problem was previously solved in (Harari and Hughes, 1991) for $ka = \pi$ on a coarse mesh of quadrilateral elements, and Galerkin and GLS results were compared. Here we examine the problem in more detail, focusing on convergence with both adaptive and uniform mesh refinement. Galerkin results and Galerkin Least-Squares results are presented below.

A brief description of the problem is now given. Referring to Figure 6.2, the interior boundary Γ represents the cross section of the cylinder. Dirichlet boundary conditions are applied to the cylinder surface—the portion $-\bar{\alpha} < \theta < +\bar{\alpha}$ is assigned a unit value, while the remaining portion is assigned a homogeneous value. The exact solution is given by

$$\phi_{exact} = \frac{2}{\pi} \sum_{n=0}^{\infty} {}' \frac{\sin n\bar{\alpha}}{n} \frac{H_n^{(1)}(kr)}{H_n^{(1)}(ka)} \cos n\theta , \quad (6.28)$$

where here we choose $\bar{\alpha} = 5\pi/32$. Highly localized features are present, namely the discontinuities in the boundary values at $\theta = \pm\bar{\alpha}$. The challenge of the adaptive analysis is to *efficiently* capture these features.

Galerkin Results

The solution was computed on a sequence of 5 meshes. Figure 6.3 shows the real part of the solution along with the corresponding mesh, for each of the 5 meshes. The number of elements and nodes in each mesh are given in Table 6.1. The computation begins with a coarse uniform mesh, while the subsequent adaptive meshes are obtained by gradually decreasing the element error tolerance, $\bar{e}_{tol}$. The value of α for each adaptive mesh, used in (6.23) to compute $\bar{e}_{tol}$, is also listed in Table 6.1.

The solution becomes highly attenuated to the left of the cylinder, and the error estimator records a very small error there. This is reflected in the second mesh, as the adaptive strategy computes a large element size. The refinement capturing the radiated wave is also evident in the second mesh. This problem contains highly localized features, namely the discontinuities in the boundary values at $\theta = \pm 5\pi/32$. It can be seen in the fourth and fifth meshes that they are efficiently captured. A high degree of refinement exists at the discontinuities, and a lesser degree of refinement resolves the radiated wave.

The mesh gradually coarsens towards the left side of the cylinder, efficiently adapting to the attenuating solution.

Table 6.1 Number of Elements and Nodes in the Sequence of Meshes Shown in Figure 6.3, for the Adaptive Solution of Nonuniform Radiation from a Rigid Infinite Circular Cylinder, $ka = 2\pi$

Mesh	n_{el}	n_{nd}	α Eq. (6.23)[a]
1	184	124	—
2	323	194	.15
3	797	458	.13
4	1812	1001	.14
5	2791	1521	.25

[a]The value of α for each adaptive mesh, used in (6.23) to compute $\bar{e}_{tol}$.

In the second mesh in Figure 6.3, it is apparent that the coarsening of the elements has caused a degradation of the geometry. This is not seen to be a problem, since the subsequent refinement restores the geometrical representation. However, if it is desired to maintain geometrical integrity throughout the refinement process, this can easily be accomplished by limiting the maximum allowable element size. In doing so only the extremely large elements are affected; thus, very few elements would be added to the mesh.

Figure 6.4 shows the final adaptive mesh and a uniform mesh with nearly the same number of elements. Alongside the full mesh is an enlargement of the region near the upper boundary condition discontinuity. The elements in the uniform mesh are extremely large in this region compared to the adaptive mesh, leading to a relatively poor resolution of the discontinuity. This contributes to a global error ($\|e\| = 3.8\%$) over 3 times greater than that in the adaptive mesh ($\|e\| = 1.2\%$).

Convergence with increasing number of elements is shown in Figure 6.5, comparing the adaptive and uniform mesh (exact) errors for the entire sequence of mesh refinements.

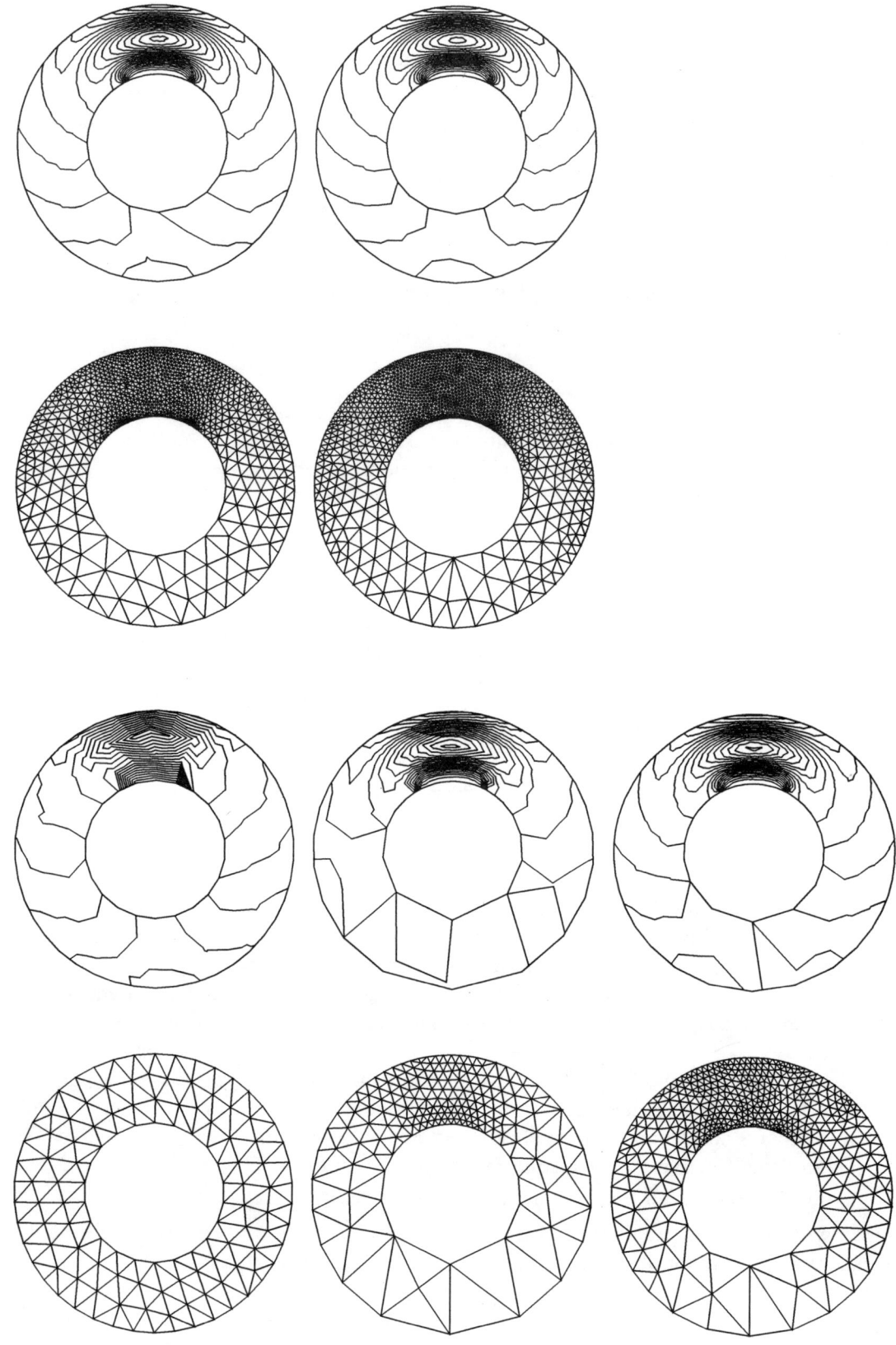

Figure 6.3 Nonuniform radiation from an infinite circular cylinder, $ka = 2\pi$: real part of solution along with corresponding mesh, showing the five meshes used in the adaptive solution. The number of elements and nodes in each mesh is given in Table 6.1.

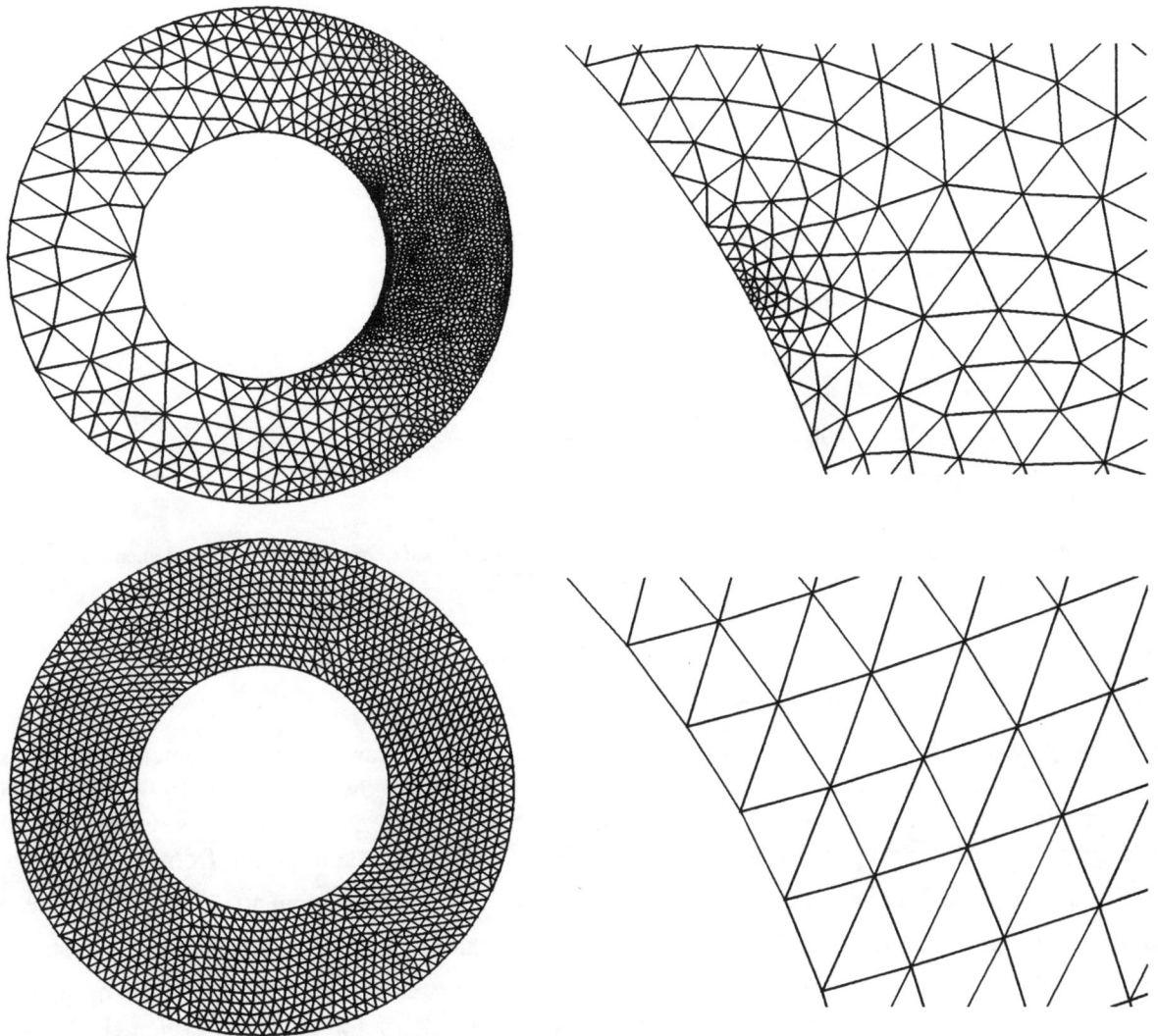

Figure 6.4 Nonuniform radiation from an infinite circular cylinder, $ka = 2\pi$: final adaptive mesh (fifth mesh in adaptive sequence; 2,791 elements, $\|e\| = 1.2\%$) and uniform mesh (2,792 elements, $\|e\| = 3.8\%$), along with detailed view of meshes in region of upper boundary condition discontinuity.

Recall that the analytical solution is given by (6.28). To get an idea of required mesh sizes for further decreases in error, linear fits of the convergence data are also shown. The slopes of these lines, indicated in the figure, show that the convergence rate of adaptive refinement is roughly the same as the convergence rate of uniform refinement. It is easily seen that the uniform mesh requires a much larger number of elements to achieve a given error tolerance, compared to the required adaptive mesh size. For example, to attain $\|e\| = 1\%$ requires just under 4,000 elements in the adaptive mesh, but over 13,000 elements in the uniform mesh.

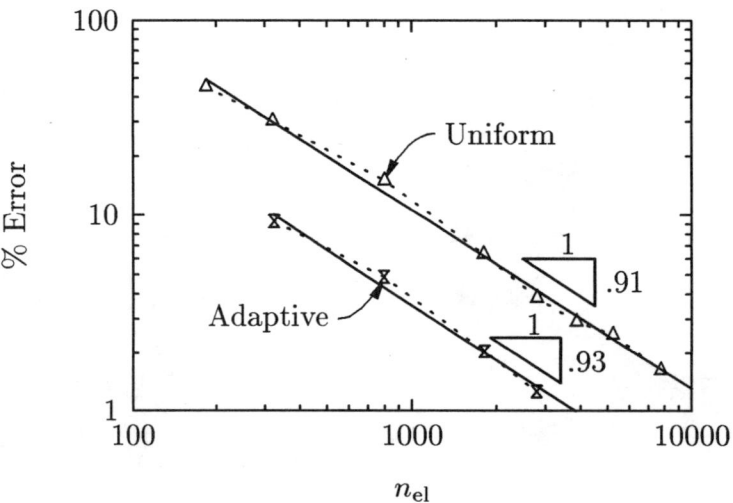

Figure 6.5 Nonuniform radiation from an infinite circular cylinder, $ka = 2\pi$: exact error vs. number of elements in the mesh. The solid lines are curve fits of the data.

Remarks:

1. Convergence studies are often presented in terms of the mesh parameter, or element size, h, as it asymptotically approaches 0. The definition of h becomes confusing, however, in adaptive meshes where h can vary considerably throughout the mesh and approach 0 at different rates. For this reason it is more convenient to present convergence studies in terms of n_{el}. An intuitive conversion from n_{el} to h can be made for uniform mesh convergence. In two dimensions, n_{el} is proportional to $1/h^2$. Therefore, the slope of -0.91 indicated in Figure 6.5 roughly translates to a slope of $+1.82$ if $\|e\|$ is plotted versus h. In other words, the error decreases at a rate slightly lower than the square of h; the latter is the optimal rate of convergence for the Galerkin method. Since the slope of $\|e\|$ vs. n_{el} for adaptive refinement is -0.93, it is concluded that nearly optimal convergence is also obtained in this case. The primary effect of h-adaptivity is to reduce the constant of proportionality between solution error and problem size.

2. For computing the exact error, 200 terms were used in the series given by (6.28). This provided a reasonable approximation of the discontinuities on the wet surface at $\theta = \pm 5\pi/32$, where the series converges to the average value $\phi = 0.5$ (note that the term "discontinuities" is used loosely—the truncated series representation of them is actually smooth but very steep). To provide consistency and meaning to the term "exact error," the truncated series solution was used as the Dirichlet boundary condition. In addition, a node was fixed at each discontinuity and assigned a value $\phi^h = 0.5$. This was found to be critical since if the discontinuities were not "captured," the convergence rate would be reduced (i.e., suboptimal convergence would occur when the mesh became refined enough for the overall error to be dominated by the error at the discontinuities).

3. Note that the boundary condition $g \notin V^h$, which violates the assumption in (6.16). Nonetheless, as seen in foregoing results, the error estimator has enough robustness to adapt to the boundary solution. As

the mesh is refined, the approximation of g improves. With respect to the analytically discontinuous problem, i.e., the problem described by computing an infinite number of terms in the analytical solution, the only error incurred by the piecewise linear representation of g would be at the two discontinuities. With respect to the truncated series form of the analytical solution, additional error (assumed to be small) is incurred away from the "discontinuities."

GLS Results

In addition to applying adaptive procedures, the efficiency of the mesh (uniform or adaptive) can be increased by improving the finite element formulation. This is precisely the goal of the Galerkin Least-Squares (GLS) formulation, given by (6.6), which was designed to enhance accuracy while maintaining stability and consistency (Harari and Hughes, 1991). The method provides nodal exactness (for a 1D model problem) for propagating solutions up to the limit of mesh resolution (4 elements per wave). In addition, GLS is trivial to implement; for details, see Harari and Hughes (1991). In this section, we compare CLS and Galerkin results for the nonuniform radiation problem.

Contours of $\text{Re}(\phi^h)$ on a coarse uniform mesh are shown in Figure 6.6, comparing Galerkin and GLS formulations. It is clear that the GLS solution is more resolved, as evidenced by a more defined wave pattern to the right of the cylinder. The exact error in the Galerkin solution is 50 percent higher.

Figure 6.7 shows the exact error as a function of n_{el}, for GLS with both uniform and adaptive refinement, as well as Galerkin with both uniform and adaptive refinement. The Galerkin convergence curves are the same ones that were presented in Figure 6.5, and are included here for comparison with GLS convergence. We consider first the GLS convergence with uniform refinement, and hereafter refer to it as GLS/uniform (analogous notation is used for the other cases). As expected, the GLS/uniform curve lies below the Galerkin/uniform curve, indicating improved accuracy using GLS.

The convergence of GLS (on uniform meshes) is now compared to that of (Galerkin) adaptive refinement. Referring again to Figure 6.7, the GLS/uniform convergence curve lies above the Galerkin/adaptive curve, indicating (at least for this problem) that adaptivity results in higher computational efficiencies relative to GLS. Finally, adaptivity is *combined* with GLS to produce the greatest gains in efficiency; this is shown by the GLS/adaptive curve in Figure 6.7. It is seen that adaptivity and GLS are complementary in that their effects are additive. It is also evident that the convergence rates of GLS mirror those of Galerkin.

The preceding conclusions are based on analysis of the *exact* error. We now focus on the *estimated* error. Figure 6.8 repeats the convergence study of Figure 6.7, this time showing the estimated error as a function of mesh refinement. The estimated error $\bar{e}$ is obtained by dividing the right-hand side of (6.16) by $C_i C_s$, viz.,

$$\bar{e} \stackrel{\text{def}}{=} \left\| h^2 r^h \right\|_{L_2(\tilde{\Omega})} + \left\| h^2 R^h \right\|_{L_2(\tilde{\Omega})}. \tag{6.29}$$

Therefore, $\bar{e}$ represents the *scaled* error:

$$\frac{\|e\|}{C_i C_S} \leq \bar{e}. \tag{6.30}$$

Note for GLS, C_{GLS} is essentially ignored in order for the scaling to be carried out. We examine, therefore, only relative magnitudes of $\bar{e}$. It is immediately evident from Figure 6.8 that $\bar{e}_{GLS}$ is nearly identical to $\bar{e}_{\text{Galerkin}}$, for both uniform refinement (top curve) and adaptive refinement (bottom curve). This is in contrast to the conclusions of Figure 6.7, where GLS significantly improves upon the accuracy of Galerkin. Since $\bar{e}$ is essentially a function of the residual, it must be concluded here that the GLS residual is nearly identical to the Galerkin residual, even when the two solutions are different.

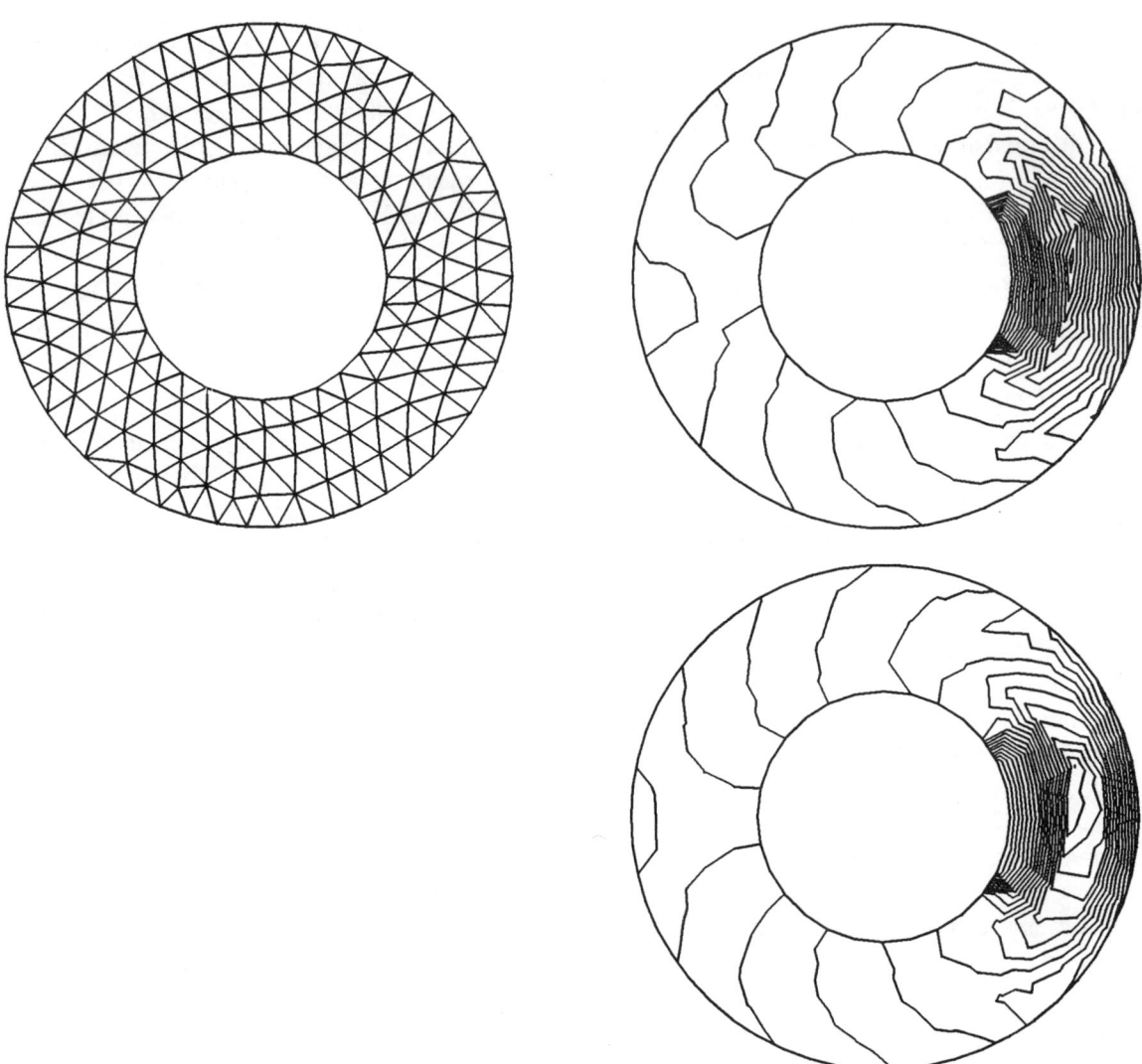

Figure 6.6 Nonuniform radiation from an infinite circular cylinder, $ka = 2\pi$: comparison of $\text{Re}(\phi^h)$ for the Galerkin (top) and GLS (bottom) formulations, on a relatively coarse uniform mesh (320 elements, 201 nodes). For the Galerkin solution, $\|e\| = 30\%$; for the GLS solution, $\|e\| = 20\%$.

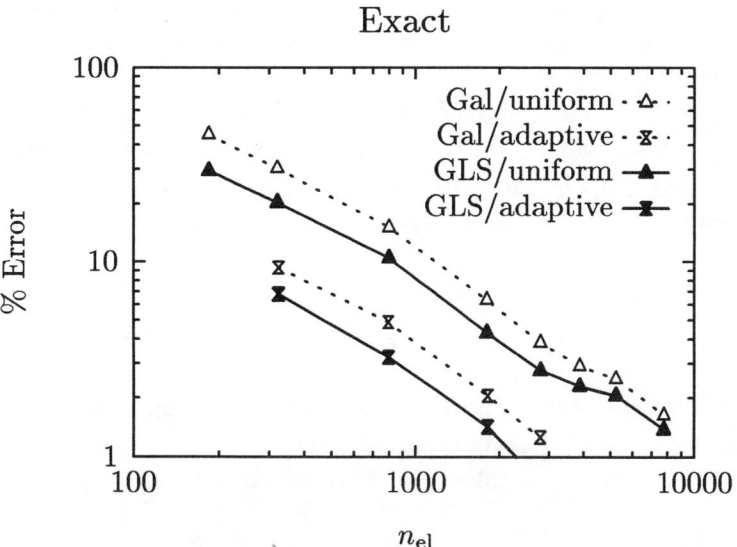

Figure 6.7 Nonuniform radiation from an infinite circular cylinder, $ka = 2\pi$: exact error vs. number of elements in mesh.

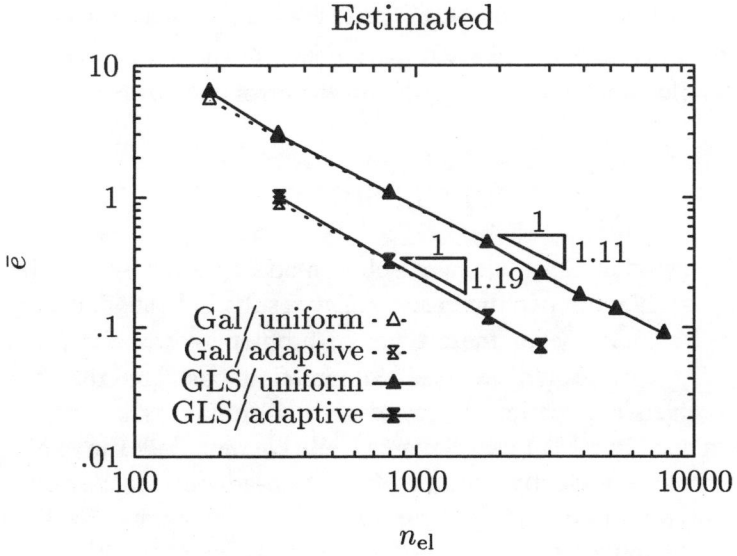

Figure 6.8 Nonuniform radiation from an infinite circular cylinder, $ka = 2\pi$: (scaled) estimated error vs. number of elements in mesh.

The convergence rates with respect to the estimated error (indicated by the slopes shown in Figure 6.8) are about 20 percent higher than the exact convergence rates (Figure 6.7). This is likely due to the error estimator being less accurate on coarser meshes, by overpredicting the error to a greater extent.[2]

From these results we can draw conclusions regarding the absolute value of the estimated GLS error. Computation of the absolute error entails including the heretofore ignored parameters $C_i C_S$ and C_{GLS} appearing in the error estimates. Comparing the GLS error expression (6.17) to the Galerkin error expression (6.16), it is clear that the estimated GLS error will be larger than the estimated Galerkin error since

$$C_i C_S + C_{GLS} > C_i C_S.$$

The overprediction of the GLS error estimator is actually greater upon noting, as discussed above, that the exact GLS error is lower than the exact Galerkin error.

A common measure of the absolute accuracy of error estimators is the global *effectivity index*, θ, given by

$$\theta = \frac{\|e_{est}\|}{\|e_{exact}\|}. \tag{6.31}$$

Thus θ_{GLS} is larger than $\theta_{Galerkin}$, since for GLS the numerator is larger and the denominator is smaller than the corresponding Galerkin values. In other words, the Galerkin a posteriori error bound is (globally) sharper than the GLS bound.

In a forthcoming paper, we will compute $C_i C_S$ and C_{GLS} as well as θ (including distributions of local, or elementwise, effectivity indices); we will also compute the *robustness index* (Babuška et al., 1994a,b), and perform a detailed analysis of the quality of the error estimator.

CONCLUSIONS

In this paper, we demonstrated finite element solution-adaptive methodologies on a model time-harmonic exterior acoustics problem in two dimensions. The results indicate that adaptivity can provide a substantial increase in mesh efficiency for these types of problems—namely, propagating solutions at moderate wave numbers. A significantly reduced computation cost can therefore be realized. The adaptive computations involved application of an advancing front mesh generator (which generates linear triangles), a Galerkin and Galerkin Least-Squares finite element code, an explicity residual-based a posteriori error estimator, and an *h*-adaptive strategy (*p* and *hp*-adaptive strategies were also presented, but computations were not performed using these types of refinement). Each of these component technologies, briefly described above, is critical in achieving an *efficient* adaptive solution.

Results were obtained for the problem of nonuniform radiation from an infinite circular cylinder, for $ka = 2\pi$. Galerkin and GLS solutions were compared, and it was shown that the GLS formulation is more efficient (i.e., accurate) on any mesh, adaptive or uniform. Thus GLS can be *combined* with adaptivity to produce the most efficient computations.

---

[2] For example, the error estimator lumps residuals on element edges onto the elements, artificially spreading the area over which these errors influence. In coarse meshes, the effect of this is exacerbated.

A specific goal in the design of the adaptive strategy was to avoid requiring knowledge of the constant appearing in the error estimator. This led to much simplification and cost savings; however, the issue of how to define a stopping criterion is perhaps made more difficult. The computation of the constant appearing in the error estimator will be discussed in a future paper.

ACKNOWLEDGMENT

We would like to thank Jaime Peraire and Ken Morgan for allowing us to use their mesh generator.

REFERENCES

Babuška, I., T. Strouboulis, C.S. Upadhyay, S.K. Gangaraj, and K. Copps, 1994a, "Validation of a posteriori error estimators by numerical approach," *Int. J. Numer. Methods Eng.* **37**, 1073-1123.

Babuška, I., T. Strouboulis, and C.S. Upadhyay, 1994b, "A model study of the quality of a posteriori error estimators for linear elliptic problems. Error estimation in the interior of patchwise uniform grids of triangles," *Comput. Methods Appl. Mech. Eng.* **114**, 307-378.

Baker, T.J., 1989, "Automatic mesh generation for complex three-dimensional regions using a constrained Delaunay triangulation," *Engineering with Computers* **5**, 161-175.

Bayliss, A., and E. Turkel, 1980, "Radiation boundary conditions for wave-like equations," *Commun. Pure Appl. Math.* **33**, 707-725.

Bettess, P., 1977, "Infinite Elements," *Int. J. Numer. Methods Eng.* **11**, 53-64.

Blacker, T.D., and M.B. Stephenson, 1991, "Paving: A new approach to automated quadrilateral mesh generation," *Int. J. Numer. Methods Eng.* **32**, 811-847.

Burnett, D., 1995, "A 3-D acoustic infinite element based on a generalized multipole expansion," submitted to *J. Acoust. Soc. Am.*

Burton, A.J., and G.F. Miller, 1971, "The application of integral equation methods to the numerical solution of some exterior boundary-value problems," *Proc. R. Soc. London Ser. A* **323**, 201-210.

Cavendish, J.C., D.A. Field, and W.H. Frey, 1985, "An approach to automatic three-dimensional finite element mesh generation," *Int. J. Numer. Methods Eng.* **21**, 329-347.

Cunefare, K.A., G. Koopman, and K. Brod, 1989, "A boundary element method for acoustic radiation valid for all wave numbers," *J. Acoust. Soc. Am.* **85**(1), 39-48.

Demkowicz, L., and J.T. Oden, 1994, "Recent progress on application of hp-adaptive BE/FE methods to elastic scattering," *Int. J. Numer. Methods Eng.* **37**, 2893-2910.

Demkowicz, L., J.T. Oden, M. Ainsworth, and P. Geng, 1991, "Solution of elastic scattering problems in linear acoustics using h-p boundary element methods," *J. Comput. Appl. Math.* **36**, 29-63.

Eriksson, K., and C. Johnson, 1988, "An adaptive finite element method for linear elliptic problems," *Math. Comp.* **50**, 361-383.

Givoli, D., 1991, "Non-reflecting boundary conditions—a review," *J. Comput. Phys.* **94**(1), 1-29.

Givoli, D., and J.B. Keller, 1989, "A finite element method for large domains," *Comput. Methods Appl. Mech. Eng.* **76**, 41-66.

Givoli, D., and J.B. Keller, 1990, "Non-reflecting boundary conditions for elastic waves," *Wave Motion* **12**, 261-279.

Harari, I., 1991, "Computational methods for problems of acoustics with particular reference to exterior domains," Ph.D. Thesis, Division of Applied Mechanics, Stanford University, Stanford, Calif.

Harari, I., and T.J.R. Hughes, 1990, "Design and analysis of finite element methods for the Helmholtz equation in exterior domains," *Appl. Mech. Rev.* **43**(2), 366-373.

Harari, I., and T.J.R. Hughes, 1991, "Finite element methods for the Helmholtz equation in an exterior domain: model problems," *Comput. Methods Appl. Mech. Eng.* **87**, 59-96.

Harari, I., and T.J.R. Hughes, 1992a, "A cost comparison of boundary element and finite element methods for problems of time-harmonic structural acoustics," *Comput. Methods Appl. Mech. Eng.* **97**, 77-102.

Harari, I., and T.J.R. Hughes, 1992b, "Analysis of continuous formulations underlying the computation of time-harmonic acoustics in exterior domains," *Comput. Methods Appl. Mech. Eng.* **97**, 103-124.

Harari, I., and T.J.R. Hughes, 1994, "Studies of domain-based formulations for computing exterior problems of acoustics," *Int. J. Numer. Methods Eng.* **37**, 2935-2950.

Johnson, C., 1990, "Adaptive finite element methods for diffusion and convection problems," *Comput. Methods Appl. Mech. Eng.* **82**, 301-322.

Johnson, C., 1992, "Finite element methods for flow problems." In: *AGARD Report 787* (AGARD, 7 Rue Ancelle, 92299 Neuilly sur Seine, France), 1.1-1.47.

Johnson, C., and P. Hansbo, 1992, "Adaptive finite element methods in computational mechanics," *Comput. Methods Appl. Mech. Eng.* **101**, 143-181.

Kirkup, S.M., 1989, "Solution of exterior acoustic problems by the boundary element method," Ph.D. Thesis, Department of Mathematical Sciences, Brighton Polytechnic, Brighton, U.K.

Kleinman, R.E., and G.F. Roach, 1974, "Boundary integral equations for the three-dimensional Helmholtz equation," *SIAM Review* **16**(2), 214-236.

Lohner, R., and P. Parikh, 1988, "Generation of three-dimensional grids by the advancing-front method," *Int. J. Numer. Methods Fluids* **8**, 1135-1149.

Peraire, J., M. Vahdati, K. Morgan, and O.C. Zienkiewicz, 1987, "Adaptive remeshing for compressible flow computations," *J. Comput. Phys.* **72**, 449-466.

Peraire, J., J. Peiro, L. Formaggia, K. Morgan, and O.C. Zienkiewicz, 1988, "Finite element Euler computations in three dimensions," *Int. J. Numer. Methods Eng.* **26**, 2135-2159.

Schroeder, W.J., and M.S. Shephard, 1989, "An $O(N)$ algorithm to automatically generate geometric triangulations satisfying the Delaunay circumsphere criteria," *Engineering with Computers* **5**, 177-193.

Schroeder, W.J., and M.S. Shephard, 1990, "A combined Octree/Delaunay method for fully automatic 3-D mesh generation," *Int. J. Numer. Methods Eng.* **29**, 37-55.

Seybert, A.F., and T.K. Rengarajan, 1987, "The use of CHIEF to obtain unique solutions for acoustic radiation using boundary integral equations," *J. Acoust. Soc. Am.* **81**, 1299-1306.

Shephard, M.S., and M.K. Georges, 1991, "Automatic three-dimensional mesh generation by the finite octree technique," *Int. J. Numer. Methods Eng.* **32**, 709-749.

Stewart, J.R., and T.J.R. Hughes, 1995, "An a posteriori error estimator and hp-adaptive strategy for finite element discretizations of the Helmholtz equation in exterior domains," to appear in *Finite Elements Anal. Des.*

van den Berg, P.M., A.P.M. Zwamborn, G.C. Hsiao, and R.E. Kleinman, 1991, "Iterative solutions of first kind integral equations." In: *Direct and Inverse Boundary Value Problems*, R.E. Kleinman, R. Kress, and E. Martensen (eds.), New York: P. Lang, 213-232.

7

Modeling of Optically "Assisted" Phased Array Radar

Alan Rolf Mickelson
University of Colorado

> Integrated optics holds great promise for being a technology that, when used synergistically with electronic and microwave technology, can lead to increased performance at decreased cost. A phased array radar system, however, is a sufficiently complex system that one must have some certainty of improvement to system performance before mocking up a complete hardware demonstration. Ideally, one needs to develop a methodology to obtain accurate system simulation from measurements made on individual devices or small subsystems. For a system of even minimal complexity, such a simulation cannot be carried out on the physical level due to time and computer power constraints. As in very large scale integrated (VLSI) simulation, we have chosen to use modeling methodology that is hierarchical. Some discussion here is given to present-day VLSI as well as microwave and optical system design strategies. Some modeling techniques developed at the Guided Wave Optics Laboratory are then discussed. Examples of some anti-intuitive yet somehow optimal strategies are given, as are some concluding remarks on time- versus frequency-domain techniques.

INTRODUCTION

Here at the Guided Wave Optics Laboratory (GWOL), we have been involved in a long-term effort to determine the efficacy of the use of optical, and more specifically integrated optical, components in phased array radars. There are various reasons why people believe that optics can lead to improved phased array implementations. The optical carrier frequency is roughly 5 orders of magnitude larger than the microwave one. This means that the achievable bandwidth of an optically fed, optically read-out system could extend all the way to the microwave carrier frequency and beyond, allowing for multiple microwave frequency antenna operation, with each beam achieving steering bandwidth near its center frequency. Also, optical waveguides, such as optical fibers, are known to be quite low dispersion guides with respect to their microwave counterparts. This feature could allow systems to be remoted, that is, to separate processing, generation, and transmission/reception, with no decrease in performance. Further, the smallness in wavelength of the optical carrier allows much reduced crosstalk between channels for a given channel packing density. However, optics has yet to become an integral part of any phased array system. Phased arrays are not especially commonplace elements either. They tend to be quite expensive in general, as hybrid implementations are painstaking, and monolithic (MIMIC) implementations of subsystems have not proven economically feasible. Interestingly enough, though, optics has been proposed as a technology that could lead to considerable cost reduction of phased arrays, possibly to the point that this technology could become viable to compete in a mass consumer market.

A major problem with applying any new technology to an already developed application area is that of convincing those already in the application area that the new technology will not cause more new problems than solve old problems. Phased array systems require high dynamic range and low crosstalk, among other attributes. Also, the performance levels that have been achieved with the standard microwave components are impressive and have been hard to come by. An engineer in such an area is loath to embrace a new technology that may increase bandwidth at, perhaps, the cost of an increased probability of false detection or, worse, that will only promise improvement while actually degrading system performance in every way. In order to "sell" such a new technology, one needs to demonstrate marked improvements in end-to-end system performance. Prototyping is not generally a good way to do this. A first prototype is generally of modest performance and possibly astronomical cost. A phased array system is expensive and can probably be made most cost effectively by the very people to whom the prototyper would like to demonstrate his improved technique. Still, simulation of noncommercial (which some engineers call imaginary) components is not enough to convince a systems engineer. Some operational characteristics, other than best-case figures for all components, are a necessity.

The approach that we have taken over the years has been a phenomenological one. We have worked on the fabrication and characterization of various integrated optical and microwave components, while simultaneously trying to make simple, physical models for these components. These models can then be calibrated from the experimental data in order to make them predictive, in the sense that the input to the model would be actual fabrication conditions, rather than physical (and perhaps indeterminable) parameters. The necessary physical parameters could, if necessary, be extracted from the model. For example, the fabrication of an integrated optical phase shifter requires one to mask off a channel, indiffuse it to obtain an index difference, and then remask and deposit electrodes. The fabrication parameters in such a case are the design dimensions of the masks, the design metal thicknesses, and times of indiffusion. The actual fabrication parameters, of course, will never be identical to the design parameters due to the basic nature of the fabrication process, and, therefore, calibrations through mask measurement, line thickness measurement, and so on are necessary. The physical parameters would be such things as the index profile and the electric field distributions in the substrate. To obtain good prediction of the modulation depth will also, in general, require calibration of the physical parameters through such measurements as m-lines, near-field profiles, and half-wave voltages. These predictive models would then be used to fabricate components to the achievable specifications. There is a limit to the amount of pure trial and error that can be carried out in an academic laboratory, and the phenomenological approach is adopted so as to minimize this trial and error. There is also a limit to the complexity of fabrication and characterization that can be carried out in an academic laboratory. If one is to work with new state-of-the-art components that must by nature be in-house fabricated, the educational process (being of finite duration per student) precludes serious experimental work in complex systems. The very existence of phenomenological models, however, forms a basis for system modeling. Simple models can be put together in packages to attempt to predict complex system behavior. The basic nature of the modeling process that follows from this phenomenological basis is a hierarchical one. Smaller models are successively grouped into larger models, which are successively grouped into system blocks, etc. Such modeling is not uncommon. As will be discussed in the next section, most of the standard computer aided design (CAD) tools employed in both the very large scale integrated (VLSI) circuit industry as well as in the microwave industry are hierarchical tools. The analysis techniques typically employed in optical communication link design, those of loss budgeting as well as of noise propagation, are also hierarchical.

Perhaps the primary problem in working from phenomenological device models toward complex system modeling is that of determining the minimum set of measurements necessary to fix the minimum necessary parameter set which will determine the desired system characteristics to the desired accuracy. The solution to this problem is more like a long-term program than a calculation or even a well-determined process. An example of this is the integrated optical phase shifter. One would like to minimize the number of calibration parameters, as the measurements necessary to determine these may have to be repeated periodically, at the least, every time that any piece of equipment, or processing step is modified. But as the effects that make such calibrations necessary are by nature unknown, it is impossible a priori to determine which calibration steps will have the greatest effects. One cannot know what is necessary until after one is well into the process of trying to find out what is necessary. Determining the importance of various factors is almost inevitably a problem of analysis of data from well-designed experiments. Discussion of this process as applied to optically assisted phased array radar is the prime motivation for this talk and will be the specific topic of the third section of this document.

Perhaps we should point out that we do not feel that the use of simple models will by any means preclude the possibility of uncovering new and interesting processes. As evidenced by the simple second-order difference equation that Feigenbaum (1980) used to discover the doubling route to chaos, even the simplest of nonlinear systems can exhibit the most complex behavior. Even the most stable of open loop linear amplifier circuits can exhibit such behavior when its loop is closed. In a large-scale structure made of simple stable pieces, even tiny spurious reflections from junctions can lead to unexpected collective behaviors. This is the wonder of complex systems. In the majority of the devices, interconnections, and systems we have investigated, we have found full-wave corrections to the models we employ to be quite small. Yet, one need look no further than a book such as *Proceedings of the First Experimental Chaos Conference* (Vohra et al., 1992) to see that even the best designed naval sensor systems, where I would also imagine that the full wave corrections in each subsection are small, can give rise to all the forms of classified mathematical chaos as well as some forms of chaos never seen before. Exotic effects in large-scale structures can be the result of the most ordinary standard, nonlinear-devices in-transmission-line designs when they are working in a multicomponent, multi-wavelength environment.

In the next section we will discuss what we perceive to be the state of the art of the industrial modeling process, first for the digital VLSI design process, next for the microwave system design process, and then for the optical link design process. The next section will then discuss results of the phenomenological design process employed at GWOL, before turning that discussion to our present attempts to extend the approach to predict end-to-end system performance of phased array radars containing "exotic" components.

STATE-OF-THE-ART INDUSTRIAL MODELING

In this section we discuss the types of modeling used in industry for design of first digital, then microwave, and last optical systems. It is important to point out that in each case the modeling is carried out in a hierarchical fashion.

VLSI Design

The VLSI design process is, probably, the process in which the tools of simulation have been most advantageously applied. Boards containing tens of chips, each containing millions of transistors, can be fabricated with an end-to-end process yield better than 50 percent. And the vast majority of the continuously occurring advances can be attributed to the use of a set of ever-evolving design tools. But although these tools are ever-evolving, there is a basic methodology behind them that lends them much of their adaptability, which is much of their power.

Whether a specific company refers to their design system as 5-level or 4-layer or as just layered in general, a common thread seems to be the hierarchical nature of the process. Figure 7.1 depicts a possible 5-level scheme. The upper levels of the design process are called logical in that the modeling at those levels contains no reference to devices, voltage levels, or waveforms. The highest of the logical levels will be one in which the system blocks are probably chips, and simulations at this level would be of protocols and perhaps line delays, to see if the pulse length is sufficiently short compared to clock window that timing errors do not occur. The next level down would be of logical blocks within a chip, i.e., ALUs, half adders, etc. At this level, simulation primarily consists of looking for critical delays. At this level and each level below it, if models for the blocks are necessary they can be pulled down from standardized libraries. And so the hierarchy proceeds. A possible hierarchy to be used in the design of a distributed processor is depicted in Figure 7.2. The productive designer would like to truncate the design process as soon as the given subsystem specifications can be met, preferably with library components that are standardly fabricated at the foundry. In fact, the chip designer in most design processes cannot even access the physical (lowest) design level. Some houses will have custom designers who have access to a time-domain simulator (all of which are similar to SPICE) that is accompanied by a myriad of device libraries, which are compiled at the foundry, most likely from circuit model fits to a mass of experimental data (amassed using the techniques of automated testing). If the existing library devices cannot be used to make the required specifications by the custom circuit designer, then the job passes to the foundry to develop new components, at a tremendous cost. Although the whole process is in essence fired by the circuit simulation stage, designers are loath to use this tool (or even banned by management from use of this tool) due to the great cost of simulation of circuits containing tens of millions of transistors. The use of libraries at every stage of the design process allows for the design system to adapt each time new designs are verified at the next lower level, often obviating the need for a designer to know very much about that level.

Logical Levels

Level 5	Chip Level
Level 4	Subsystem Level
Level 3	Blocks of Logic Elements Level
Level 2	Logic Element Level

Physical Level

| Level 1 | Within Logic Elements |

Figure 7.1 Schematic depiction of a possible 5-layer CAD system hierarchy.

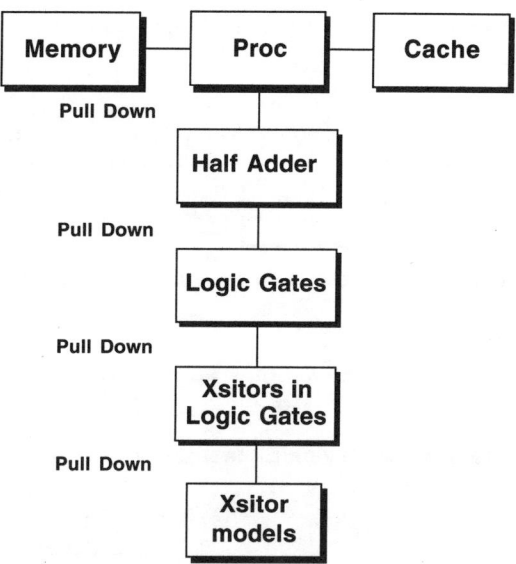

Figure 7.2 A possible hierarchical path that might be traversed during the design of a distributed processor.

An interesting point is that the digital simulations are performed in the time domain. Regeneration in digital systems is performed by a thresholding process in which, when the input voltage level passes a given value, the output shoots up to a given value, significantly above the threshold level, as fast as the circuit allows for. This technique of regeneration allows for inexpensive signal reconstruction, but is inelegant in the sense that it generates a multitude of harmonics with noise envelopes surrounding each. To simulate such a situation in the frequency domain would be, at the least, computer intensive. Further, time-domain circuit models for the nonlinear elements (generally MOSFETs) are well known and run efficiently in the time domain. An ever-growing problem in the digital simulation world, however, is the inclusion of dispersive elements, dispersively coupled lines, and other such artifacts that arise as clock rates increase. Dispersion, in general, is much easier to treat in the frequency domain than in the time domain. One could argue that it is always possible to Fourier transform frequency domain formulations but, unfortunately, the fast Fourier transform (FFT) is discrete, and windowing often precludes fidelity. Recent work on combining finite-difference time domain (FDTD) with SPICE holds promise for providing insight into simple cases of dispersive propagation, but is still too computer intensive a technique to provide a general solution to the modeling problem. The frequency-domain time-domain dichotomy is alive and well in digital modeling.

Microwave System Design

Much as in the digital world, the microwave system design process is also hierarchical, probably by example, as microwave CAD is a follow-on to digital CAD. A possible hierarchy for a microwave CAD tool may appear as that depicted in Figure 7.3. If for no other reason than its follow-on status, microwave CAD has achieved nowhere near the maturity of digital CAD. It is mature enough, however, that there do exist a number of commercial CAD tools. The popularity of some of these tools has produced a situation in that there are de facto standard microwave design procedures that are more or less followed by most of the major manufacturers of microwave systems.

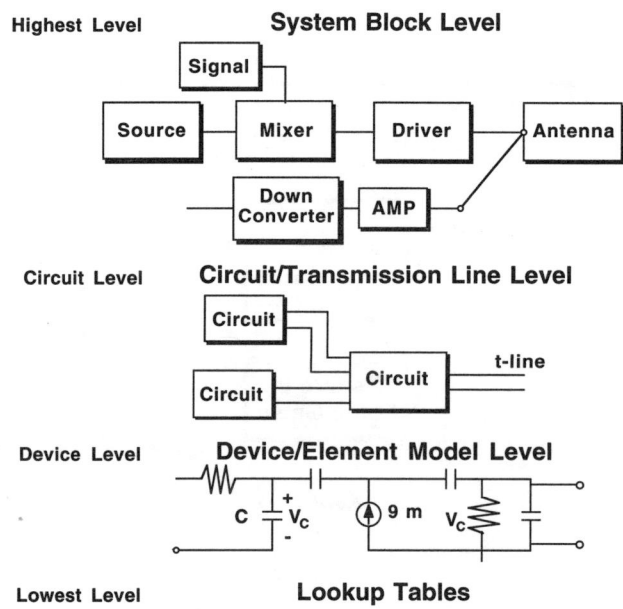

Figure 7.3 Schematic depiction of a possible hierarchy for a microwave CAD system.

Typically, a microwave system will contain a microwave source, a modulation subsystem with digital control, a transmission system, at-element amplifiers, radiating/receiving antenna elements, a bank of mixers, some RF processing elements, and some analog to digital (A/D) converters leading into some digital processing electronics. A mixer subsystem may appear as the one depicted in Figure 7.4. The highest level of design is one in which the major functional design blocks are the simulation blocks. A major difference, however, between digital and analog CAD is that in analog CAD there can be no logical levels. Simulation at all levels must be carried out on waveforms. One is primarily interested in system characteristics such as frequency response and noise propagation through the system. It is the signal to noise ratio at a given frequency that determines all of the important system parameters such as dynamical range and bit error rate (false detection probability in radar) of the external digital processing circuit. At the highest level, the blocks are essentially characterized by transfer functions and noise figures, as blocks were characterized in the pre-CAD days. Indeed, if the radar were single frequency, one could easily obtain analytical results for frequency response and signal-to-noise ratio. The reason for CAD at this level, more than all else, is that systems are no longer so simple. Waveforms can be complicated as can coding and, for that matter, functionality. By having an automated system of analysis, many different schemes can be rapidly and easily analyzed.

In microwave CAD, the next level below the system (highest) level can be physical, in the sense that this next level will employ frequency-domain circuit and device models linked together by transmission lines (i.e., refer again to Figure 7.4). The circuit and device models at this level, for the most part, will exist in pull-down libraries. The libraries for the active elements more than likely will come from the foundry. Models for the passive elements likely will come from a combination of extensive fabrication and characterization as well as possibly full or nearly full wave modeling of the interconnecting, coupling, and filtering structures. It has been in this arena of passive element modeling that we have seen the most interaction between industrial designers and academic researchers. It is also in this arena that there are still

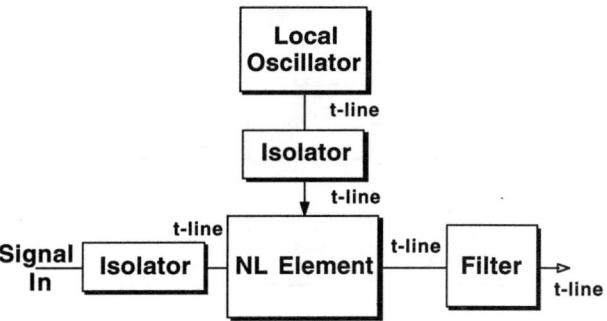

Figure 7.4 A possible realization of a microwave mixer subsystem.

extant problems with the microwave modeling process. Parasitic couplings between lines are generally small, and it is often not serious that they do not show up in the transmission line based models. This is not true when the couplings are near active elements where they can provide output to input control, system-like coupling. Yet these effects will not show up at all in the modeling process unless put in by hand by a wise design engineer who knows exactly what areas on a chip or board really need full wave modeling and which do not, and who further needs to know how to carry out the partitioning. In this sense, CAD for microwave systems is exactly that—computer aided design and nothing remotely like automated design.

An interesting point is that the microwave simulations are frequency-domain simulations. Due to propagation dispersion, one generally wants to modulate about a center carrier frequency. At a receiver, there will generally be a local oscillator that will beat down this carrier to an intermediate frequency. Power that goes into harmonics of the carrier frequency will then simply be lost in the receiver end filtering process. One, therefore, strives in a microwave system to minimize any nonlinear behaviour. In the vicinity of necessary nonlinearities, such as those inherent in mixers, liberal use of filtering techniques are employed to keep the unwanted harmonics from forming. For this reason, frequency-domain analysis, together with harmonic balance, are acceptable analysis tools. This situation is quite compatible with the transmission line models employed. Use of the frequency domain allows for inclusion of dispersion in lines, circuits, and devices. The use of the frequency domain also allows for the models to be verified. The high center frequency of microwave carriers pretty much requires the use of frequency-domain (network analyzer) based measurement techniques. Transient effects in the microwave world can only be measured in narrow bands around the center frequency and its (unwanted) harmonics. Microwave CAD is strongly grounded in the frequency domain.

Optical System Analysis

The field of optical networking is still a young field. The growth of the fiber optic network, driven by the need for a transmission medium that would fit into existing right of way, was explosive, even when compared to the ever-increasing pace of development at this point in the twentieth century. An outcropping of this pace of growth was that the point-to-point system architecture and loss budgeting techniques became pretty much frozen in from day one. A typical block diagram for an optical transmission system may appear as in Figure 7.5 and indeed loss budgeting methods are based on associating a loss factor with each of the stages and transitions appearing in this model. This is not to say that there has not been a large amount of

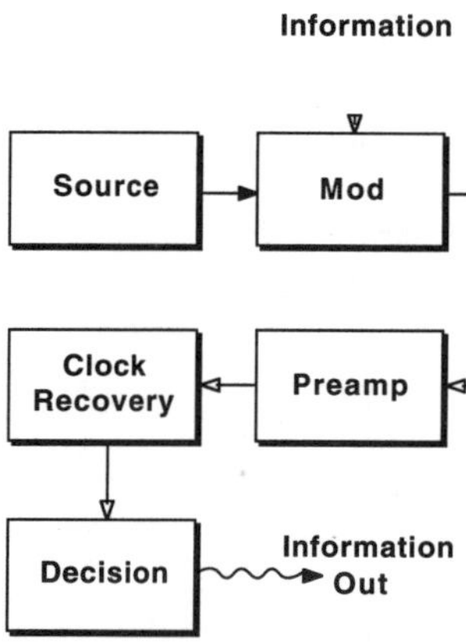

Figure 7.5 A block diagram for an archetypical optical communication system.

advanced modeling work on various components (especially lasers), nor extensive work into more advanced data communications architectures, just that the practice of link design seems to be based on some quite simple, yet formalized, design rules. (The literature indicates that in-house design programs at various optical communication groups are beginning to exist, but are not yet commercial). These rules are quite similar to those used in point-to-point RF and microwave link design (pre-CAD) and essentially track the total signal and noise power through the system (i.e., again refer to Figure 7.5). Frequency response in an optical link generally means response to the signal modulation as parameterized by the optical carrier, and also tends to be handled as the signal and noise power. The budgeting processes require loss, noise and dispersion figures for various components and channels in the system, and in this sense the design process is hierarchical, much as is the microwave design process.

RECENT MODELING RESULTS FROM GWOL

In the present section, we discuss first some of the historical reasons for our having adopted some of the simple models we have adopted for physical devices, as well as the original "optical communications" model (as in Figure 7.5) that we adopted for prediction of salient system characteristics. Now, a typical phased array radar system might be represented by a block diagram such as that displayed in Figure 7.6. An optically assisted phased array radar system block diagram appears in Figure 7.7. A salient feature of that diagram is that there needs to be digital, optical and microwave modeling involved in any prediction of the operational characteristics of this microwave optical radar. As with many other groups, our first attempts to come up with accurate system modeling were aimed at simulation of as complete a subsystem as we could. To start with, rather than relying on pure computer power, we tried to use simplified models we thought could rapidly simulate large areas of our microwave optical circuits. Such a model is contained in the paper by Radisic et al. (1993). We soon found that our microwave (electrical) model was far from optimal in that it had much too much accuracy over the majority of the layout area, yet too little in certain neighborhoods (a problem that seems to be shared by many of the techniques presented by participants of this workshop). For example, the effect of parasitic coupling on passive lines is a small effect, unless one is near an active element. Various full wave as well as quasi-static methods can be used to simulate the passive lines, taking

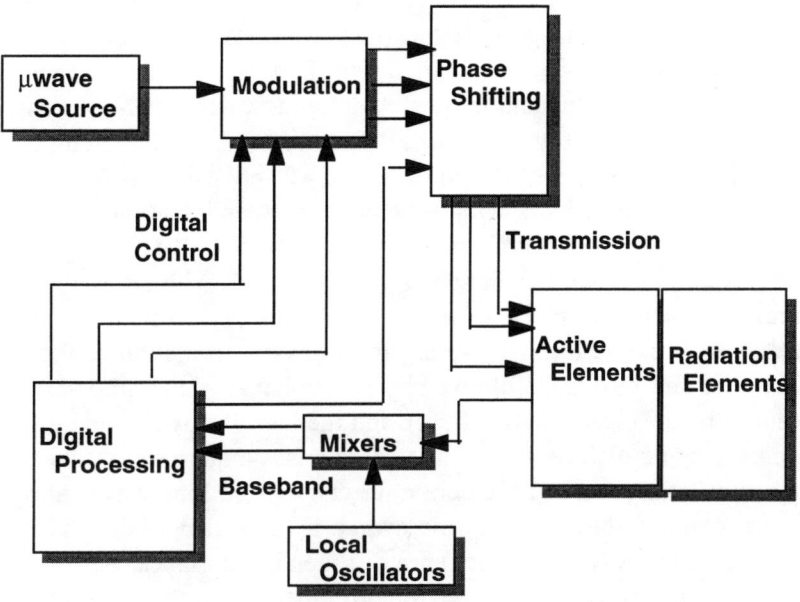

Figure 7.6 A block diagram for a rather typical phased array radar system.

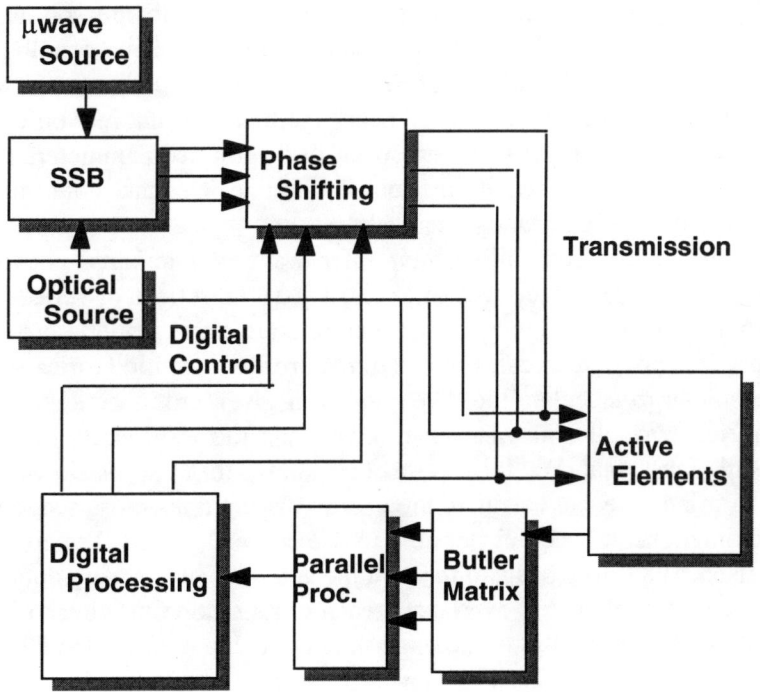

Figure 7.7 A block diagram for a phased array radar architecture that employs optical components for phase shifting as well as phase decoding of incoming signals. SSB, single sideband modulator.

full account of parasitics. Full wave models, however, are of no real efficacy in active regions, where the important effects are those of charge transport. To try to make a transistor model that fits into the simulation is a bit pretentious, as transistor (active device) modeling is a mature subject, and there exist excellent models for such devices, both in the time and frequency domain, which do not necessarily easily mesh with electromagnetic simulation. At some point we realized that, perhaps, the hierarchical techniques that were so ubiquitous in industrial CAD may well be the ones most applicable to our problems. We will then discuss some of our initial attempts to work with these new ideas, and then indicate some of the pitfalls we have fallen into, while emphasizing some of the potential advantages of the approach.

Integrated optical devices, in general, consist of optical channels with electrodes running over them, to effect electrical control of optical stream. To solve exactly for the optical field exiting such a device, given the input optical field, would require a full wave electromagnetic simulation of the electrical field, which involves both the electrodes and feedlines, followed by a calculation of the time varying dielectric constant in the substrate (as caused by the electro-optic effect), and then a full wave simulation of the propagation of an optical field in a time varying dielectric medium. There arises a question of what information we really need from the model and what accuracy of prediction we can expect. In a phased array, one is interested in transforming signals back and forth from the microwave to optical domains. The efficiencies of these transformations (at a given microwave frequency) are of crucial importance. Also important are deviations from pure harmonic behavior whether they be deterministic (harmonic distortion due to nonlinear modulation transfer function) or stochastic in nature. The accuracy with which we can determine such parameters as the mask parameters, due to fabrication tolerance, or diffusion parameters, due to oven ramping or simply imprecise knowledge of the diffusion coefficient of something of finite purity into something else of finite purity, cannot ever be as good as 1 percent. For this reason, from early on we have adopted phenomenological modeling techniques. A block diagram of a phenomenological tool may have a representation such as that of the block diagram of Figure 7.8. The idea is that one uses a model that is as accurate as possible (includes as much physics as one can) but that does not rely on exact knowledge of all parameters. Instead one sets up the model with somewhat ill-determined parameters that can be calibrated on a specific fabrication line in order to specify the tool to that line. This takes out much of the uncertainty induced in the fabrication process if it is done judiciously.

Now, in integrated optical devices, unlike their microwave counterparts, one can generally ignore reflections except at discrete points (interfaces, mirrors) where the effect of reflection can generally be calculated by hand. From the fabrication models and a calibration step involving propagation constant and near field measurements, it is possible to calculate an index profile to within fabrication tolerance. One can therefore use some technique to calculate the eigenmodes in given cross sections. Further, the electrode structure is, in general, much smaller in transverse extent than the wavelength of the highest frequency component of the modulation signal. For this reason one knows that the quasi-static approximation is a good one. With these in mind, one can calculate the charge distributions cross section by cross section as well. With these approximations, it becomes possible to use coupled modes (Weierholt et al., 1987, 1988; Charczenko and Mickelson, 1989) to propagate forward the continuously varying optical mode distribution, needing only to calculate electrical/optical overlaps at each cross section in order to integrate the resulting set of first-order equations. Both simulations and experiments (Vohra et al., 1989; Charczenko et al., 1991; Januar et al., 1992; Januar and Mickelson, 1993) have shown this technique to be as accurate as any other, given the measurement uncertainties that one began with.

To illustrate the technique, we could well revisit an example that was touched on in the introduction, the example of the modeling of an integrated optical phase modulator being driven at microwave frequencies. Figure 7.9 illustrates a block diagram for a system which supplies two optical signals whose beat note is a

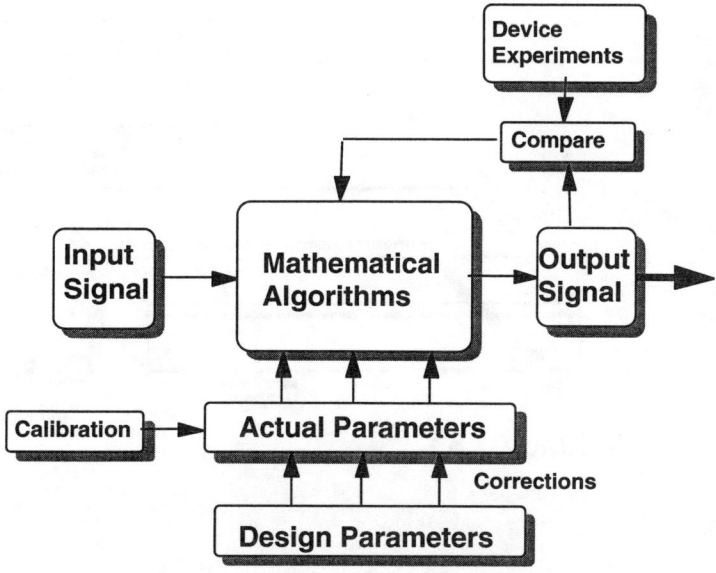

- **Calibrated Parameter Sets**
- **Verification Using Actual Devices**
- **Yields Efficient and Accurate CAD**

Figure 7.8 A block diagram for a phenomenological CAD tool which includes in it the possible calibration paths.

typical radar signal, a microwave carrier with phase information impressed. The system piece of interest to us here is the optical phase modulator. As was mentioned in the preceding paragraph, uncertainties in the fabrication process preclude the use of models that require sharp parameter definition. The primary processing steps necessary for fabrication of a polymeric integrated optical phase modulator are indicated in Figure 7.10. The uncertainties in this fabrication process will include polymer thickness, both due to initial spin speed as well as due to processing induced shrinkage, photolithographic errors in line definition and development, metallization thickness, and quality. All of the above are of course compounded by the amorphous nature of the polymeric material, which may cause numerous inhomogeneities. Now, the primary technique for a priori determining the behavior of the device will be an application of the coupled mode equations, which in turn requires various input data of both the electrical and optical behavior of the pieces of the device. The process is illustrated in the block diagram of Figure 7.11. The figure includes as blocks both the calibration input and the comparison with actual device blocks. The primary input block, that directly to the left of the coupled mode block, requires the input from these lower level blocks. The optical and electrical lower level blocks are schematically depicted in Figures 7.12 and 7.13. Again the calibration and comparison blocks have been included in these diagrams.

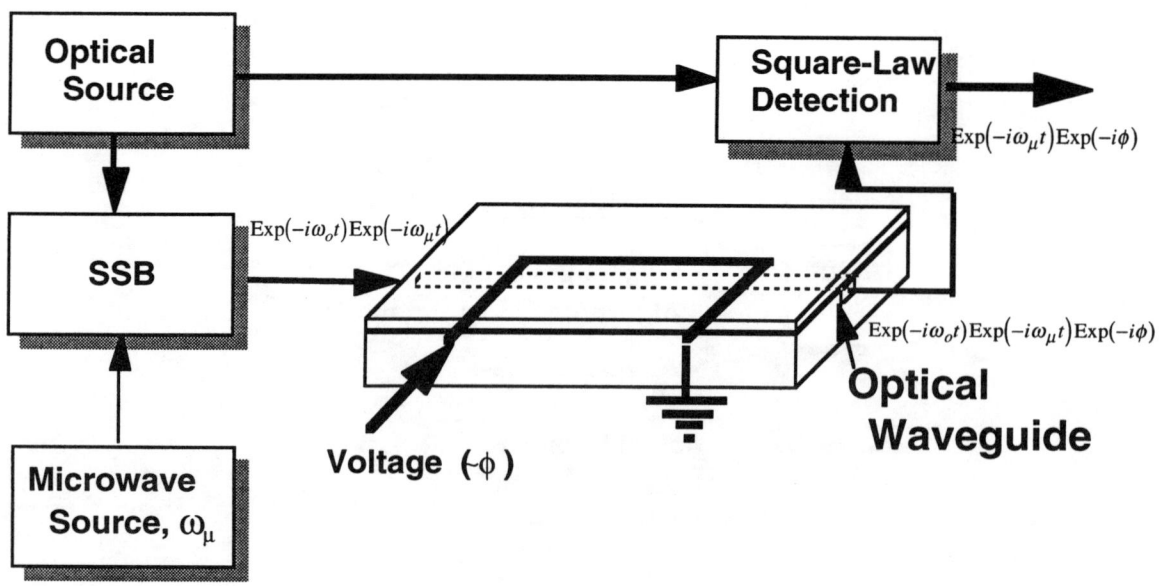

Figure 7.9 Block diagram of a system that generates a microwave heterodyne beat note between two optical signals. SSB, single sideband modulator.

- **Substrate Preparation**

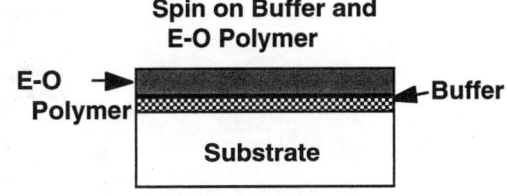

- **Optical Waveguide Formation**

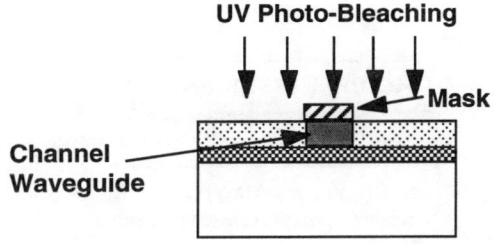

- **Microwave Electrode Formation**

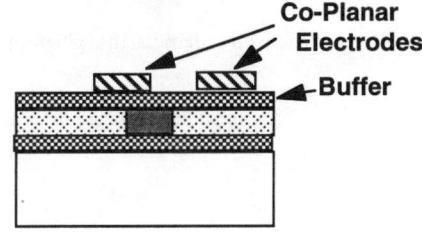

Figure 7.10 Schematic of the processing steps involved in the fabrication of an optical phase modulator. EO, electro-optical polymer.

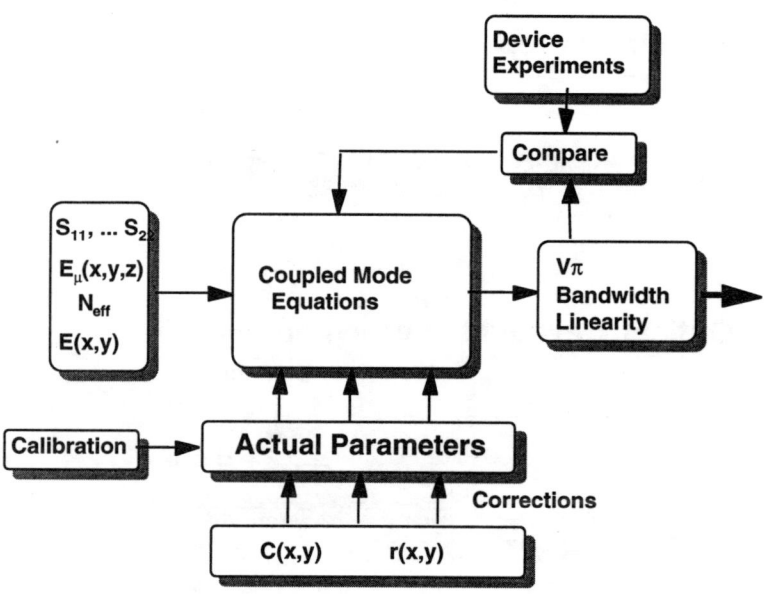

- **Design Parameters**
  - $C(x,y)$: Chromophore Concentration
  - $r(x,y)$: E-O Coefficient

Figure 7.11 A block diagram that schematically depicts the phenomenological CAD system used to design optical phase modulators at GWOL.

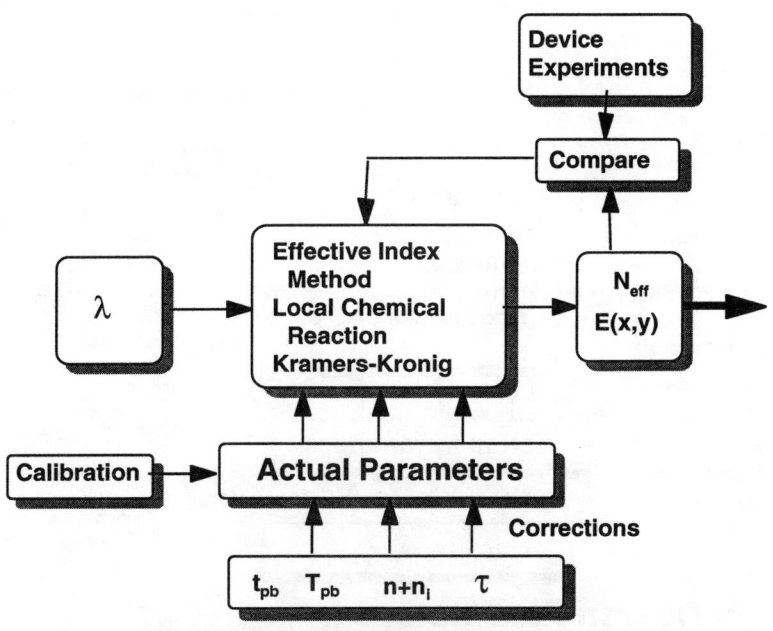

Figure 7.12 A block diagram of a lower level of the phase modulator design system of Figure 7.11, that level which is used to design the optical waveguide channels.

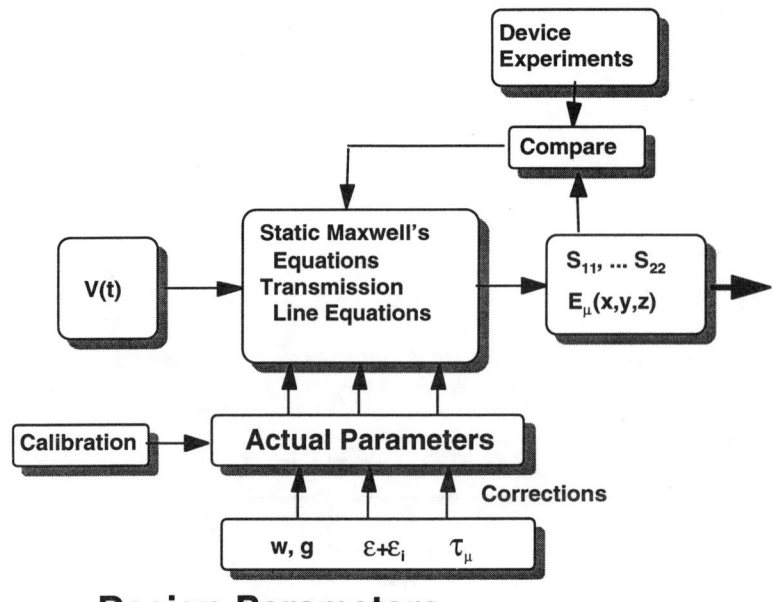

- **Design Parameters**
  - $\varepsilon+\varepsilon_i$: Polymer dielectric constant
  - τ_μ: Metallization thickness
  - w, g: Electrode width and gap

Figure 7.13 A block diagram of a second lower-level design subsystem of the design system of Figure 7.11, the subsystem for designing the electrode structure of the optical phase modulator.

At some point during the effort to employ integrated optics into phased array, it became clear that it was important to be able to model both the microwave and microwave optical versions of the system on roughly the same footing, for purposes of comparison. As was mentioned above, optical loss budgeting and noise propagation modeling were really taken from earlier microwave system modeling techniques. Comparing microwave Butler matrices with optical ones therefore could be performed relatively straightforwardly (Charczenko et al., 1990). It is in places where the components become dense that better modeling becomes necessary. Most joint academic-industrial efforts have concentrated on analysis of dense passive circuits where parasitics become important. Our original answer to this problem was to develop a two step approach (which we called Zoom; Radisic et al., 1993) that could replicate most of the results of full wave approaches but at a much greater computation speed. The basic idea was to first identify a parasitically coupled region, then use a static solution to find the quasi-static charge distribution that can be used to define an effective transmission line circuit that takes into account the parasitics. Integration of the first-order equations of propagation along this transmission line model could then be used to find the S-parameters that then, together with the charge distribution, can fully describe the fields throughout the substrate and surrounding space. The hope was that this technique would be sufficiently fast that it could be applied to a large area of a microwave or microwave optical circuit. A problem that arose with this idea was that the corrections on interconnect lines turned out to be rather insignificant compared to the S-parameters that would come out from a straight transmission line analysis (no added elements to represent the parasitics). This is not to say that the effects could not be dramatic. Filters can show significant frequency shifts due to parasitics as has been well illustrated by the double stub tuner (Goldfarb and Platzker, 1991). Unfortunately, filter design is a rather mature field, one of those that would show up in the pull down menus in a VLSI system if microwave filters were a part of a VLSI system. There is not much need to simulate such things as that. Parasitic effects in the neighborhoods of active components, be they amplifiers, oscillators, mixers or what have you, could also exhibit dramatic and unpredictable effects. These are certainly worthy of simulation and study. Unfortunately, there is no easy way to take active regions of transistors into account in an electromagnetic simulation.

It was the realization that any number of commercial simulation tools could perform better active element analysis than we could practically do that really brought us to our current philosophy. It has always been clear that graphical interfaces are best purchased from a supplier. In the past, this was not so clear concerning analysis programs unless one really believed in the models included. But at this point, any number of commercial tools allow for one to define their own component models. This allows for one to develop a hierarchical modeling structure well in line with what was the original philosophy behind Zoom, and still looks quite a lot like the hierarchical techniques employed in VLSI CAD and microwave system CAD. In essence, a microwave circuit simulator (e.g., MDS, Libra, etc.) is only a tool for solving the transmission line equations that one could as well have used the first piece of Zoom (call it, for example, the circuit parameter definition part) to determine. For example, let's consider an optical link that consists of a CW laser diode, a modulator that inserts the microwave signal onto the optical carrier, a short optical propagation path, and then a receiver that consists of an optical fiber radiating into an FET that is the active element of the radiating structure, assumed to be an active antenna. The optical system block diagram of Figure 7.7 could well be applied to this system as well as a number of others. A system level model would show the blocks as the microwave source, the modulator, the link, and the active antenna. A good system level simulation package (at least two of which are commercially available) would allow one to drop to lower levels (which will presently be described) to define the system blocks, and then could perform all kinds of quite complex waveform and noise analyses quite simply. To define the blocks would require dropping first one level, to a transmission line modeling level in which the simulation, at least for the standard microwave style software, would be a frequency-domain circuit analysis. In this simulation, active components would be pulled down from standard menus, but other circuit parameters would be determined from a true physical level simulation as performed by a circuit parameter definition routine. There still is some level of art in this approach—as in areas where parasitics are important, one must still have a good idea of what areas to focus on—but at least this approach harnesses some of the power inherent in the commercial software packages. From other work presented at this workshop, the focusing problem seems to be an inherent one in modeling that cannot be attributed to hierarchical, full wave, or any other modeling technique. If there exists an inherent problem with the hierarchical scheme, it may lie in the original partitioning.

SOME CONCLUDING REMARKS

We did not want to entitle this section conclusions as this work is not concluded but only beginning. However, in addition to some differences, there are various common points that have shown up in our work and much of the other work presented at this symposium, and we deemed it appropriate to point out these commonalities (while necessarily coloring the discussion with our own bias).

A first point we would like to reiterate is that we have found that the simplest models that we have found to be reliable have been the ones that we have found to be most amenable to the phenomenological approach. For example, coupled modes together with an effective transmission line model allow one to propagate solutions forward through the system, using only the overlap of the optical and microwave fields as a parameter in solving the first-order differential equations. Such a technique can be calibrated with physical fabrication parameter data as well as with the simplest device data of m-lines, near fields, and switching voltages. The technique that we have referred to as Zoom (which we discovered without enough study of the literature to determine its origination, but we would imagine it was with Lord Rayleigh (1945) fits well into both phenomenology and hierarchy, as it employs physical level calculations to feed a transmission line model. Models we have developed for optical injection into MESFETs have also been physical simulations to determine changes in effective circuit parameters of transistor models. We feel that it is more complex to determine the effective circuit parameters of full wave models where the circuit interpretation of the results is not so immediately clear.

The problem of choice of, as well as interfacing of, time-domain and frequency-domain tools, is a real and serious one. In the majority of the modeling work I have discussed in this talk, I have concentrated on the frequency-domain modeling we are carrying out on phased arrays. This is mainly due to the fact that we started several years ago working in the frequency domain and now that we have decided to go hierarchical, the commercial tools available are all frequency-domain tools, a characteristic of present-day microwave systems modeling. This is not to say that this is the only effort. We are also looking into digital interconnects using integrated optics. The frequency regimes of interest in the two problems are now becoming comparable, yet the tools and techniques are incompatible. This is more than a modeling problem; it is a real problem. People who model microwave systems have solutions to high-frequency propagation problems, yet work in such a different world from the digital modelers that their solutions cannot be applied to the digital world. It would be nice to see a bridge built.

A common thread in many of the works presented here has been that one must find ways to partition a large structure in order to analyze it. The point is that some regions may require totally different techniques of analysis than others. A pitfall in making the decision as to where to partition, especially strongly seen in hierarchical techniques, is that the boundary characteristics can seriously affect the system performance in a real sense, as well as the system analysis.

REFERENCES

Charczenko, W., and A.R. Mickelson, 1989, "Symmetric and asymmetric perturbations of the index of refraction in three-waveguide optical planar couplers," *J. Opt. Soc. Am.* **6**, 202-212.

Charczenko, W., M. Surette, P. Mathews, H. Klotz, and A.R. Mickelson, 1990, "Integrated optical butler matrices for beam forming in phased array antennas," *SPIE Proc. 1217*, Los Angeles.

Charczenko, W., P.S. Weitzman, H. Klotz, M. Surette, J.M. Dunn, and A.R. Mickelson, 1991, "Characterization and simulation of proton exchanged integrated optical modulators with various dielectric buffer layers," *J. Lightwave Technol.* **9**, 92-100.

Feigenbaum, M.J., 1980, "Universal behavior in nonlinear systems," *Los Alamos Sci.* **1**, 4-27. This paper is reprinted along with a number of other relevant papers in the collection *Universality in Chaos*, P. Cvitanovic (ed.), 1984, Bristol: Adam Hilger Ltd.

Goldfarb, M., and A. Platzker, 1991, "The effects of electromagnetic coupling on MMIC design," *Microwave and Millimeter Wave CAE* **1**, 38-47.

Januar, I.P., and A.R. Mickelson, 1993, "Dual wavelength (X = 1300-1650 nm) directional coupler multiplexer/demultiplexer by the annealed proton exchange process in LiNbO3," *Opt. Lett.* **18**, 6, 1-3.

Januar, I.P., R.J. Feuerstein, A.R. Mickelson, and J.R. Sauer, 1992, "Wavelength sensitivity in directional couplers," *J. Lightwave Technol.* **10**, 1202-1209.

Radisic, V., D. Hjelme, A. Horrigan, Z. Popovic, and A. Mickelson, 1993, "Experimentally verifiable modeling of coplanar waveguide discontinuities," Special Issue on Modeling and Design of Coplanar Monolithic Microwave and Millimeter-Wave Integrated Circuits, *IEEE Trans. Microwave Theory Tech.* **41**, 1524-1533.

Rayleigh, J.W.S., 1945, *The Theory of Sound*, New York: Dover Publications. This two-volume set is actually a reprint of an 1896 tome.

Vohra, S.T., A.R. Mickelson, and S.E. Asher, 1989, "Diffusion characteristics and waveguiding properties of proton exchanged and annealed LiNbO3 channel waveguides," *J. Appl. Phys.* **66**, 5161-5174.

Vohra, S.T., M. Spano, M. Shlesinger, L. Pecora, and W. Ditto, 1992, *Proceedings of the First Experimental Chaos Conference*, River Edge, N.J.: World Scientific.

Weierholt, A.J., A.R. Mickelson, and S. Neegard, 1987, "Eigenmode analysis of parallel waveguide couplers," *IEEE J. Quantum Electron.* **JQE-23,** 1689-1700. See also ELAB report STF44 A8611.

Weierholt, A.J., S. Neegard, and A.R. Mickelson, 1988, "Eigenmode analysis of optical switches in LiNbO3: Theory and experiment," *IEEE J. Quantum Electron.* **JQE-24,** 2477-2490.

8

Synthesis and Analysis of Large-Scale Integrated Photonic Devices and Circuits

Lakshman S. Tamil
Arthur K. Jordan
University of Texas at Dallas

> Optical devices and circuits can be synthesized from specified transmission characteristics using the methods of inverse scattering. Both analytical and numerical inverse scattering techniques that have been developed to synthesize optical devices and circuits are discussed. Large-scale guided-wave structures such as optical logic gates and optical interconnects can be synthesized using the techniques discussed here. Finite difference-based frequency-domain analysis technique has been used to verify the results obtained by these inverse scattering techniques.

INTRODUCTION

The conventional method of designing optical guided-wave devices or structures is to assume a refractive index profile and solve the governing differential equation to find the various propagating modes and their propagation characteristics. If the propagation characteristics do not meet the expected behavior, the refractive index is changed and the propagation characteristics are again evaluated; this is repeated until the expected propagation behavior of the modes is obtained. This being an iterative procedure, it is time consuming. Also, to obtain certain arbitrary transmission characteristics, one may not be able to guess the correct initial refractive index profile.

The procedure discussed in this paper, as opposed to the direct method, starts with the required propagation characteristics of the guided-wave device and obtains the refractive index profile as the end result. We achieved this by transforming the wave equation for both the TE and TM modes in the planar waveguide to a Schrodinger-type equation and then applying the inverse scattering theory as formulated by Gelfand, Levitan, and Marchenko (Gelfand and Levitan, 1955; Marchenko, 1950). The inverse scattering problem encountered here has a direct analogy to the inverse scattering problem of the quantum mechanics. The refractive index profile of the planar waveguide is contained in the potential of the Schrodinger-type equation and the propagating modes are the bound states of the quantum mechanics (Marcuse, 1972). In general the guided-wave devices are based on channel waveguides; however, we have considered here planar waveguides for mathematical simplicity. The theory presented here can be extended to channel waveguide structures, though it is nontrivial.

PHYSICAL MODEL OF A PLANAR WAVEGUIDE

The wave equations for the inhomogeneous planar optical waveguides can be derived from the Maxwell's equations. If we take z as the propagation direction and let ω represent the frequency of laser radiation, we have the following wave equations for one-dimensional inhomogeneous planar waveguides (Tamir, 1990)

$$\frac{d^2}{dx^2}E_y(x) + \left[k_0^2\varepsilon(x) - \beta^2\right]E_y(x) = 0 \tag{8.1}$$

for TE modes and

$$\frac{d^2}{dx^2}E_x(x) + \frac{d}{dx}\left[\frac{1}{\varepsilon(x)}\frac{d\varepsilon(x)}{dx}E_x(x)\right] + \left[k_0^2\varepsilon(x) - \beta^2\right]E_x(x) = 0 \tag{8.2}$$

for TM modes. The planar waveguide we are considering here has a refractive index that varies continuously in the x direction. For the planar optical waveguide shown in Figure 8.1, our problem is to find the refractive index profile function in the core for a set of specified propagation constants.

We assume that this planar waveguide has a refractive index profile guiding N modes. The propagation constants $\{\beta_n\}$ are $k_0 n_1 > \beta_1 > \beta_2 > \ldots \beta_N \geq k_0 n_\infty$, in which n_∞ is the value of $n(x)$ as $x \to \infty$ and $n_1 = \sup n(x)$. Designing an optical waveguide is analogous to the inverse problem encountered in quantum mechanics. We are trying to get the potential function from the given bound states and scattering data. The wave equation for the TE modes can be easily transformed to an equivalent Schrodinger equation

$$\frac{d^2}{dx^2}E_y(x) + \left[k^2 - V(x)\right]E_y(x) = 0 \tag{8.3}$$

by letting

$$V(x) = -k_0^2\left[n^2(x) - n_\infty^2\right] \tag{8.4}$$

and

$$k^2 = -\kappa_n^2 = -\left(\beta_n^2 - k_0^2 n_\infty^2\right). \tag{8.5}$$

We can see in our case the potential function $V(x)$ is continuous and $V(x) \to 0$ as $|x| \to \infty$. The TE mode cases have been solved by Yukon and Bendow (1980) and Jordan and Lakshmanasamy (1989).

We now need to transfer the wave equation for the TM modes to Schrodinger-type equation to apply the inverse scattering method. In (8.2), the first derivative of E_x can be eliminated if we let $E_x(x) = \varepsilon^{-1/2}(x)\Phi(x)$. The wave equation then becomes

$$\frac{d\Phi^2}{dx^2} + \left[\frac{1}{2\varepsilon(x)}\frac{d^2\varepsilon(x)}{dx^2} - \frac{3}{4\varepsilon^2(x)}\left(\frac{d\varepsilon(x)}{dx}\right)^2\right]\Phi + \left[k_0^2\varepsilon(x) - \beta^2\right]\Phi = 0. \tag{8.6}$$

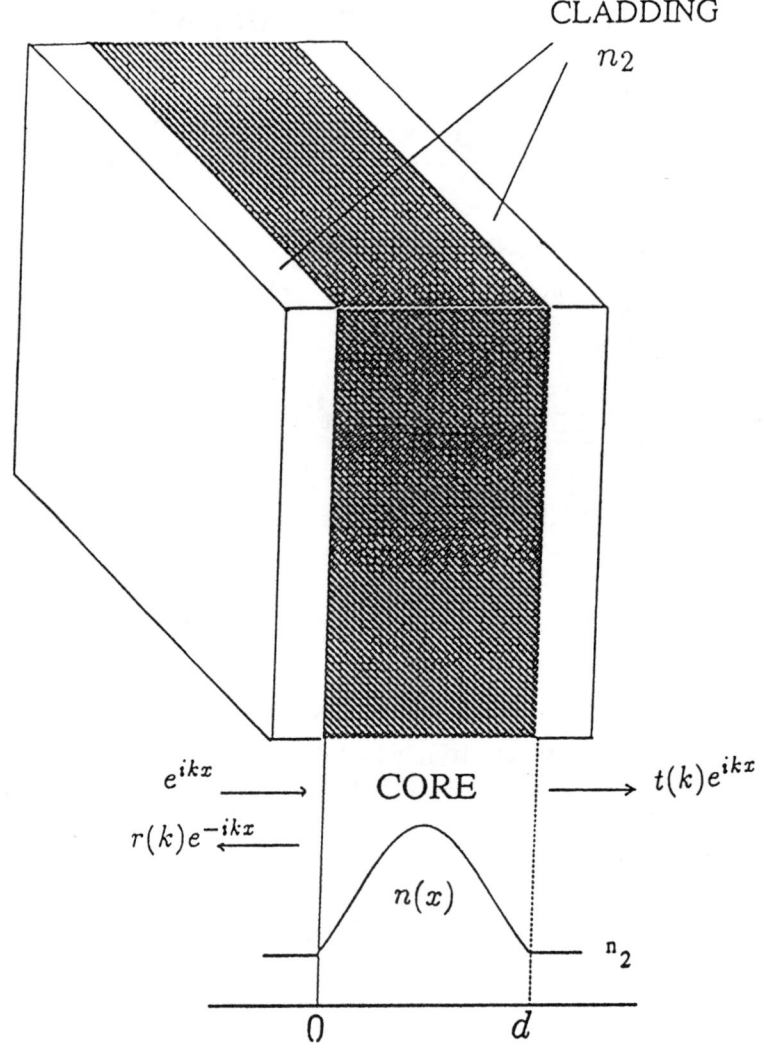

Figure 8.1 The physical structure of an inhomogeneous symmetrical planar optical waveguide showing the reflection and transmission of electromagnetic wave.

We are now able to obtain the equivalent Schrodinger equation

$$\frac{d^2\Phi(x)}{dx^2} + \left[k^2 - V(x)\right]\Phi(x) = 0 \tag{8.7}$$

by setting the potential function as

$$V(x) = \frac{3}{4\varepsilon^2(x)}\left(\frac{d\varepsilon(x)}{dx}\right)^2 - \frac{1}{2\varepsilon(x)}\frac{d^2\varepsilon(x)}{dx^2} - k_0^2\left(\varepsilon(x) - n_\infty^2\right) \tag{8.8}$$

and letting

$$k^2 = -\kappa_n^2 = k_0^2 n_\infty^2 - \beta_n^2. \tag{8.9}$$

The TM mode case has been solved by Tamil and Lin (1993).

INVERSE SCATTERING THEORY

The inverse scattering theory of Kay and Moses (Kay, 1955) provides us with a way to obtain the potential from the reflection coefficient that characterizes the propagation properties of the planar waveguide. As the potential we defined vanishes at infinity, we can apply the Gelfand-Levitan-Marchenko (G-L-M) equation to solve our problem. Let us consider a time-dependent formulation of the scattering. We take the Fourier transform of (8.7) (the transform pairs are $\Phi(x,k) \Leftrightarrow \Psi(x,t)$ and $k \Leftrightarrow t$) to obtain

$$\frac{\partial^2}{\partial x^2}\Psi(x,t) - \frac{\partial^2}{\partial t^2}\Psi(x,t) - V(x)\Psi(x,t) = 0, \tag{8.10}$$

in which t is the time variable with the velocity of light $c \equiv 1$. The incident plane wave is represented by the unit impulse

$$\Psi(x,t) = \delta(x-t), \quad x < 0, \quad t < 0, \tag{8.11}$$

which will produce the reflected transient wave function

$$R(x+t) = \frac{1}{2\pi}\int_{-\infty}^{\infty} r(k)e^{-ik(x+t)}dk + \sum_{n=1}^{N} A_n e^{-i\kappa_n(x+t)}, \tag{8.12}$$

where $k^2 = -\kappa_n^2$ are the discrete eigenvalues of Schrodinger-type (8.7), $r(k)$ is the complex reflection coefficient, and A_n are arbitrary constants normalizing the wave equation such that

$$\int_{-\infty}^{+\infty} \Phi(x)\Phi^*(x)dx = 1. \tag{8.13}$$

The reflected transient is produced only after the incident unit impulse has interacted with the inhomogeneous core of the optical waveguide and therefore

$$R(x+t) = 0 \quad \text{for } x+t \leq 0. \tag{8.14}$$

A linear transform independent of k can now relate the wave amplitude $\Psi(x,t)$ in the core region with the wave amplitude $\Psi_0(x,t)$ in the exterior region

$$\Psi(x,t) = \begin{cases} \Psi_0(x,t) + \int_{-x}^{x} K(x,\xi')\Psi_0(\xi',t)d\xi' & x > 0 \\ \Psi_0(x,t) & x \leq 0 \end{cases}. \tag{8.15}$$

Here the exterior field is

$$\Psi_0(x,t) = \delta(x-t) + R(x+t). \tag{8.16}$$

From physical consideration, since $\Psi(x,t)$ is a rightward moving transient

$$\Psi(x,t) = 0 \quad \text{for } t < x. \tag{8.17}$$

Thus the kernel $K(x,t) = 0$, for $t > x$ and $K(x,t) = 0$ for $t \leq -x$. We substitute (8.16) into (8.15) and using (8.14) and (8.17) yield the integral equation

$$K(x,t) + R(x+t) + \int_{-x}^{x} K(x,\xi')R(\xi'+t)d\xi' = 0 \quad t < x. \tag{8.18}$$

By substituting (8.15) into (8.10) the kernel $K(x,t)$ satisfies a differential equation of the same form as (8.10) provided the following conditions are imposed

$$K(x,-x) = 0, \tag{8.19}$$

and

$$2\frac{d}{dx}K(x,x) = V(x). \tag{8.20}$$

We now can see how the solution of the integral (8.18) for the function $K(x,t)$ can lead to the synthesis of optical waveguides.

DESIGN EXAMPLE 1:
ZERO REFLECTION COEFFICIENT

The reflection coefficient characterizes the propagation properties of the guided-wave optical devices. The zero reflection coefficient characterizes a system with propagating modes only, whereas a non-zero reflection coefficient characterizes a system with both guided and nonguided modes. Let us first consider the special case of zero reflection coefficient (Kay and Moses, 1956). We substitute (8.12) for $r(k) = 0$ in Gelfand-Levitan-Marchenko (8.18)

$$K(x,t) + \sum_{n=1}^{N} A_n e^{\kappa_n(x+t)} + \sum_{n=1}^{N} A_n \int_{-\infty}^{x} K(x,\xi)e^{\kappa_n(t+\xi)}d\xi = 0. \tag{8.21}$$

It is clear from the above equation that the solution for $K(x,t)$ should have the form (Kay and Moses, 1956)

$$K(x,t) = \sum_{n=1}^{N} f_n(x) e^{\kappa_n t}. \tag{8.22}$$

Substituting (8.22) into (8.21) produces a system of equations for $f_n(x)$:

$$A_n \sum_{\nu=1}^{N} \left(\frac{e^{(\kappa_\nu + \kappa_n)x}}{\kappa_n + \kappa_\nu} \right) f_\nu(x) + f_n(x) + A_n e^{\kappa_n x} = 0 \tag{8.23}$$

where $n = 1, 2, \ldots N$. This system can be conveniently written as

$$[\mathbf{A}][\mathbf{f}] + [\mathbf{B}] = 0, \tag{8.24}$$

where $[\mathbf{f}]$ and $[\mathbf{B}]$ are column vectors with f_n, and $B_n = A_n \exp(\kappa_n x)$ respectively, and $[\mathbf{A}]$ is a square matrix with elements

$$A_{\nu n} = \delta_{\nu n} + A_\nu \left(\frac{e^{(\kappa_\nu + \kappa_n)x}}{\kappa_\nu + \kappa_n} \right), \tag{8.25}$$

in which $\delta_{\nu n}$ is a Kronecker delta. The solution for $\mathbf{f}$ is $\mathbf{f} = -\mathbf{A}^{-1}\mathbf{B}$ and then from (8.22) $K(x,x) = \mathbf{E}^T \mathbf{f}$ where $\mathbf{E}$ is the column vector with element $E_n = \exp(\kappa_n x)$ and T denotes transpose. Now,

$$\frac{d}{dx} A_{\nu n} = A_\nu e^{(\kappa_\nu + \kappa_n)x} = B_n E_n, \tag{8.26}$$

and so

$$K(x,x) = E_n f_n = -E_n A_{\nu n}^{-1} B_\nu = A_{\nu n}^{-1} \frac{d}{dx} A_{n\nu}, \tag{8.27}$$

when written with subscript notation and the summation convention. The $K(x,x)$ given by (8.22) can be recognized in the form

$$K(x,x) = tr\left(\mathbf{A}^{-1} \frac{d\mathbf{A}}{dx} \right) = \frac{d}{dx} ln(det\,\mathbf{A}), \tag{8.28}$$

and therefore the potential $V(x)$ according to (8.20) is

$$V(x) = -2 \frac{d^2}{dx^2} ln(det\,\mathbf{A}). \tag{8.29}$$

Given N modes with desired propagation constants, we can obtain a potential function as given by (8.29). Here we have N degrees of freedom due to N arbitrary constants $\{A_n | n = 1, 2 \ldots N\}$.

For TE modes the refractive index profiles is simply given by

$$n^2(x) = n_\infty^2 - \frac{V(x)}{k_0^2}, \qquad (8.30)$$

in which k_0 is the free space wave number. Whereas for TM modes, obtaining the refractive index profile is more complicated because it is a solution to a nonlinear differential equation [(8.8)]. The nonlinear differential equation can only be solved numerically. First we transform (8.8) into a convenient form by setting $\varepsilon(x) = e^{y(x)}$. We then obtain

$$\frac{1}{2}\frac{d^2 y(x)}{dx^2} - \frac{1}{4}\left[\frac{dy(x)}{dx}\right]^2 + k_0 e^{y(x)} + \left[V(x) - k_0^2 n_\infty^2\right] = 0. \qquad (8.31)$$

This is a constant coefficient equation that yields the refractive index profile $\sqrt{\varepsilon(x)}$ provided the potential $V(x)$ is given.

DESIGN EXAMPLE 2:
NON-ZERO REFLECTION COEFFICIENT

In the previous section, we took advantage of assuming that the reflection coefficient is zero, which simplified the problem a lot. Now we are going to solve the problem with non-zero reflection coefficient.

We take the rational function approximation for our scattering data. We represent our reflection coefficient using a three-pole rational function of transverse wave number k (Jordan and Lakshmanasamy, 1989). The three poles are as follows: one pole on the upper imaginary axis of the complex k plane, which represents discrete spectrum of function $R(x+t)$ [(8.12)] characterizing the propagating mode; two symmetric poles lie in the lower half of the k plane, which represent the continuous spectrum of $R(x+t)$ characterizing the unguided modes; the three-pole reflection coefficient can be written as

$$r(k) = \frac{r_0}{(k - k_1)(k - k_2)(k - k_3)}, \qquad (8.32)$$

where r_0 can be determined by the normalization condition $r(0) = -1$. This condition ensures total reflection at $k = 0$. k_1, k_2 have the following forms: $k_1 = -c_1 - \iota c_2$ and $k_2 = -c_1 - \iota c_2$. The third pole on the positive imaginary axis is $k_3 = \iota a$.

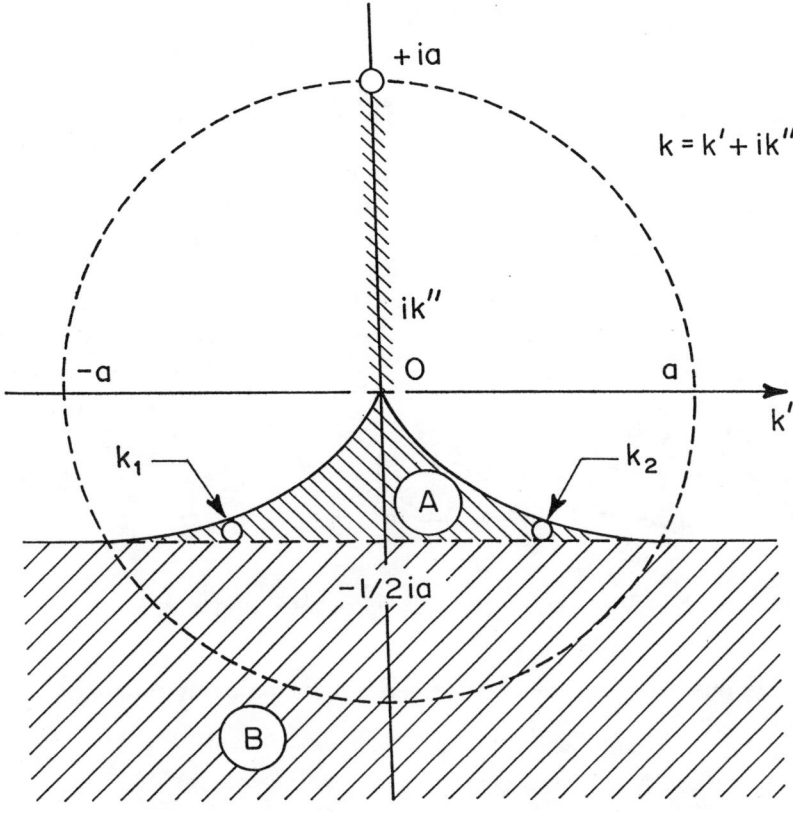

Figure 8.2 Permitted regions of the complex k plane for the pole positions in a three-pole reflection coefficient.

The pole positions are confined to certain "allowed regions" that are determined by the law of conservation of energy, which can be represented by $|r(k)|^2 \leq 1$ for all real k; see Figure 8.2 and refer to Jordan and Lakshmanasamy (1989) for details.

It has been shown that the reconstructed potential function $V(x)$ has following form:

$$V(x) = 2\left[\frac{d(\mathbf{a}^T(x))}{dx} - \mathbf{a}^T(x)\mathbf{A}^{-1}(x)\frac{d(\mathbf{A}(x))}{dx}\right]\mathbf{A}^{-1}(x)\mathbf{b}, \qquad (8.33)$$

in which **a** and **b** are column vectors, and are given by

$$\mathbf{a}^T(x) = \begin{bmatrix} 1 & x & e^{\eta_1 x} & e^{-\eta_1 x} & e^{\eta_2 x} & e^{-\eta_2 x} \end{bmatrix} \qquad (8.34)$$

$$\mathbf{b}^T = \begin{bmatrix} 0 & 0 & 0 & 0 & 0 & -a(c_1^2 + c_2^2) \end{bmatrix}, \qquad (8.35)$$

where

$$\eta_1 = \left[\frac{1}{2}a^2 + c_2^2 - c_1^2 + \frac{1}{2}\left(a^2 - 4c_2^2\right)^{1/2}\left(a^2 + 4c_1^2\right)^{1/2}\right]^{1/2} \tag{8.36}$$

$$\eta_2 = \left[\frac{1}{2}a^2 + c_2^2 - c_1^2 - \frac{1}{2}\left(a^2 - 4c_2^2\right)^{1/2}\left(a^2 + 4c_1^2\right)^{1/2}\right]^{1/2}. \tag{8.37}$$

Matrix $\mathbf{A}(x)$ is given by

$$\begin{bmatrix} 0 & 1 & 0 & 0 & 0 & 0 \\ 0 & 0 & f(\eta_1) & a(c_1^2 + c_2^2) & 0 & 0 \\ 0 & 0 & 0 & 0 & f(\eta_2) & a(c_1^2 + c_2^2) \\ 1 & -x & e^{-\eta_1 x} & e^{\eta_1 x} & e^{-\eta_2 x} & e^{\eta_2 x} \\ 0 & -1 & -\eta_1 e^{-\eta_1 x} & \eta_1 e^{\eta_1 x} & -\eta_2 e^{-\eta_2 x} & \eta_2 e^{\eta_2 x} \\ 0 & 0 & \eta_1^2 e^{-\eta_1 x} & \eta_1^2 e^{\eta_1 x} & \eta_2^2 e^{-\eta_2 x} & \eta_2^2 e^{\eta_2 x} \end{bmatrix}, \tag{8.38}$$

where

$$f(x) = x^3 + (2c_2 - a)x^2 + \left[c_1^2 + c_2^2 - 2ac_2\right]x - a\left(c_1^2 + c_2^2\right). \tag{8.39}$$

So, it is possible to construct the potential from the three poles of reflection coefficient using the above equations.

DESIGN EXAMPLE 3: NONRATIONAL REFLECTION COEFFICIENT

The refractive index profiles reconstructed for the cases discussed above go to zero asymptotically and they approximately model the actual refractive index profiles used in practice. The refractive index profiles used in practice are truncated and the truncations form the core-cladding boundary. For a doubly truncated refractive index profile modeling a planar optical waveguide, the reflection coefficient is not a rational function of the complex wavenumber, but a more complicated form (Mills and Tamil, 1991, 1992; Lamb, 1980). Reconstructing refractive index profiles for nonrational reflection coefficients is not possible in analytical closed forms and so numerical techniques must be used.

Discretization of the G-L-M Equation

To solve the G-L-M (8.18) by numerical methods, the space-time diagram is discretized into square grids rotated by 45° with respect to abscissa, as shown in Figure 8.3. The interval $\Delta t = 2\Delta x$, and $x = m\Delta x$, $m = 0, 1, \ldots N$, where N is the total number of grid points along the x direction, and $t = n\Delta t - (m/2)\Delta t$, $n = 0, 1, \ldots, m$. The G-L-M integral equation can then be discretized as

$$K_m(n) + R_m(n) + \sum_{l=m-n}^{m} C(l) K_m(l) R_m(l) \Delta t = 0, \quad n = 0, 1, \ldots, m, \tag{8.40}$$

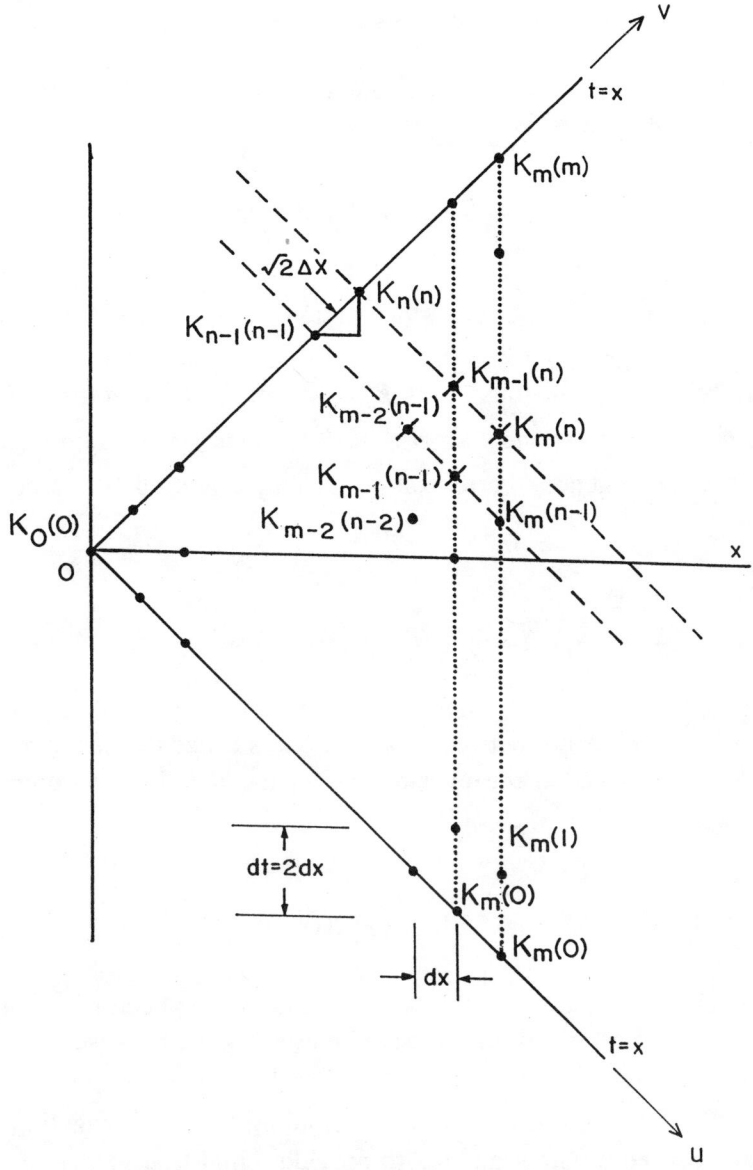

Figure 8.3 Discretized grid diagram in a space-time plane for numerical reconstruction.

where $y = l\Delta t - (m/2)\Delta t$. The subscript m in $K_m(n)$ represents the grid position along the x direction and the argument n represents the grid position along the t direction. $C(l)$ is the coefficient for numerical integration; if the trapezoidal rule is used,

$$C(l) = \begin{cases} 1/2 & l = m-n \text{ and } m \\ 1 & \text{elsewhere} \end{cases}. \tag{8.41}$$

Iteration Scheme with Relaxation

The Gelfand-Levitan-Marchenko equation is an integral equation of the second kind and can be solved numerically in an iterative manner. We rewrite (8.40) as

$$K_m^i(n) = -R_m(n) - \sum_{l=m-n}^{m} C(l) K_m^{i-1}(l) R_m(l) \Delta t, \qquad (8.42)$$

where the superscript i in $K_m^i(n)$ represents the i-th iteration result. It is worth pointing out that the iterative process involves only the grid points on the m-th column in Figure 8.3.

In (8.12) the poles on the positive imaginary axis $k_n = i\kappa_n$, $\kappa_n > 0$ are in the discrete spectrum and correspond to the guided modes. The exponential term in (8.12) then grows rapidly as $(x+t)$ increases and in order to improve the convergence, the relaxation technique is used (Press et al., 1989), so that (8.42) is revised as

$$\begin{aligned} K_m^i(n) &= (1-\omega) K_m^{i-1}(n) \\ &+ \omega \left[-R_m(n) - \sum_{l=m-n}^{m} C(l) K_m^{i-1}(l) R_m(l) \Delta t \right], \end{aligned} \qquad (8.43)$$

where ω is the relaxation factor. If ω lies between 0 and 1.0, it is called the under-relaxation method; if ω lies between 1.0 and 2.0, it is called the over-relaxation method. In our computations, $\omega < 0.7$ provides the desired results.

Initial Values for K(x, t)

The convergence of analytical solutions to the G-L-M equation has been proved (Szu et al., 1976). However, the convergence of its discretized form cannot be ascertained, because of the additional errors due to truncation and discretization. Good initial values for $K(x,t)$ are important for the numerical iterative scheme, in particular when a bound state corresponding to the propagating mode exists. The Born approximation has been used by other authors to provide initial trial values for $K(x,t)$ for cases where there are no bound states. However, for cases discussed here, where there are bound states, the Born approximation when used to provide the initial values for $K(x,t)$ fails to reconstruct the potential correctly. Although for shorter lengths of the potential the reconstruction is in agreement with the actual value, the method fails for larger lengths. The leapfrogging algorithm (Jordan and Ladoucer, 1987) provides an effective initial value for $K(x,t)$.

To obtain the leapfrogging algorithm, we substitute (8.20) into (8.10) where $\Psi(x,t)$ has been replaced by $K(x,t)$ yielding

$$\frac{\partial^2 K(x,t)}{\partial x^2} - \frac{\partial^2 K(x,t)}{\partial t^2} - 2 \frac{dK(x,x)}{dx} K(x,t) = 0, \qquad (8.44)$$

and introduce new variables u and v, defined as

$$u = \frac{x-t}{\sqrt{2}} \tag{8.45}$$

and

$$v = \frac{x+t}{\sqrt{2}} \tag{8.46}$$

(see Figure 8.3). With this coordinate transformation, the partial differential equation (8.44) can be rewritten as

$$\frac{\partial^2 K(u,v)}{\partial u \partial v} - \sqrt{2}\frac{\partial K(u,v)}{\partial v} - K(u,v) = 0 \tag{8.47}$$

so that its discretization gives the following equation (Jordan and Ladoucer, 1987),

$$\begin{aligned}K_m(n) &= K_{m-1}(n) + K_{m-1}(n-1) \\ &+ \{2\Delta x[K_n(n) - K_{n-1}(n-1)] - 1\}K_{m-2}(n-1); \; n>0, \, m>0, \, n=m,\end{aligned} \tag{8.48}$$

which relates the grid point $K_m(n)$ with the other five grid points, as shown in Figure 8.3. Note that $K_m(n)$ on the LHS of (8.48) is at the "current" reconstruction column m, while the remaining five grid points on the RHS are all located within its left region, which are either on the boundary whose values are provided by $K(x,-x) = -R(0)$ or grid points that have already been reconstructed by the step-by-step marching algorithm marching in the x direction. (8.48) does not provide values for $K_m(m)$, $m = 1, 2, \ldots N$ and a different procedure should be adopted to find those values.

Solving (8.42) for $K_m(m)$ yields

$$K_m(m) = \frac{-R_m(m) - \sum_{l=0}^{m-1} C(l) K_m(l) R_m(l) \Delta t}{1 + C(m) R_m(m) \Delta t}, \quad m > 0, \, m = n, \tag{8.49}$$

which provides initial trial values for $K_m(m)$. Furthermore, we obtain

$$K_m(0) = -R(0). \tag{8.50}$$

To summarize, (8.48), (8.49), and (8.50) can provide the initial trial values for $K(x,t)$ necessary for the iterative numerical solution of the G-L-M equation.

Reconstruction of the Potential v(x)

The potential in its discretized form can be expressed using (8.20) as

$$v(m-1) = \frac{K_m(m) - K_{m-2}(m-2)}{\Delta x}, \quad m \geq 2. \tag{8.51}$$

This expression can be used to reconstruct the potential when the values of $K_m(m)$, $m = 0, 1, \ldots N$ are already evaluated. This reconstructs the potential at every point in x except at $x = 0$, corresponding to the grid point $m = 0$.

To evaluate the potential at the origin we substitute (8.18) into (8.20), yielding

$$v(x) = -2\frac{dR(2x)}{dx} - K(x,x)R(2x) + \int_{-t}^{x}\frac{\partial}{\partial x}[K(x,y)R(y+x)]dy. \tag{8.52}$$

Because $R(t) = 0$ for $t \leq 0$, we obtain at the origin

$$v(0) = -2\frac{dR(2x)}{dx}\bigg|_{x=0} = -4R'(0), \tag{8.53}$$

which is an exact formula for recovering the potential at the origin. It is interesting to note that the perturbation expansion theory derives the approximate solutions (Kritikos et al., 1982)

$$v(x) = -2\frac{dR(2x)}{dx} \tag{8.54}$$

and

$$v(x) = -2\frac{dR(2x)}{dx} + 4[R(2x)]^2, \tag{8.55}$$

which are called the Born and the modified Born approximation, respectively. At the origin, the Born expression provides an approximate reconstruction, even though there exists a discontinuity at the boundary.

The numerical inverse scattering theory can now be summarized in the following steps:

(a) Compute the potential at the origin, $v(0)$ using (8.53);
(b) Set the initial trial values for $K_m(n)$, $n = 0, 1, \ldots m$ on the current column m using (8.48), (8.49) and (8.50);
(c) Iteratively calculate $K_m(n)$, $n = 1, 2, \ldots m$ for each value of n on the current column m using (8.42) with an appropriate choice of relaxation factor ω;
(d) Evaluate potential $v(m-1)$ using (8.51);
(e) Move the current column from m to $m+1$, and repeat the steps (b) to (d).

DISCUSSION

We have developed a method based on inverse scattering theory that can be used to design planar optical waveguides that transmit a prescribed number of TE or TM modes with prescribed propagation constants. To demonstrate some practical examples for the zero reflection case, let us compute the refractive index profiles for two cases: the single mode case and the N mode case.

For the single mode case, (8.23) becomes

$$A_1 e^{\kappa_1 x} + f_1(x) + \left(\frac{A_1 e^{2\kappa_1 x}}{2\kappa_1}\right) f_1(x) = 0. \tag{8.56}$$

Then, the potential has the form

$$V(x) = \frac{-4\kappa_1 A_1 e^{2\kappa_1 x}}{\left(1 + A_1 e^{2\kappa_1 x}/2\kappa_1\right)^2}, \tag{8.57}$$

where A_1 is an arbitrary constant, and note that κ_1 can be obtained from

$$\kappa_1^2 = \beta_1^2 - k_0^2 n_\infty^2. \tag{8.58}$$

For a desired propagation constant β_1, we can get a set of refractive index profiles corresponding to different arbitrary choice of A_1; see Figure 8.4. We use the following data relating to waveguide: $n(\infty) = n_S = 2.177$, wavelength $\lambda = 0.8\,\mu m$ and $\beta_1 = 17.20(\mu m)^{-1}$. We obtained the refractive index profiles by solving (8.31) using the potential $V(x)$ obtained from (8.57). Runge-Kutta's fourth-order approximation is applied in solving the differential equation (8.31) (Levy and Baggott, 1976). We can see from Figure 8.4 that the maximum value of refractive index lies on the positive side of $x = 0$ when $A_1 < 2\kappa_1$, on the negative side of $x = 0$ when $A_1 > 2\kappa_1$, and at $x = 0$ when $A_1 = 2\kappa_1$.

Substituting $A_1 = 2\kappa_1$ into (8.57) yields

$$V(x) = -2\kappa_1^2 \text{sech}^2 \kappa_1 x. \tag{8.59}$$

This potential is everywhere negative and goes to 0 as x goes to infinity. Also the potential is symmetric about its minimum point. We can truncate the potential at the point where the potential is 1 percent of its maximum value to find the width of the core d. The refractive index profile corresponding to this potential is shown by the continuous line in Figure 8.4.

Similarly, for the N mode case, we need to construct the potential first using (8.25) and then solve the nonlinear differential (8.31) for the refractive index profiles. For a set of prescribed propagation constants, every arbitrary choice of normalization constants will produce a different potential and a corresponding refractive index profile. In order to construct a symmetric refractive index profile with single peak, we found that the normalization constants $\{A_n | n = 1, 2 \ldots N\}$ must satisfy the following equation (Deift and Trubowitz, 1979):

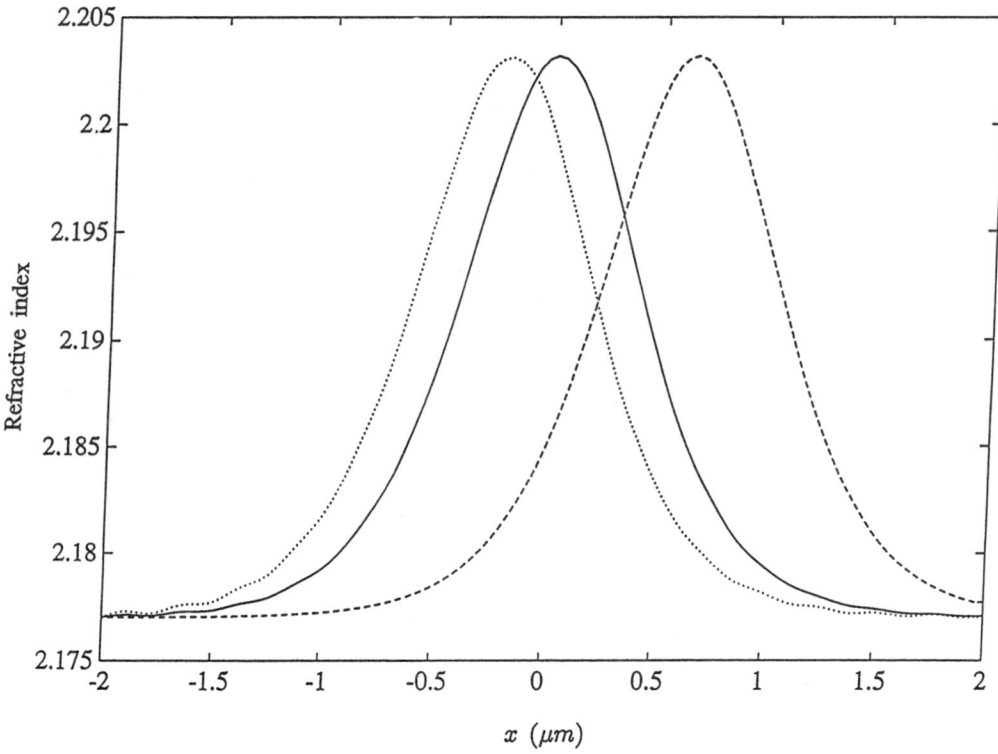

Figure 8.4 The reconstructed refractive index profiles for a single prescribed TM mode with $\beta_1 = 17.2$ and various $A_1 = 2\kappa_1 = 3.7386, 0.4,$ and 0.7 corresponding to the solid, dashed, and dotted curves, respectively.

$$A_n = \sqrt{2\kappa_n P_n}, \qquad (8.60)$$

where

$$P_n = (-1)^{n-1} \prod_{v=1(v \neq n)}^{N} \frac{\kappa_v + \kappa_n}{\kappa_v - \kappa_n} \qquad n = 1, 2, \ldots N \qquad (8.61)$$

for the reflectionless case. Here N is the number of guided modes in the planar waveguide. For the case $N = 5$, using sets of arbitrary normalization constants $\{A_n | n = 1, 2 \ldots N\}$ we have computed the refractive index profiles and these are shown in Figure 8.5. The symmetric profile obtained using the condition (8.60) is shown by a continuous line in the figure.

To demonstrate the reconstruction of the potential from a three-pole reflection coefficient (a case of non-zero rational reflection coefficients) we have chosen here two examples. In example 1, the poles are determined by the following parameters: $a = 1.0$, $c_1 = 0.8$, and $c_2 = 0.499$; example 2 has different unguided modes characterized by $c_1 = 0.05$, $c_2 = 0.1$, and the same propagating mode characterized by $a = 1.0$. Figure 8.6 shows the plots of potential functions for examples 1 and 2. In example 2, we see that the potential is everywhere negative.

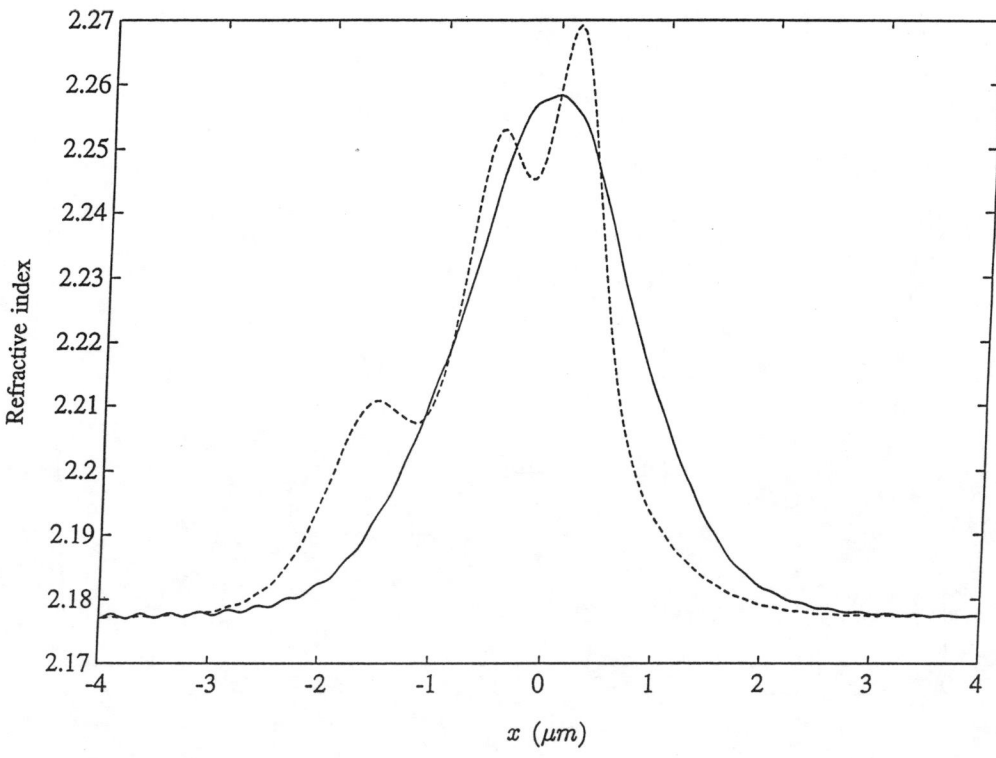

Figure 8.5 Reconstructed refractive index profiles for five prescribed TM modes with correspondence to $A_n = \{1, 2, 3, 3, 1\}$ (dashed curve) and for A_n satisfying (8.60) (solid curve).

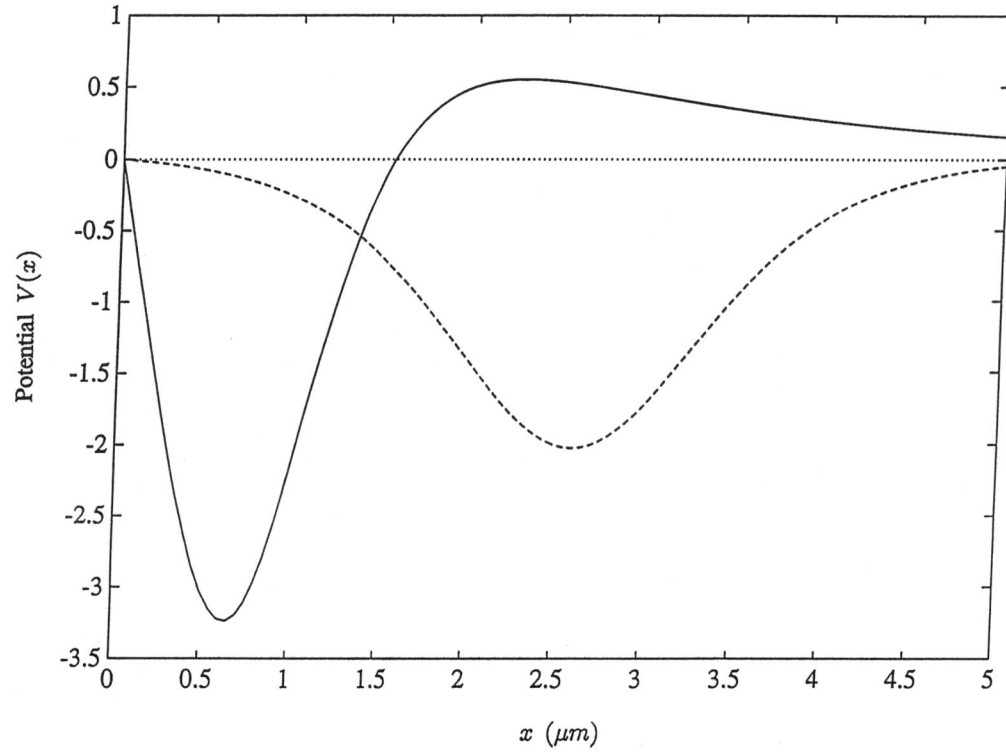

Figure 8.6 Potentials of a waveguide characterized by a three-pole rational reflection coefficient. The solid curve corresponds to $a = 1.0$, $c_1 = 0.8$, $c_2 = 0.499$; the dashed curve corresponds to $a = 1.0$, $c_1 = 0.05$, $c_2 = 0.1$.

Figure 8.7 shows the refractive index profiles for TM mode in both the above discussed examples obtained by substituting the potentials into the nonlinear differential equation (8.31) and solving for $\sqrt{\varepsilon(x)}$. We notice that a depressed cladding is obtained in example 1, and we also see that the profiles we found here resemble the profiles we normally find in practical optical waveguides (Okoshi, 1976).

Introducing a truncated potential to model the planar waveguide (Mills and Tamil, 1992), it can be shown that both propagating and non-propagating modes appear when the reflectionless potential $v(x) = -2\,\mathrm{sech}^2(x)$ is truncated at a point on the left $x = x_1$. Based on the Jost solutions corresponding to the untruncated potential $v_0 = -2\,\mathrm{sech}^2(x)$, the reflection coefficient from the left for the truncated potential can be derived (Mills and Tamil, 1992) as

$$r(k) = -\exp(i2kx_1)\frac{\mathrm{sech}^2(x_1)}{k^2 + 1 + (k + i\tanh(x_1))^2}, \tag{8.62}$$

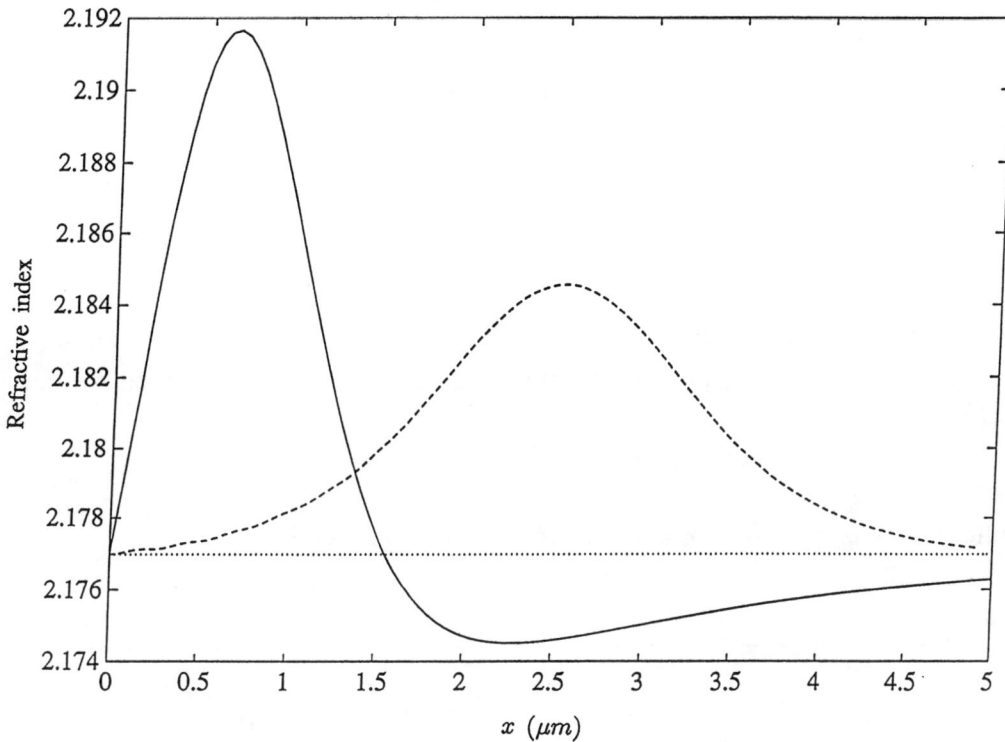

Figure 8.7 Reconstructed refractive index profiles corresponding to the potentials shown in Figure 8.6.

which has two poles in the complex k-plane located at

$$k_1 = -\frac{i}{2}\left[\sqrt{2-\tanh^2(x_1)} + \tanh(x_1)\right] \tag{8.63}$$

and

$$k_2 = \frac{i}{2}\left[\sqrt{2-\tanh^2(x_1)} - \tanh(x_1)\right]. \tag{8.64}$$

Since $\sqrt{2-\tanh^2(x_1)} > |\tanh(x_1)|$, both poles are located on the imaginary axis, so that $k_2 = i\kappa$ corresponds to the guided mode. The exponential factor $\exp(i2kx_1)$ in (8.62) represents a shift x_1 on the x axis relative to the corresponding untruncated potential. Equation (8.62) can then be rewritten as

$$r_0(k) = \frac{\text{sech}^2(x_1)}{k^2 + 1 + (k + i\tanh(x_1))^2}, \tag{8.65}$$

in which the phase shift factor has been excluded. The characteristic function is

$$R_0(t) = \frac{\text{sech}^2(x_1)}{2\sqrt{2-\tanh^2(x_1)}} \left\{ -\exp\left[-0.5\left(\sqrt{2-\tanh^2(x_1)} + \tanh(x_1)\right)t\right] \right.$$
$$\left. + \exp\left[0.5\left(\sqrt{2-\tanh^2(x_1)} - \tanh(x_1)\right)t\right] \right\}. \tag{8.66}$$

Using (8.53), the potential at the truncation location is

$$v_0(0) = -2\text{sech}^2(x_1). \tag{8.67}$$

This is a case of nonrational reflection coefficient. Figure 8.8 gives the potential, assuming $x_1 = -1.0$, where the asterisks show the potential obtained by numerical reconstruction, and the exact potential

$$v_0(x) = -2\text{sech}^2(x + x_1) \tag{8.68}$$

is plotted in solid line for comparison. Again good agreement is achieved.

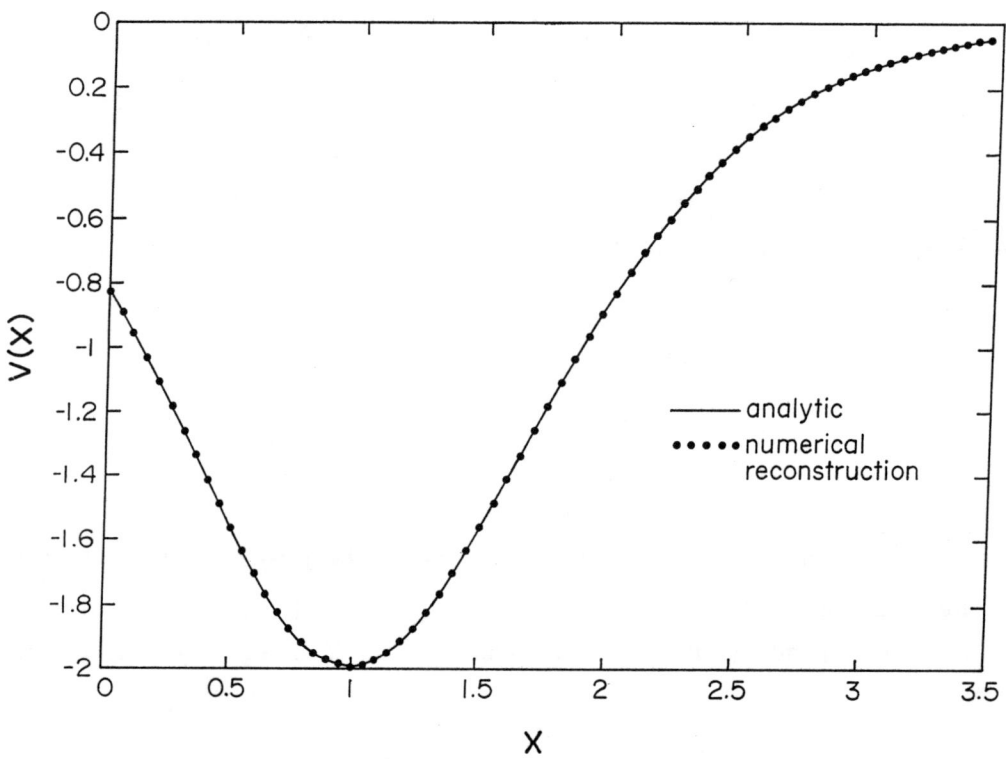

Figure 8.8 Potential $-2\text{sech}^2(x)$ truncated at the left, $x_1 = -1.0$. Solid curve, exact potential; circles, numerical reconstruction.

The results obtained by inverse scattering theory can be verified by a finite difference-based analysis scheme. Using this method we find the propagation constants of guided modes of an optical waveguide with arbitrary refractive index profile. Owing to its simplicity and flexibility, this method is proved to be very effective. For demonstration purposes we consider here a symmetric planar waveguide. We have compared in Table 8.1 the propagation constants of various modes that we used in reconstructing the refractive index profile of the waveguide against the propagation constants obtained by analysis for the normalized frequency at which the propagation constants are prescribed. We see that the last two columns of the table agree very well. This shows that the inverse technique outlined here can be used to synthesize waveguides with prescribed modes.

Table 8.1 Prescribed TM Mode Spectra Used in Reconstructing Refractive Index of Planar Waveguide and Spectra Obtained by Analysis Using Finite Difference Scheme

Number of Modes	Mode Number γ	Prescribed Mode Spectra β_γ / k_0	β_γ / k_0 Obtained by Authors' Analysis
N = 1	0	2.18997	2.18995
N = 2	0	2.20556	2.20553
	1	2.18417	2.18398
N = 3	0	2.20926	2.20916
	1	2.19140	2.19100
	2	2.18061	2.18036
N = 5	0	2.21288	2.21266
	1	2.20003	2.19968
	2	2.18998	2.18968
	3	2.18278	2.18254
	4	2.17845	2.17797
N = 7	0	2.21466	2.21452
	1	2.20473	2.20449
	2	2.19630	2.19606
	3	2.18927	2.18915
	4	2.18397	2.18379
	5	2.18010	2.17997
	6	2.17778	2.17753

The method demonstrated here can be extended to the synthesis of optical devices (Mills and Tamil, 1993, 1994) with specified transmission characteristics.

REFERENCES

Deift, P., and E. Trubowitz, 1979, "Inverse scattering on the line," *Commun. Pure Appl. Math.* **32**, 121-251.

Gelfand, I.M., and B.M. Levitan, 1955, "On the determination of a differential equation by its spectral function," *Trans. Am. Math. Soc. Ser.* **21**, 253-304.

Jordan, A.K., and H. Ladoucer, 1987, "A renormalized inverse scattering theory for discontinuous profiles," *Phy. Rev. A* **36**, 4245-4253.

Jordan A.K., and S. Lakshmanasamy, 1989, "Inverse scattering theory applied to the design of single-mode planar optical waveguides," *J. Opt. Soc. Am. A* **6**, 1206-1212.

Kay, I., 1955, *The Inverse Scattering Problem*, Rep. **EM-74**, New York, N.Y.: New York University.

Kay, I., and H. Moses, 1956, "Reflectionless transmission through dielectrics and scattering potentials," *J. Appl. Phys.* **27**, 1503-1508.

Kritikos, H.N., D.L. Jaggard, and D.B. Ge, 1982, "Numeric reconstruction of smooth dielectric profiles," *Proc. IEEE* **70**, 295-297.

Lamb, Jr., G.L., 1980, *Elements of Soliton Theory*, New York: Wiley and Sons.

Levy, H., and E.A. Baggott, 1976, *Numerical Solution of Differential Equations*, New York: Springer-Verlag.

Marchenko, V.A., 1950, "Concerning the theory of a differential operator of second order," *Dokl. Akad. Nauk SSSR T2*, 457-463.

Marcuse, D., 1972, *Light Transmission Optics,* Princeton, N.J.: Van Nostrand Reinhold.

Mills, D.W., and L.S. Tamil, 1991, "A new approach to the design of graded-index guided-wave devices," *IEEE Microwave Guided Wave Lett.* **1**, 87-88.

Mills, D.W., and L.S. Tamil, 1992, "Analysis of planar waveguides using scattering data," *J. Opt. Soc. Am. A* **9**, 1769-1778.

Mills, D.W., and L.S. Tamil, 1993, "Synthesis of guided-wave optical interconnects," *IEEE J. Quantum Electron.* **29**, 2825-2834.

Mills, D.W., and L.S. Tamil, 1994, "Coupling in multilayer optical waveguides: An approach based on scattering data," *J. Lightwave Technol.* **9**, 1560-1568.

Okoshi, T., 1976, *Optical Fibers*, New York: Academic Press.

Press, H.H., B.P. Flannery, S.A. Teukolsky, and W.T. Vetterling, 1989, *Numerical Recipes,* New York: Cambridge University Press.

Szu, H.H., C.E. Carroll, C.C. Yang, and S. Ahn, 1976, "A new functional equation in the plasma inverse scattering problem and its analytical properties," *J. Math. Phys.* **17**, 1236-1247.

Tamil, L., and Y. Lin, 1993, "Synthesis and analysis of optical planar waveguides with prescribed TM modes," *J. Opt. Soc. Am. A* **9**, 1953-1962.

Tamir, T., 1990, *Guided-Wave Optoelectronics,* New York: Springer-Verlag, Chap. 2.

Yukon, S.P., and B. Bendow, 1980, "Design of waveguides with prescribed propagation constants," *J. Opt. Soc. Am.* **70**, 172-179.

9

Design and Analysis of Finite Element Methods for Transient and Time-Harmonic Structural Acoustics

Peter M. Pinsky
Stanford University
M. Malhotra
Stanford University
Lonny L. Thompson
Clemson University

INTRODUCTION

In this paper the development of efficient computational methods for the exterior structural acoustics problem is considered from two standpoints. The first is a space-time finite element method for transient structural acoustics. The second is the development of efficient and robust iterative solution methods for large-scale problems in the frequency domain.

In the first part of this paper, a new space-time finite element approach for solving the coupled structural acoustics problem involving the interaction of vibrating structures submerged in an infinite acoustic fluid is described. The proposed method employs the simultaneous discretization of the spatial and temporal domains and is based on a new time-discontinuous variational formulation for the coupled fluid-structure system. In the proposed approach, the concept of space-time slabs is employed, which allows for discretizations that are discontinuous in time and offers greater discretization flexibility, e.g., through the use of space-time meshes oriented along space-time characteristics. In the space-time approach, increased algorithmic stability is obtained through the introduction of temporal jump operators, which give rise to a natural high-frequency dissipation required for the accurate resolution of sharp gradients in the physical solution. Additional stability is obtained by a least-squares modification. The order of accuracy of the solution depends directly on the order of the finite element spatial and temporal basis functions, which can be chosen to any accuracy for general unstructured discretizations in space and time. The resulting space-time approach, therefore, permits the development of a finite element method for transient structural acoustics with the desired combination of increased stability and high accuracy. Additionally, the proposed time-discontinuous Galerkin space-time method provides a natural variational setting for the incorporation of high-order accurate nonreflecting boundary conditions that are local in time. Since the temporal and spatial domains are treated in a consistent manner in the space-time variational equations, the method inherits a firm mathematical foundation from which rigorous a posteriori error estimates useful for reliable and efficient adaptive schemes may be established, (see, e.g., Johnson, 1990, 1993).

In the second part of this paper, we address issues relating to the efficient solution of large-scale matrix problems arising in steady state acoustics. Although considerable progress has been made in finite element methods for exterior problems in the frequency domain, efficient numerical algorithms for robust and accurate solution of large-scale acoustics problems still need to be developed. Although gradient-type iterative methods are an attractive alternative to direct methods for efficient solution of large sparse matrix problems arising from finite element discretizations, the convergence of these methods deteriorates with increasing mesh density and increasing frequency of analysis. In such cases effective preconditioning becomes essential in order to accelerate iterative convergence. In the second part, we investigate a multilevel preconditioning approach that is based on the *h*-version of the hierarchical finite element method.

THE TRANSIENT STRUCTURAL ACOUSTICS PROBLEM

Consider the coupled system illustrated in Figure 9.1, consisting of the computational domain $\Omega = \Omega_f \cup \Omega_s$, composed of a fluid domain Ω_f, and structural domain Ω_s. The fluid boundary $\partial \Omega_f$ is divided into the fluid-structure interface boundary Γ_i and the artificial boundary Γ_∞. The structural boundary $\partial \Omega_s$ is composed of the fluid-structure interface boundary Γ_i and a traction boundary Γ_σ. The infinite domain outside the artificial boundary is denoted by Ω_∞. The temporal interval of interest is $I =]0, T[$ and the number of spatial dimensions is d.

The structure is governed by the equations of elastodynamics, while the fluid equations are derived under the usual linear acoustic assumptions of an inviscid, compressible fluid with small disturbance. The strong form of the fluid-structure initial boundary-value problem is given by:

Find $\boldsymbol{u}: \overline{\Omega}_s \times [0, T] \mapsto \mathbb{R}^d$, and $\phi: \overline{\Omega}_f \times [0, T] \mapsto \mathbb{R}^1$, such that

$$\nabla \cdot \boldsymbol{\sigma} - \rho_s \ddot{\boldsymbol{u}} = \boldsymbol{0} \quad \text{in } Q_s \equiv \Omega_s \times I \tag{9.1}$$

$$\boldsymbol{\sigma} = C : \nabla^s \boldsymbol{u} \quad \text{in } Q_s \equiv \Omega_s \times I \tag{9.2}$$

$$\nabla^2 \phi - a^2 \ddot{\phi} = f \quad \text{in } Q_f \equiv \Omega_f \times I \tag{9.3}$$

$$\dot{\boldsymbol{u}} \cdot \boldsymbol{n} = \nabla \phi \cdot \boldsymbol{n} \quad \text{on } \Psi_i \equiv \Gamma_i \times I \tag{9.4}$$

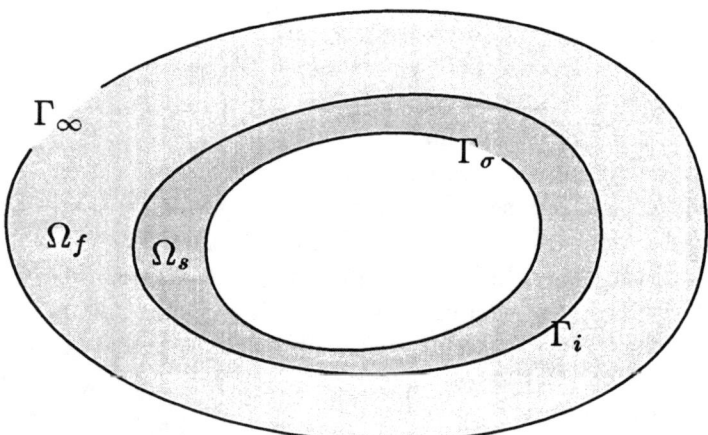

Figure 9.1 Coupled system for the exterior fluid-structure interaction problem, with artificial boundary Γ_∞ enclosing the finite computational domain $\Omega = \Omega_f \cup \Omega_s$.

$$\boldsymbol{\sigma} \cdot \boldsymbol{n} = \rho_f \dot{\phi} \boldsymbol{n} \qquad \text{on } \Psi_i \equiv \Gamma_i \times I \qquad (9.5)$$

$$\boldsymbol{\sigma} \cdot \boldsymbol{n} = \bar{\boldsymbol{t}} \qquad \text{on } \Psi_\sigma \equiv \Gamma_\sigma \times I \qquad (9.6)$$

$$\nabla \phi \cdot \boldsymbol{n} = -S_m \phi \qquad \text{on } \Psi_\infty \equiv \Gamma_\infty \times I \qquad (9.7)$$

In the above, $u(x,t)$ with $x \in \Omega_s$ is the structural displacement vector, $\boldsymbol{\sigma}$ is the symmetric Cauchy stress tensor, and $\phi(x,t)$ with $x \in \Omega_f$ is the scalar acoustic velocity potential. The phase velocity of acoustic wave propagation is denoted by c; $\rho_s > 0$ and $\rho_f > 0$ are the reference densities of the structure and fluid, respectively. A superposed dot indicates partial differentiation with respect to time t, and ∇^s refers to the symmetric gradient. The acoustic pressure, p, and the acoustic velocity, $\boldsymbol{v}$, are related to the velocity potential by $p = -\rho_f \dot{\phi}$ and $\boldsymbol{v} = \nabla \phi$. Equation (9.7) is the radiation boundary condition imposed on the artificial boundary Γ_∞, which approximates the asymptotic behavior of the solution at infinity, as described by the Sommerfeld radiation condition. Also, appropriate initial conditions are assumed corresponding to the above coupled second-order system of hyperbolic equations.

SPACE-TIME FINITE ELEMENT FORMULATION

Finite Element Discretization

The development of the space-time method proceeds by considering a partition of the time interval, $I =]0,T[$, of the form $0 = t_0 < t_1 < ... < t_N = T$, with $I_n =]t_n, t_{n+1}[$. Using this notation, $Q_n^s = \Omega_s \times I_n$ and $Q_n^f = \Omega_f \times I_n$ are the nth space-time slabs for the structure and fluid respectively. For the nth space-time slab, the spatial domain is subdivided into $(n_{el})_n$ elements, and the interior of the eth element is defined as Q_n^e. Figure 9.2 shows an illustration of two consecutive space-time slabs Q_{n-1} and Q_n for the fluid where the superscript is omitted for clarity.

Within each space-time element, the trial solution and weighting function are approximated by pth-order polynomials in x and t. These functions are assumed $C^0(Q_n)$ continuous throughout each space-time slab, but are allowed to be discontinuous across the interfaces of the slabs. This feature of the time-discontinuous method allows for the general use of high-order elements and spectral-type interpolations in both space and time. The collections of finite element interpolation functions are given by the spaces as follows:

Trial structural displacements

$$\mathcal{S}^h = \bigcup_{n=0}^{N-1} \mathcal{S}_n^h, \quad \mathcal{S}_n^h = \left\{ \boldsymbol{u}^h(x,t) \Big| \boldsymbol{u}^h \in (C^0(Q_n^s))^d, \boldsymbol{u}^h \Big|_{Q_n^{se}} \in (\mathcal{P}^p(Q_n^{se}))^d \right\}$$

Trial fluid potential

$$\mathcal{T}^h = \bigcup_{n=0}^{N-1} \mathcal{T}_n^h, \quad \mathcal{T}_n^h = \left\{ \phi^h(\mathbf{x},t) \Big| \phi^h \in C^0(Q_n^f), \phi^h \Big|_{Q_n^{fe}} \in \mathcal{P}^p(Q_n^{fe}) \right\},$$

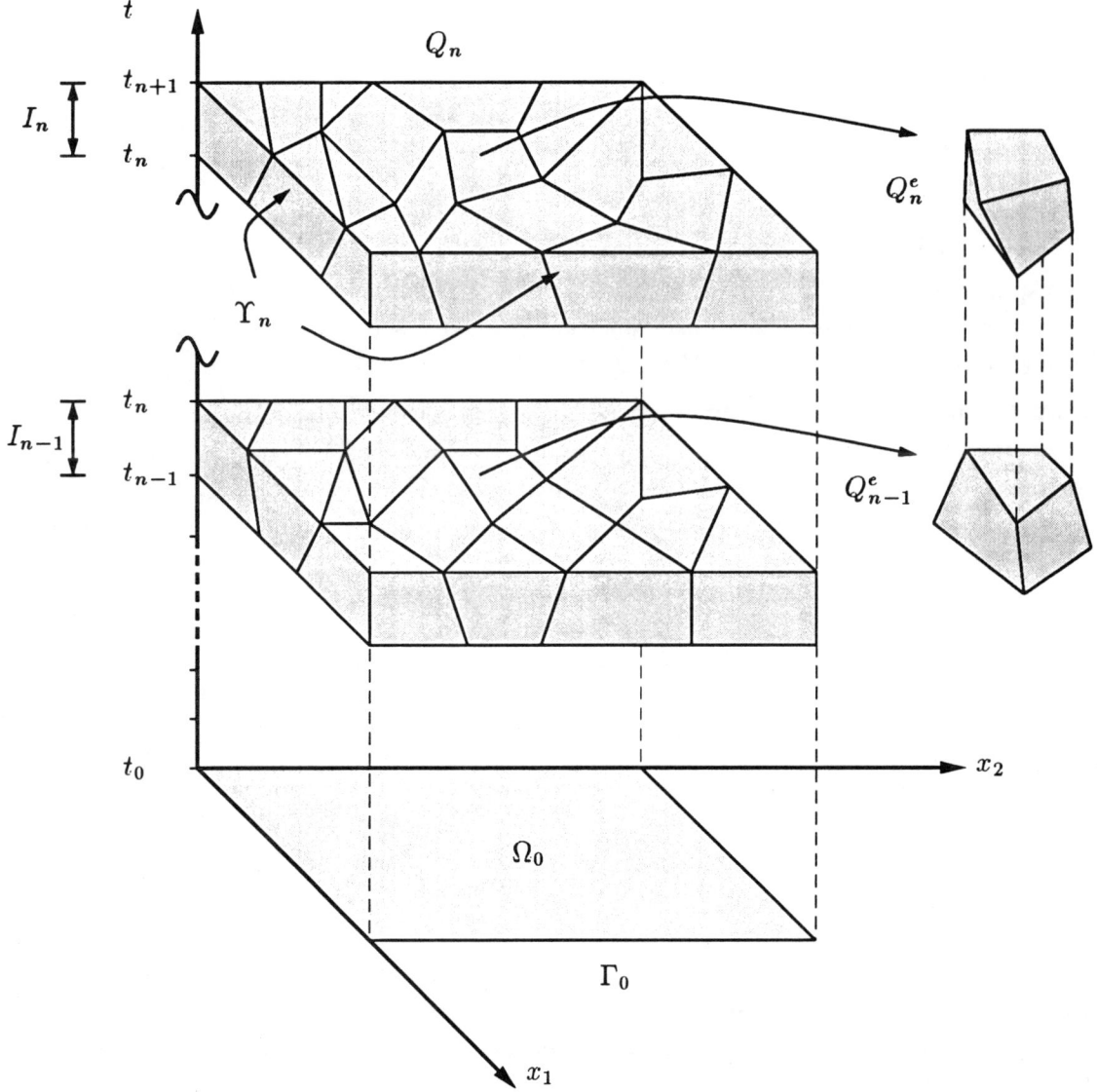

Figure 9.2 Illustration of two consecutive space-time slabs with unstructured finite element meshes in space-time.

where $\mathcal{P}^p$ denotes the space of pth-order polynomials and C^0 denotes the space of continuous functions. For clarity, only natural boundary conditions are employed. As a result the trial function spaces and weighting spaces are identical, i.e., $\mathcal{S}^h = \mathcal{V}^h$ and $\mathcal{T}^h = \mathcal{W}^h$. Before stating the space-time variational equations, it is useful to introduce the following notation:

$$(w^h, u^h)_{\Omega_S} = \int_{\Omega_s} w^h \cdot u^h d\Omega$$
$$a(w^h, u^h)_{\Omega_S} = \int_{\Omega_s} \nabla w^h \cdot \sigma(\nabla u^h) d\Omega$$
$$(w^h, \phi^h)_{Q_n} = \int_{t_n}^{t_{n+1}} (w^h, \phi^h)_\Omega dt.$$

The interpretation of other similar terms may be inferred from these. The L_2 norm is denoted by $\|\phi\|_\Omega = (\phi,\phi)_\Omega$. A natural measure of stability for the coupled structural acoustics problem is the total energy for the system:

$$\mathsf{E}(u,\phi) := \mathcal{E}_s(u) + \mathcal{E}_f(\phi) \tag{9.8}$$

$$\mathcal{E}_s(u) = \frac{1}{2}(\dot{u},\rho_s\dot{u})_{\Omega_s} + \frac{1}{2}a(u,u)_{\Omega_s} \tag{9.9}$$

$$\mathcal{E}_f(\phi^h) = \frac{1}{2}\rho_f\|a\dot{\phi}^h\|_{\Omega_f}^2 + \frac{1}{2}\rho_f\|\nabla\phi^h\|_{\Omega_f}^2, \tag{9.10}$$

where $\mathcal{E}_s$ and $\mathcal{E}_f$ denote energy of the elastic structure and of the acoustic fluid, respectively.

SPACE-TIME VARIATIONAL EQUATIONS

The space-time variational formulation is obtained from a weighted residual of the governing equations and incorporates time-discontinuous jump terms. The specific form of this new formulation is designed such that unconditional stability for arbitrary space-time finite element discretizations can be proved through a functional analysis of the method. The time-discontinuous Galerkin formulation can be stated formally:

Within each space-time slab, $n = 0, 1, \ldots, N-1$, find $(u^h,\phi^h) \in \mathcal{S}_n^h \times \mathcal{T}_n^h$, such that when $\forall (w^h,w^h) \in \mathcal{V}_n^h \times \mathcal{W}^h$, the following coupled variational equations are satisfied:

$$G_s(w^h,u^h,\phi^h)_n = 0 \tag{9.11}$$
$$G_f(w^h,u^h,\phi^h)_n = 0, \tag{9.12}$$

where

$$G_s(w^h,u^h,\phi^h)_n = \int_{t_n}^{t_{n+1}}(\dot{w}^h,\rho_s\ddot{u}^h)_{\Omega_s}dt + \int_{t_n}^{t_{n+1}}a(\dot{w}^h,u^h)_{\Omega_s}dt \tag{9.13}$$
$$-\int_{t_n}^{t_{n+1}}(\dot{w}^h\cdot n,\rho_f\dot{\phi}^h)_{\Gamma_i}dt - \int_{t_n}^{t_{n+1}}(\dot{w}^h,\bar{t})_{\Gamma_\sigma}dt$$
$$+(\dot{w}^h(t_n^+),\rho_s[\dot{u}^h(t_n)]_{\Omega_s}) + a(w^h(t_n^+),[u^h(t_n)]_{\Omega_s})$$

$$G_f(w^h,u^h,\phi^h)_n = \int_{t_n}^{t_{n+1}}(\dot{w}^h,\rho_f a^2\ddot{\phi}^h)_{\Omega_f}dt + \int_{t_n}^{t_{n+1}}(\nabla\dot{w}^h,\rho_f\nabla\phi^h)_{\Omega_f}dt \tag{9.14}$$
$$+\int_{t_n}^{t_n}(\dot{w}^h,\rho_f\dot{u}^h\cdot n)_{\Gamma_i}dt - \int_{t_n}^{t_{n+1}}(\dot{w}^h,\rho_f f)_{\Omega_f}dt$$
$$+(\dot{w}^h(t_n^+),a^2\rho_f[\dot{\phi}^h(t_n)]_{\Omega_f}) + (\nabla w^h(t_n^+),\rho_f[\nabla\phi^h(t_n)]_{\Omega_f})$$
$$+G_\infty(w^h,\phi^h)_n.$$

In the operator G_s, the terms evaluated over $\Omega_s \times I_n$ weakly enforce the momentum balance in the structure while in G_f, the terms evaluated over $\Omega_f \times I_n$ weakly enforce the scalar wave equation over the interior domain of the space-time slab. Fluid-structure interaction is accomplished through the coupling operators defined on the fluid-structure interface $\Gamma_i \times I_n$. The operator G_∞ incorporates the time-dependent radiation boundary conditions on the boundary Γ_∞, and will be described later in the next section (on Exact Nonreflecting Boundary Conditions).

An important component in the success of the space-time method is the incorporation of discontinuous temporal jump terms, $[w^h(t_n)] = w^h(x,t_n^+) - w^h(x,t_n^-)$, at each space-time slab interface. These jump operators weakly enforce initial conditions across time slabs and are crucial for obtaining an unconditionally stable algorithm for unstructured space-time finite element discretizations with high-order interpolations. The specific form of these jump operators is designed such that a natural norm emanates from the variational equation and satisfies a strong coercivity condition. From a Fourier analysis, it can be shown that the jump operators introduce beneficial numerical dissipation for frequencies above the spatial resolution limit.

The method is applied in one space-time slab at a time; data from the end of the previous slab are employed as initial conditions for the current slab. Matrix equations are obtained by introducing space-time finite element approximations for $u^h(x,t) = N_s(x,t)d$ and $\phi^h(x,t) = N_f(x,t)\phi$. In these expressions $N_s \in Q_n^s$ and $N_f \in Q_n^f$ are basis functions defined over a space-time slab, and d and ϕ are global solution vectors. Inserting into the variational (9.13) and (9.14) leads to the coupled system of algebraic equations to be solved in sequence for each time interval $I_n =]t_n, t_{n+1}[$, $n = 0, 1, \ldots, N-1$:

$$\begin{bmatrix} K_S & A \\ A^T & K_f \end{bmatrix} \begin{Bmatrix} d \\ \phi \end{Bmatrix} = \begin{Bmatrix} f_S \\ f_f \end{Bmatrix} \tag{9.15}$$

where K_s is the matrix emanating from the structural operator G_s, K_f is the matrix emanating from the fluid operator G_f, and A is the fluid-structure coupling matrix:

$$A = \int_{t_n}^{t_{n+1}} \int_{\Gamma_i} \rho_f N_{s,t}^T n N_{f,t} d\Gamma dt, \tag{9.16}$$

where $N_s(x,t) \in Q_n^s$ and $N_f(x,t) \in Q_n^f$ shape functions for the structure and fluid respectively.

Exact Nonreflecting Boundary Conditions

New time-dependent nonreflecting boundary conditions, which are exact for the first N spherical wave harmonics on Γ_∞, have been derived recently (Thompson and Pinsky, 1994, 1995a; Thompson, 1994). Two alternative sequences of time-dependent nonreflecting boundary conditions have been obtained; the first involves both time and spatial derivatives (local in time and local in space version), and the second involves time derivatives while retaining a spatial integral (local in time and nonlocal in space version).

In the first version, a local in time counterpart to the spatially nonlocal DtN map is derived in Thompson (1994) and Thompson and Pinsky (1995a), which exactly represents the first N spherical wave harmonics. This new sequence of boundary conditions retains the nonlocal spatial integral, yet replaces the time-convoluted DtN map with higher-order local time derivatives. This form of time-dependent boundary conditions has the advantage that, when implemented in the time-discontinuous finite element formulation, standard $C^0(\Gamma_\infty \times I_n)$ interpolation functions may be used for both the space and time variables.

The second version starts from the localization of the acoustic impedance relation, the Dirichlet-to-Neumann (DtN) map in the frequency domain. In Thompson (1994) and Thompson and Pinsky (1995b), it is shown that when the solution on the boundary Γ_∞ contains only a finite number of spherical harmonics, then such a transformation gives an exact condition that is local in both x and t. The sequence of local boundary conditions is obtained by truncating the DtN map. Time-dependent boundary conditions are then obtained through the application of an inverse Fourier transform. This new sequence of local time-dependent boundary conditions provides increasing accuracy with order N, which, however, is also a measure of the difficulty of implementation. In general, the Nth-order condition contains all the even tangential and temporal derivatives up to order $2(N-1)$. Because the time-discontinuous formulation allows for the use of C^0 interpolations to represent the high-order time derivatives, it is possible to implement this sequence of time-dependent absorbing boundary conditions up to any order desired. However, for high-order operators in the sequence extending beyond $N \geq 3$, the lowest possible order of spatial continuity on the artificial boundary that can be achieved after integration by parts is C^{N-2}. For these high-order operators, a layer of boundary elements adjacent to Γ_∞ possessing high-order tangential continuity on Γ_∞ is needed. Further details on the finite element implementation of these boundary conditions using the space-time formulation are described in Thompson (1994).

ANALYSIS OF THE SPACE-TIME FORMULATION

Stability Analysis

In this section, results are summarized from a stability analysis of the space-time finite element formulation for the exterior structural acoustics problem. In the absence of forcing terms, i.e., $\bar{t}=0$ and $f=0$ and for $\mathbb{S}_1$, it has been proved in Thompson (1994) and Thompson and Pinsky (1995b) that the following energy decay inequality holds for the coupled space-time formulation:

$$\mathcal{E}_s(u^h(t_{n+1}^-)) + \mathcal{E}_f(\phi^h(t_{n+1}^-)) + \frac{1}{2R}\|\phi^h(t_{n+1}^-)\|_{\Gamma_\infty}^2 + \frac{1}{c}\int_0^{t_{n+1}}\|\dot{\phi}^h(t)\|_{\Gamma_\infty}^2 dt$$
$$\leq \mathcal{E}_s(u_0) + \mathcal{E}_f(\phi_0) \tag{9.17}$$

for $n = 0, 1, 2, \ldots, N-1$. In the above inequality, $\mathcal{E}_s$ and $\mathcal{E}_f$ were defined earlier in Yserentant (1986) and Zienkiewicz et al. (1981) and denote the energy for the elastic structure and acoustic fluid respectively. Equation (9.17) states that the total energy in the fluid-structure system, plus the energy absorbed through the radiation boundary, is always less than or equal to the initial energy in the system. A corollary to this estimate is that the computed total energy for the system plus the radiation energy absorbed through the artificial boundary at the end of a time step is always less than or equal to the total energy at the previous time step for arbitrary step sizes, i.e.,

$$\begin{aligned}
&\mathcal{E}_s(\mathbf{u}^h(t_{n+1}^-)) + \mathcal{E}_f(\phi^h(t_{n+1}^-)) + \frac{1}{2R}\left\|\phi^h(t_{n+1}^-)\right\|_{\Gamma_\infty}^2 + \frac{1}{c}\int_{t_n}^{t_{n+1}}\left\|\dot\phi^h(t)\right\|_{\Gamma_\infty}^2 dt \\
&\leq \mathcal{E}_s(\mathbf{u}^h(t_n^-)) + \mathcal{E}_f(\phi^h(t_n^-)) + \frac{1}{2R}\left\|\phi^h(t_n^-)\right\|_{\Gamma_\infty}^2
\end{aligned} \qquad (9.18)$$

for $n = 0, 1, 2, \ldots, N-1$. Results (9.17) and (9.18) both imply that *the space-time formulation presented is unconditionally stable*. See Thompson (1994) for an analogous result for the interior problem where no radiation boundary conditions are present.

Galerkin Least-Squares Stabilization

For additional stability, local residuals of the governing differential equations in the form of least-squares may be added to the Galerkin variational equations. The Galerkin/Least-Squares addition to the variational equation for the fluid (9.14) is

$$\begin{aligned}
G_{GLS}^f(w^h, \mathbf{u}^h, \phi^h)_n &= G_f(w^h, \mathbf{u}^h, \phi^h)_n + (\rho_f c^2 \tau \mathcal{L}_1 w^h, (\mathcal{L}_1 \phi^h - f))_{\tilde{Q}_n^f} \\
&\quad + (\rho_f c^2 s \mathcal{L}_2 w^h, \mathcal{L}_2 \phi^h)_{(\tilde\Psi_\infty)_n} + (\rho_f c^2 s \mathcal{L}_3 w^h, \mathcal{L}_3 \phi^h)_{(\tilde\Psi_i)_n} \\
&\quad + (\rho_f c^2 s [\![w_{,n}^h(x)]\!])_{(\tilde\Psi_e)_n},
\end{aligned}$$

where $(\mathcal{L}_1 w^h - f = \nabla^2 w^h - a^2 \ddot w^h - f)$ is the residual for the wave equation, $(\mathcal{L}_2 w^h = w_{,n}^h + S_m w^h)$ is the radiation boundary residual, and $(\mathcal{L}_3 w^h = w_{,n}^h - \dot{\mathbf{u}} \cdot \mathbf{n})$ is the interface boundary residual. In the above expressions, a tilde refers to integration over element interiors and τ and s are local mesh parameters designed to improve desirable high-frequency numerical dissipation without degrading the accuracy of the underlying time-discontinuous Galerkin method. For the structural equation (9.13), similar least-squares terms are added (see Thompson, 1994). Consistency of both the underlying method and the least-squares addition is clear from the fact that a sufficiently smooth exact solution of the coupled initial/boundary-value problem as stated in (9.1)-(9.7), satisfies the variational equations (9.13), (9.14), and (9.19) identically.

Convergence Analysis

To study the convergence rates of the space-time finite element formulation for the exterior structural acoustics problem the following space-time mesh size parameters are introduced. For the structural domain Ω_s, $h_s = max\{c_L \Delta t, \Delta x\}$ where c_L is the dilatational wave speed and Δx and Δt are maximum element diameters in space and time, respectively. For the fluid domain Ω_f, $h_f = max\{c \Delta t, \Delta x\}$ where c is the acoustic wave speed. Assuming that the exact solution to the strong form of the initial boundary value problem with S_1 is sufficiently smooth in the sense that

$$\mathbf{u} \in (H^{k+1}(Q_s))^d \text{ and } \phi \in H^{p+1}(Q_f) \qquad (9.19)$$

and assuming standard finite element interpolation estimates hold, then it has been proved in Thompson (1994) and Thompson and Pinsky (1995b) that the following error estimate holds for the time-discontinuous Galerkin Least-Squares formulation,

$$|||E|||^2 \leq c(\mathbf{u})h_s^{2k-1} + c(\phi)h_f^{2p-1}, \tag{9.20}$$

where k and p are the finite element interpolation orders for the structure and fluid respectively. In the above, the error is defined as

$$E = \{e, e\} \quad \text{where} \quad e = \mathbf{u}^h - \mathbf{u} \quad \text{and} \quad e = \phi^h - \phi, \tag{9.21}$$

and $c(\mathbf{u})$ and $c(\phi)$ are values that are independent of the mesh size parameters h_s and h_f. The norm in which convergence is measured is given by

$$\begin{aligned}
|||E|||^2 =\ & \mathcal{E}_s(e(0^+)) + \sum_{n=1}^{N-1} \mathcal{E}_s([e(t_n)]) + \mathcal{E}_s e(T^-)) \\
& + \mathcal{E}_f(e(0^+)) + \sum_{n=1}^{N-1} \mathcal{E}_f([e(t_n)]) + \mathcal{E}_f(e(T^-)) \\
& + \frac{1}{2R}\|e(0^+)\|_{\Gamma_\infty}^2 + \sum_{n=1}^{N-1} \frac{1}{2R}\|[e(t_n)]\|_{\Gamma_\infty}^2 + \frac{1}{2R}\|e(T^-)\|_{\Gamma_\infty}^2 \\
& + \sum_{n=0}^{N-1} \frac{1}{c} \int_{t_n}^{t_{n+1}} (\dot{e}(t), \dot{e}(t))_{\Gamma_\infty} dt \\
& + \sum_{n=0}^{N-1} \{(\mathcal{L}_s e, \rho_s^{-1} \tau \mathcal{L}_s e)_{\tilde{Q}_n^s} \\
& + ([\sigma(\nabla e))(x)] \cdot \mathbf{n}, \rho_s^{-1} s [\sigma(\nabla e)(x)] \cdot \mathbf{n})_{(\Psi_s)_n} \\
& + (\sigma(\nabla e) \cdot \mathbf{n}, \rho_s^{-1} s \sigma(\nabla e), \mathbf{n})_{(\Psi_\sigma)_n} \\
& + (\sigma(\nabla e) \cdot \mathbf{n} - \rho_f \dot{e}\mathbf{n}, \rho_s^{-1} s \sigma(\nabla e) \cdot \mathbf{n} - \rho_f \dot{e}\mathbf{n})_{(\Psi_i)_n} \\
& + \|c\tau^{1/2} \mathcal{L}_1 e\|_{\tilde{Q}_n^f}^2 + \|cs^{1/2} \mathcal{L}_2 e\|_{(\tilde{\Psi}_\infty)_n}^2 \\
& + \|cs^{1/2} [e_{,n}(x)]\|_{(\tilde{\Psi}_f)_n}^2 \\
& + (\dot{e}\mathbf{n} - e_{,n}, c^2 s \dot{e} \cdot \mathbf{n} - e_{,n})_{(\Psi_i)_n} \}.
\end{aligned} \tag{9.22}$$

In the above expression $\mathcal{L}_s \mathbf{u}^h$ is the residual for the structure. This norm emanates naturally from the coupled fluid-structure variational equations (9.13) and (9.14) together with the least-squares operators discussed above. The error estimate is optimal in the sense that the finite element error converges at the same rate as the interpolate. This result indicates that the error for the coupled system is controlled by the convergence rates in both the structure and the fluid. In other words, for an accurate solution to the coupled fluid-structure problem, discretizations for both the structural domain and the fluid domain must be adequately resolved.

Convergence rates are verified numerically by calculating the response of the one-dimensional wave equation in the interval $0 \leq x < 4$. On the boundary $L = 4$, the exact nonreflecting boundary condition

$$\phi_{,x}(L,t) = -(1/c)\dot{\phi}(L,t) \qquad (9.23)$$

is imposed. The left end ($x = 0$) is fixed. A transient pulse of the form

$$\phi_0(x) = \frac{1}{4}(1 - \cos\frac{2\pi}{\lambda}(x - x_0))^2 \qquad (9.24)$$

is initiated at $x_0 = 2.4$ with wavelength $\lambda = 0.8$. The response was calculated for the time interval $0 \leq t \leq T = 1.8$. Each space-time slab was discretized with a uniform mesh of 160 biquadratic elements. Figure 9.3 shows the error computed using the Galerkin Least-Squares formulation at time T = 1.8. This result confirms that a cubic rate of convergence is obtained as predicted by (9.20).

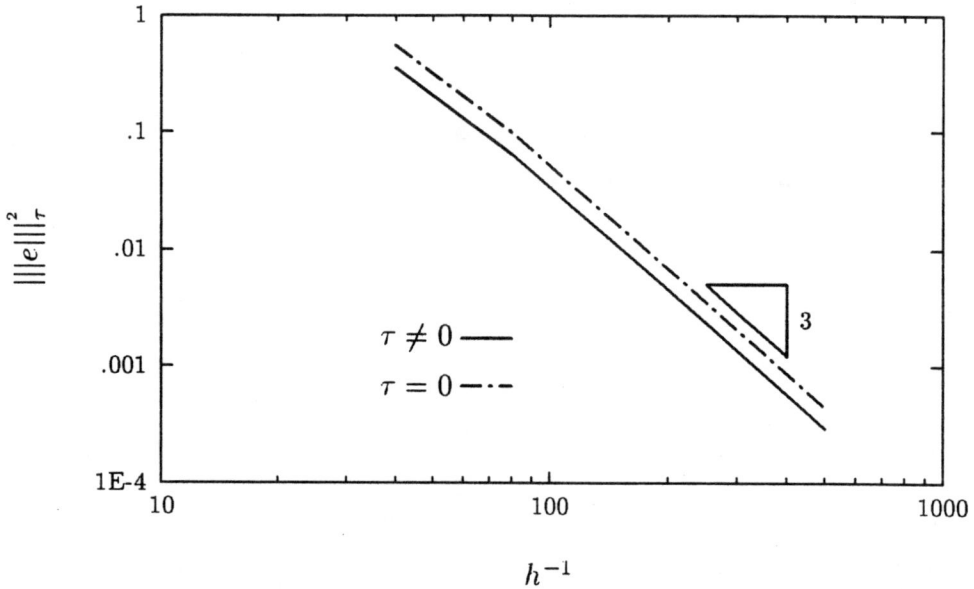

Figure 9.3 Convergence of the numerical error employing the $Q2$ element; $h = h_f$ is the element mesh size parameter. Results confirm the high-order convergence rate $(2p - 1) = 3$ for $p = 2$.

REPRESENTATIVE NUMERICAL EXAMPLE

A numerical example is presented to demonstrate the effectiveness of the time-discontinuous space-time finite element method to accurately model transient scattering from geometrically complex structures. The problem investigated also assesses the performance of the local nonreflecting boundary conditions. For all the numerical results presented, the GLS mesh parameters are set to zero and standard quadratic finite element shape functions are used in both the time and space dimensions.

Consider the time-dependent scattering from a rigid cylinder with conical-to-spherical end caps and a large length to diameter ratio. Figure 9.4 illustrates the finite element spatial discretization of the computational domain bounded internally by the lateral projection of the benchmark cylinder and externally by a circular artificial boundary. A total of 1,600 space-time elements are used for this example. For this two-dimensional problem, we consider the the following second-order local time-dependent boundary operator (Pinsky et al., 1992):

$$\mathbb{S}_2 \phi = \frac{1}{2R}\left(\frac{3}{4} - \frac{\partial^2}{\partial \theta^2}\right)\phi + \frac{3}{2c}\dot{\phi} + \frac{R}{c}\frac{\partial \dot{\phi}}{\partial r} + \frac{R}{c^2}\ddot{\phi}. \qquad (9.25)$$

The pulse $f = \delta(x_0, y_0)\sin \omega t$ and $t \in [0,3]$ is positioned inside the computational domain, simulating an oblique incident wave during a short time period. The numerical simulation is continued until just prior to reaching the practical disappearance of the signal from the domain. This example represents a challenging problem where the multiple-scales involving the ratio of the wavelength to cylinder diameter and cylinder length dimension play a critical role in the complexity of the resulting scattered wave field.

The numerical simulation starts with the initial pulse shown in Figure 9.4 at $t = 3$. The accompanying figures show the contours of the scattering phenomena from the cylinder with homogeneous Neumann boundary conditions on the wet surface, i.e., "rigid" boundary conditions. At $t = 6$ the incident pulse has expanded in a cylindrical wave and has just reached the boundaries of the rigid cylinder. At the artificial boundary, the wave front is allowed to pass through the boundary with no reflection. At $t = 9$, the wave has begun to reflect off the rigid boundary, creating a complicated backscattered wave. As time passes, the originally cylindrical incident wave has been scattered into a part that travels along the upper part of the cylinder, and a part that diffracts around the backside.

ITERATIVE SOLUTION OF LARGE-SCALE TIME-HARMONIC PROBLEMS

In the second part of this paper, we consider efficient methods for the solution of large-scale matrix problems that arise from finite element methods for structural acoustics. Direct solution techniques become expensive for solving large systems of linear equations due to excessive growth in computational and storage requirements. Iterative solution strategies are an attractive alternative in these situations. Also, unlike direct methods, characteristic computational kernels associated with iterative solution methods parallelize very efficiently, making them even more attractive for use on modern vector and parallel computers. However, for matrix problems arising from finite element discretizations in acoustics, the convergence of gradient-type iterative methods deteriorates under increasing mesh density and increasing frequency. Under these circumstances, effective preconditioning becomes essential. For

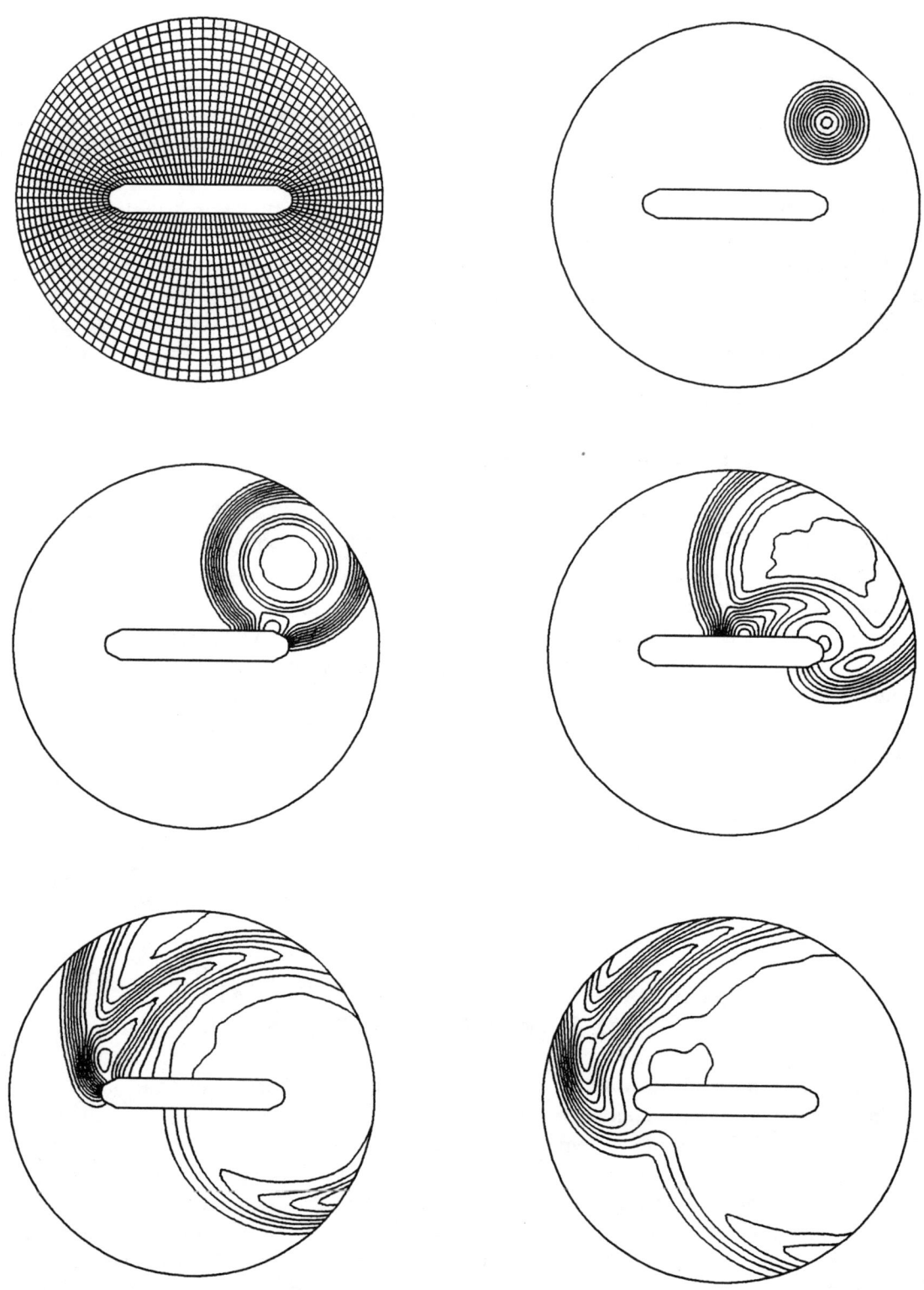

Figure 9.4 Scattering from a geometrically complex rigid structure due to point source. Solution contours shown for initial pulse at $t = 3$ and later times $t = 6, 9, 15, 18$.

effective preconditioning of the Helmholtz operator we examine a novel approach that is based on an understanding of the properties of finite element discretization of a second-order elliptic differential operator.

Model Boundary Value Problem

As a model problem we consider a linear, second-order, elliptic partial differential equation along with specified boundary conditions,

$$\mathcal{L}u = f \text{ in } \Omega \tag{9.26}$$

$$u = g \text{ on } \Gamma_g, \quad \frac{\partial u}{\partial n} = h \text{ on } \Gamma_h, \tag{9.27}$$

with the boundary of the domain $\partial\Omega = \Gamma_g \cup \Gamma_h$. The variational form of the problem is expressed as follows: Find $u \in H$ satisfying $u = g$ on Γ_g such that $a(u,v) = f(v)$ holds $\forall v \in H$ satisfying $v = 0$ on Γ_g. The bilinear functional $a(.,.)$ is defined on the Sobolev space H and corresponds to the differential operator $\mathcal{L}$; $f(.)$ is a linear functional corresponding to the forcing function and specified flux boundary conditions. Introducing a finite element discretization of the domain and choosing a basis on this discretization, the approximate statement of the variational problem becomes: Find $u^h \in H^h$ satisfying $u^h = g$ on Γ_g^h such that

$$a(u^h, v^h) = f(v^h) \tag{9.28}$$

holds $\forall v^h \in H^h$ satisfying $v^h = 0$ on Γ_g^h. By employing the Galerkin approach, the discrete system of equations to be solved is $Kd = f$, where $K_{ij} = a(\phi_i, \phi_j)$ and $f_j = f(\phi_j)$. If $a(.,.)$ is evaluated using the usual nodal basis consisting of Lagrange polynomials, shown in Figure 9.5b, then the spectral condition number of K grows as $O(h^{-2})$, where h is the characteristic mesh size (see, e.g., Axelsson and Barker, 1984). However, for an appropriate set of h-hierarchical basis functions the condition number grows as $O(\log h^{-1})^2$ (Yserentant, 1986). In the h-hierarchical approach, basis functions of the same polynomial order are introduced in a hierarchic fashion corresponding to nodes in the interior of an element (see Figure 9.5a). Although not necessary, it may be useful to consider such interior nodes as those arising from refinement of an initial coarse mesh. Alternatively, it is useful to think of these nodes as being associated with a "multilevel splitting" of a given mesh (Yserentant, 1986).

Using these multiple levels, the hierarchical basis can be defined as follows: (1) basis functions at the coarsest level, level 1, consist of the nodal basis functions, and (2) the hierarchical basis at any level $j > 1$ consists of nodal basis functions corresponding to nodes in that level which are not present in any of the coarser levels, together with the hierarchical basis for level $j - 1$. This recursive definition leads to a set of basis functions that is complete and unique. We now describe transformation matrices that relate the nodal basis with an equivalent set of h-hierarchical basis functions.

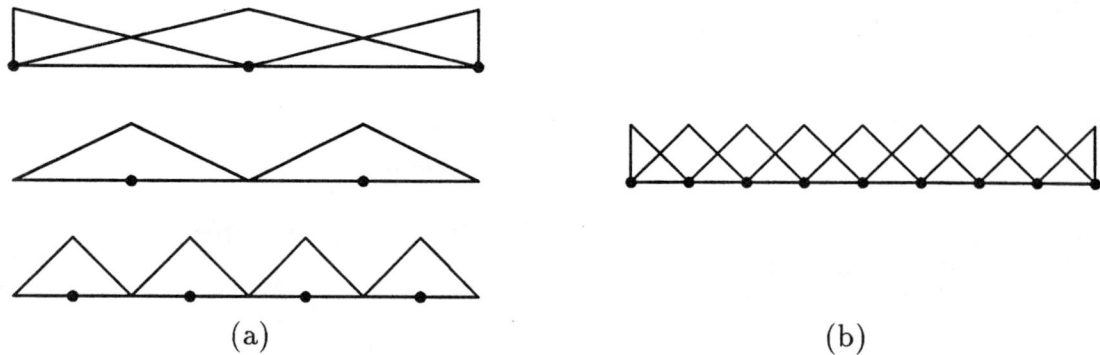

Figure 9.5 (a) *h*-Hierarchical linear shape functions introduced with two refinement levels. (b) An equivalent nodal basis for the final mesh with linear Lagrange polynomials.

h-HIERARCHICAL BASIS PRECONDITIONER

Formulation of the Preconditioner

Consider a function $u^h(x) \in H^h$ approximated on a given finite element mesh with n unknowns (or degrees of freedom) using nodal and hierarchical basis functions as

$$u^h(x) = \sum_{i=1}^{n} \alpha_i \phi_i(x) \quad \text{and} \quad u^h(x) = \sum_{i=1}^{n} \beta_i \psi_i(x), \tag{9.29}$$

respectively. Let $\Phi = \{\phi_1(x), \phi_2(x), ..., \phi_n(x)\}^T$ and $\Psi = \{\psi_1(x), \psi_2(x), ..., \psi_n(x)\}^T$ denote column vectors comprising nodal and hierarchical basis functions respectively. We are interested in the linear transformation matrix P such that $\Psi = P\Phi$. For the case of two levels, it is easy to show that the nodal and hierarchical basis functions are related as

$$\phi_i(x) = \psi_i(x) - \sum_{j=n_1+1}^{n} \psi_i(x_j) \psi_j(x), \quad i = 1, ..., n_1 \tag{9.30}$$

$$\phi_i(x) = \psi_i(x), \quad i = n_1 + 1, ..., n \tag{9.31}$$

$$\text{or} \quad \Psi = \begin{bmatrix} I_{n_1} & -F \\ 0 & I_{n-n_1} \end{bmatrix}^{-1} \Phi, \tag{9.32}$$

where n_1 is the number of nodes in Level 1, and nodes in Level 1 are assumed to be numbered before those in Level 2; I_{n_1} and I_{n-n_1} are square identity matrices of size n_1 and $n - n_1$; and F is a sparse matrix containing the interaction of hierarchical shape functions between the two levels. To generalize (9.32) for the case of multiple levels, recall the definition of the hierarchical basis functions. Due to the recursive construction of the hierarchical basis, the transformation matrix in the case of $j > 2$ levels can be stated as

$$\psi = P_1^{-1} P_2^{-1} \dots P_{j-1}^{-1} \Phi, \tag{9.33}$$

where P_k is the transformation between the nodal basis on Level $k+1$, the hierarchical basis on nodes in Level k and Level $k+1$. Each P_k essentially has the same structure as P_1 (shown in (9.32)).

Now consider the solution of our nodal basis finite element equations $Kd = f$. In order to achieve improved conditioning of the matrix equations, we start by employing hierarchical basis functions in the bilinear form $a(.,.)$ of (9.28). Now, using (9.33), introduce nodal basis functions Φ in place of Ψ to get a relation between K and its equivalent representation in the hierarchical basis $\hat{K}$,

$$\hat{K} = a(\Phi^T P^T, \Phi^T P^T) \tag{9.34}$$
$$= PKP^T. \tag{9.35}$$

If we choose P^{-1} and P^{-T} as left and right preconditioners, the hierarchical basis preconditioner can be stated as $M_{HB} = (P^T P)^{-1}$. Such a construction of the hierarchical basis preconditioner from transformation matrices was obtained by Yserentant (1986). It is interesting to note that the matrix entries in M_{HB} are independent of the coefficients that appear in the differential operator $\mathcal{L}$. In order improve this situation, we combine diagonal scaling with the preconditioner M_{HB}. Although $diag(\hat{K})$ is not easily available, it is frequently useful (Greenbaum et al., 1989) to employ instead $diag(K)$ in the following way:

$$M_{HBDS} = (P^T [diag(K)]^{-1} P)^{-1}. \tag{9.36}$$

Another approach in enhancing M_{HB} becomes evident if we consider the block form of $\hat{K}$. Ordering all nodes in coarsest level first, we form the preconditioner

$$M_{HBCS} = \left(P^T \begin{bmatrix} ILU(\hat{K}_{11}) & 0 \\ 0 & I \end{bmatrix}^{-1} P \right)^{-1}, \tag{9.37}$$

where the operator $\{ILU(\hat{K}_{11})\}^{-1}$ represents an approximate solution of unknowns associated with nodes in the coarsest level, and the block matrix $\hat{K}_{11} = a(\phi_i, \phi_j)$, with ϕ_i and ϕ_j being the basis functions corresponding to nodes in Level 1. We compare the performance of preconditioners (9.36) and (9.37) in our numerical examples.

Implementation Issues

The use of a preconditioner in conjunction with a gradient-type iterative method requires solving a system of equations, of the form $Mz = r$, at each step of the iteration. An attractive feature of the preconditioners M_{HB} or M_{HBDS} is that solving for z doesn't require a matrix inversion, and only matrix-vector products of the form

$$z = P^T y \quad \text{and} \quad y = Pr \tag{9.38}$$

need to be evaluated. Since the preconditioner appears in the above operations only, significant reductions in storage and computational costs can be realized if these matrix-vector products are computed directly and the explicit calculation of P and P^T is avoided. This is a key algorithmic feature that was also exploited in Yserentant (1986). On unstructured grids, the preconditioning steps in equation (9.38) can be performed using nearest-neighbor type connectivity data between successive hierarchical levels. Such a procedure entails only a limited amount of additional storage of $O(n_{np})$ words, and a computational complexity of $O(n_{np})$ flops. Here n_{np} denotes total number of mesh points. See Malhotra and Pinsky (1995) for a description of data-structures and algorithms for efficient implementations on serial and distributed memory computers.

Selection of Multiple Levels

An important ingredient in the construction of hierarchical basis preconditioners is the choice of a multilevel splitting associated with the finite element mesh on which the problem needs to be solved. Multiple grid levels inherent in such a splitting arise quite naturally in h-adaptive finite element computations if nested mesh refinement is employed. However, in applying the preconditioning approach to indefinite problems, a more careful selection of grid levels is required. Results, based on both a priori convergence analyses (Schatz, 1974) as well as dispersion analyses (Thompson, 1994) of finite element discretizations, indicate that for indefinite problems it is essential that the mesh size be sufficiently small in order to maintain accuracy of finite element approximations. For the Helmholtz equation, results of one-dimensional dispersion analysis indicate a limit of resolution given by $h_o < \sqrt{12}/k$ where k is the wavenumber for linear basis functions. In our numerical tests on two-dimensional scattering problems using multilevel hierarchical grids, the coarse grid was chosen to satisfy this limit of resolution.

Matrix-Free Computations

It is noteworthy that the preconditioning approach described here is not directly based on a "splitting" or "factorization" of the large sparse coefficient matrix K. The hierarchical basis preconditioners, therefore, can also be employed in conjunction with highly storage-efficient matrix-free implementations of a gradient-type iterative algorithm. Matrix-free iterative approaches involve calculating characteristic computational kernels required in the iterative method without the explicit assembly of any global matrix, which enables substantial storage reductions. This is particularly useful in the context of acoustics, where a matrix-free representation of the non-local DtN contribution allows the use of this highly accurate boundary condition without any storage penalties associated with its nonlocal nature (Malhotra and Pinsky, 1995).

REPRESENTATIVE NUMERICAL EXAMPLES

A canonical problem in structural acoustics is the scattering of plane waves from a rigid body submerged in a fluid of infinite extent. Here we consider two-dimensional plane-wave scattering from a rigid cylinder with conical-to-spherical end caps and a large length to diameter ratio as shown in Figure 9.6. The governing equations are:

$$-\nabla^2 p - k^2 = 0 \quad \text{in } \mathcal{D} \quad (9.39)$$
$$\nabla p \cdot n = -\nabla p_{inc} \cdot n \quad \text{on } \Gamma_i \quad (9.40)$$
$$\nabla p \cdot n = -\mathcal{S}(p) \quad \text{on } \Gamma_\infty \quad (9.41)$$

In the above equations, p is the unknown acoustic pressure in the fluid domain $\mathcal{D}$, k is the wavenumber, and n denotes unit outward normals from respective boundaries. Equation (9.41) represents the impedance of the infinite acoustic medium, exterior to Γ_∞, through the boundary operator $\mathcal{S}(p)$. We employ an exact representation of the radiation impedance by choosing $\mathcal{S}(p)$ to be the Dirichlet-to-Neumann (DtN) map (Keller and Givoli, 1989); for details on finite element modeling of (9.39)-(9.14), see Harari (1991). We consider plane waves incident along the length of the cylinder such that $p_{inc} = e^{ikx}$, where $i = \sqrt{-1}$ and x is the coordinate along the length of the cylinder. As our basic iterative method, we employ the complex-symmetric preconditioned quasi-minimal residual (QMR) iterative method (Freund and Nachtigal, 1991; Freund, 1992). Iterations are stopped when the true residual at iteration k satisfies $\|b - Ax_k\| / \|b\| \leq 10^{-7}$.

The effect of mesh refinement is examined first by fixing the frequency of analysis at $k = \pi / 6$ and solving the problem on three successively refined meshes. In order to construct M_{HBDS}, an initial coarse

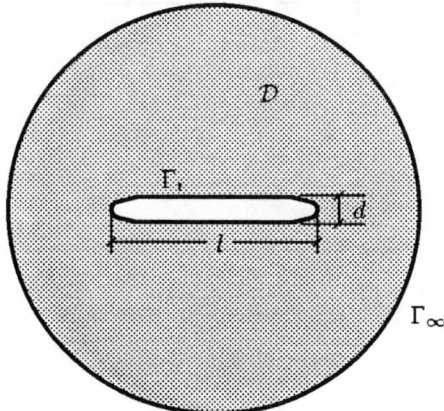

Figure 9.6 Two-dimensional trace of a rigid cylinder with end caps, $l / d = 8$.

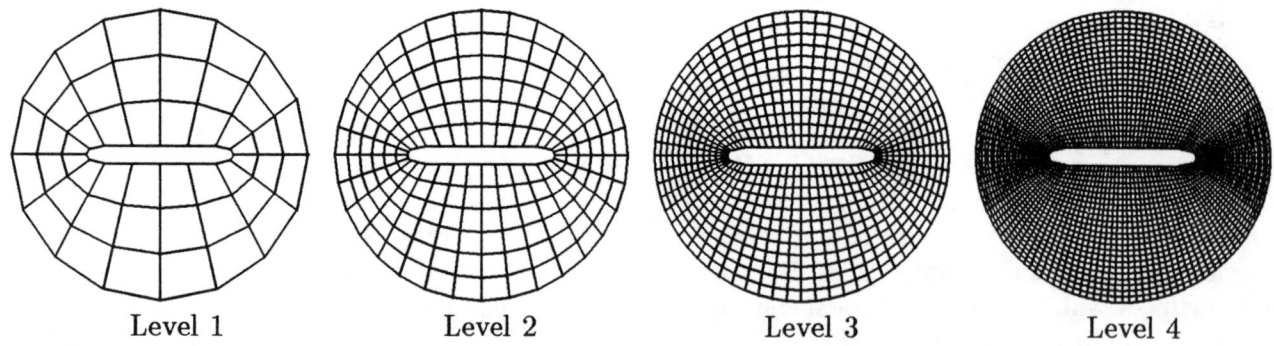

Figure 9.7 Multiple grid levels for the scattering example: solution on mesh with 3,200 unknowns, $kd = \pi / 6$.

mesh of 3×16 elements (radial × circumferential elements) was used and intermediate mesh levels were obtained by successive uniform refinement (Figure 9.7). We denote by "HB Levels" the total number of hierarchical levels used to construct the transformations from the nodal to the hierarchical basis.

Table 9.1 summarizes the number of iterations required for convergence using various preconditioners. Observe that the unpreconditioned algorithm, denoted by M_I, suffers substantial deterioration in iteration count as the mesh size decreases. Diagonal preconditioning, M_D, offers little improvement. The performance of M_{HBDS} appears to be almost mesh-independent. Similar behavior is observed for analysis at $kd = \pi / 2$ and $kd = \pi$ illustrated in Figure 9.8.

Table 9.1 Iteration Counts for Convergence of Scattering Problem

Mesh	(Unknowns)	kd	kl	M_I	M_D	HB Levels	M_{HBDS}
24×128	(3200)	$\pi/6$	$8\pi/6$	286	244	4	70
48×256	(12544)			592	468	5	83
96×512	(49664)			1372	1083	6	102

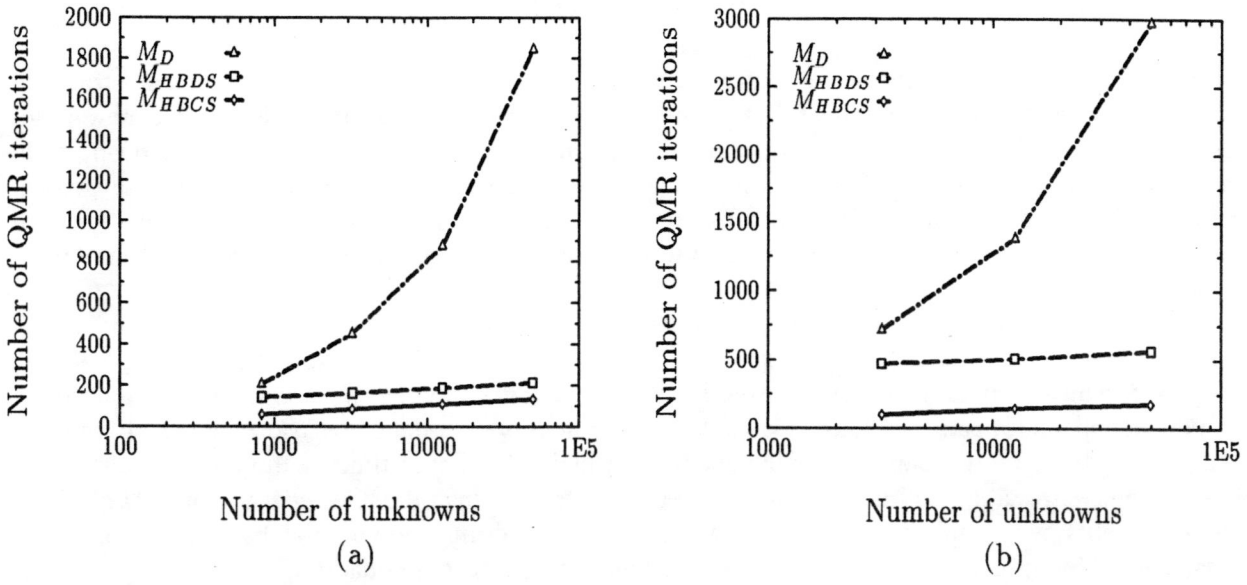

Figure 9.8 Convergence with mesh refinement for (a) $kd = \pi / 2$, (b) $kd = \pi$.

Next, we examine the effect of increasing frequency on the performance of various preconditioners. Figure 9.9 shows the number of iterations needed for convergence as the wavenumber is increased to $kd = \pi / 2 +$ and then $kd = \pi$. Although M_{HBDS} performs better than diagonal scaling, the iteration counts for M_{HBDS} do not remain as favorable when the wavenumber increases to $kd = \pi$. However, the preconditioner M_{HBDS} accounts for the deterioration in convergence with frequency increase and leads to a rate of convergence that is almost both mesh- and frequency-independent.

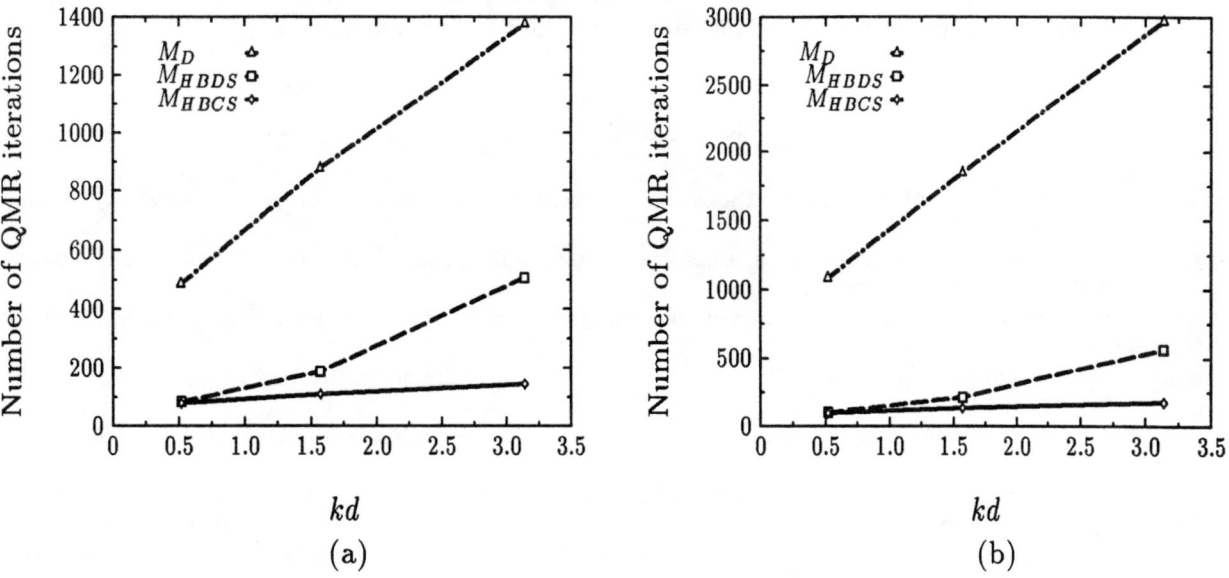

Figure 9.9 Convergence with frequency increase for meshes with unknowns (a) $n = 12544$, (b) $n = 49664$.

CONCLUSION

In this paper, a new space-time finite element method for solution of the transient structural acoustics problem in exterior domains has been presented. The formulation is based on a new time-discontinuous Galerkin Least-Squares variational equation for both the structure and the acoustic fluid together with their interaction. The resulting space-time algorithm gives for the first time a general solution to the fundamental problem of constructing a finite element method for transient structural acoustics with unstructured meshes in space-time and the desired combination of increased numerical stability and high accuracy.

Desirable attributes of the new computational method for transient structural acoustics include a natural framework for the design of rigorous a posteriori error estimates for self-adaptive solution strategies for unstructured space-time discretizations, and the implementation of high-order accurate time-dependent nonreflecting boundary conditions. Furthermore, through the use of acoustic velocity potential and structural displacement as the solution variables, the space-time method is unconditionally stable and converges at an optimal rate in a norm that is stronger than the total energy norm. High-order accuracy is obtained simply by raising the order of the space-time polynomial basis functions; both standard nodal interpolation and hierarchical shape functions are accommodated.

In the second part of this paper, a preconditioning approach based on exploiting the properties of h-hierarchical basis functions associated with a multilevel splitting of a given finite element mesh was considered. Numerical tests were performed for the solution of two-dimensional scattering problems in acoustics in order to examine iterative convergence rates obtained with various preconditioners. These tests indicate that convergence rates predicted by theoretical asymptotic bounds, which motivate the formation of h-hierarchical basis preconditioners, are actually realized on practical discretizations for problems in acoustics. Moreover, by combining hierarchical basis transformations with the solution of a coarse problem, we achieve iterative convergence rates for the Helmholtz equation that are almost frequency- and discretization-independent. The basis transformations inherent in these preconditioners can also be applied efficiently in conjunction with matrix-free gradient-type iterative algorithms. We have found such an approach to be very efficient and quite promising for solving very large-scale problems in exterior structural acoustics on distributed-memory parallel computers.

REFERENCES

Axelsson, O., and V.A. Barker, 1984, *Finite Element Solution of Boundary Value Problems: Theory and Computation*, Orlando, Fla.: Academic Press.

Freund, R.W., 1992, "Conjugate gradient-type methods for linear systems with complex symmetric coefficient matrices," *SIAM J. Sci. Statist. Comput.* **13**, 425-448.

Freund, R.W., and N.M. Nachtigal, 1991, "QMR: A quasi-minimal residual method for non-Hermitian linear systems," *Numer. Math.* **60**, 315-339.

Greenbaum, A., C. Li, and H.Z. Chao, 1989, "Parallelizing preconditioned conjugate gradient algorithms," *Comput. Phys. Commun.* **53**, 295-309.

Harari, I., 1991, *Computational Methods for Problems of Acoustics with Particular Reference to Exterior Domains*, PhD Thesis, Stanford University.

Johnson, C., 1990, "Adaptive finite element methods for diffusion and convection problems," *Comput. Methods Appl. Mech. Eng.* **82**, 301-322.

Johnson, C., 1993, "Discontinuous Galerkin finite element methods for second order hyperbolic problems," *Comput. Methods Appl. Mech. Eng.* **107**, 117-129.

Keller, J.B., and D. Givoli, 1989, "Exact non-reflecting boundary conditions," *J. Comput. Phys.* **82**(1), 172-192.

Malhotra, M., and P.M. Pinsky, 1995, "Parallel preconditioning based on h-hierarchic finite elements with application to acoustics," in preparation.

Pinsky, P.M., L.L. Thompson, and N.N. Abboud, 1992, "Local high order radiation boundary conditions for the two-dimensional time-dependent structural acoustics problem," *J. Acoust. Soc. Am.* **91**(3), 1320-1335.

Schatz, A.H., 1974, "An observation concerning Ritz-Galerkin methods with indefinite bilinear forms," *Math. Comput.* **28**, 959-962.

Thompson, L.L., 1994, *Design and Analysis of Space-Time and Galerkin Least-Squares Finite Element Methods for Fluid-Structure Interaction in Exterior Domains*, PhD Thesis, Stanford University.

Thompson, L.L., and P.M. Pinsky, 1994, "New space-time finite element methods for fluid-structure interaction in exterior domains," *Computational Methods for Fluid/Structure Interaction* **178**, 101-120.

Thompson, L.L., and P.M. Pinsky, 1995a, "A space-time finite element method for structural acoustics in infinite domains, part i: Formulation, stability, and convergence," *Comput. Methods Appl. Mech. Eng.*, submitted.

Thompson, L.L., and P.M. Pinsky, 1995b, "A space-time finite element method for structural acoustics in infinite domains, part ii: Exact time-dependent non-reflecting boundary conditions," *Comput. Methods Appl. Mech. Eng.*, submitted.

Yserentant, Harry, 1986, "On the multi-level splitting of finite element spaces," *Numer. Math.* **49**, 379-412.

Zienkiewicz, O.C., D.W. Kelly, J. Gago, and I. Babuška, 1982, "Hierarchical finite element approaches, error estimation and adaptive refinement." In: *The Mathematics of Finite Elements and Applications IV*, J.R. Whiteman (ed.), pp. 314-346, New York: Academic Press.

10

An Overview of the Application of the Method of Moments to Large Bodies in Electromagnetics

Edward H. Newman
I. Tekin
Ohio State University

> This paper presents an overview of recently developed techniques that improve the efficiency of the method-of-moments (MM) solution of electromagnetic radiation and scattering problems and thus allow it to be applied to electrically larger bodies. The techniques reviewed include fast iterative methods, recursive methods, sparse matrix methods, and the use of different unknowns or expansions. It is concluded that only with a hybrid method that combines a "low-frequency" technique, such as the MM, with an asymptotic "high-frequency" technique can problems of arbitrary size and complexity be treated.

INTRODUCTION

The frequency-domain method of moments (MM) is possibly the most well developed and widely used of the numerical methods for the analysis of electromagnetic radiation and scattering problems (Harrington, 1982, 1987; Hansen, 1990; Miller, 1988; Miller et al., 1991; Wang, 1991). As it is typically applied in electromagnetics, the MM is used to solve an integral equation for the currents on or in a body, by converting the integral equation to a matrix equation. For this reason, the MM is often referred to as an integral equation method. An advantage of integral equation methods is that the unknowns of the problem are limited to the surface or volume of the radiating or scattering body, as compared to differential equation methods (such as the finite element or finite difference methods), in which the unknowns can fill all space. A second advantage of integral equation methods is that the elements in the MM matrix equation are all expressed as *integrals* of fields, as opposed to *derivatives* of fields in the differential methods. The averaging process of integration tends to make integral equation methods more stable and reliable than differential equation methods. Finally, the MM is a method that is applicable to almost arbitrary geometries, and thus it has been implemented in a number of user-oriented general purpose computer codes. A disadvantage of the standard MM is that it generates full matrices. Thus, for an N unknown MM solution, the matrix fill time and storage requirements are $O(N^2)$, while the matrix

---

Note: This work was sponsored by the Joint Services Electronics Program under Contract N00014-78-C-0049 with the Ohio State University Research Foundation.

solution time (by direct methods) is $O(N^3)$ and dominates for large N. Recently, Canning (1991) has pointed out that this solution time can be reduced by a factor of 2 for symmetric matrices. Also, Cohoon (1980) has shown that group theory can be used to exploit problem symmetries and reduce computational expense.

The main limitation of the MM is that N is proportional to some power of frequency, and thus the required computer storage and CPU time increase dramatically as the frequency (and thus the electrical size of the body) increases. For this reason, the MM has always been viewed as a "low-frequency" method applicable when the radiating or scattering body is not too large in terms of the operating wavelength. Over the last several years, a number of methods have been studied to improve the efficiency of the MM and thus allow it to be applicable to electrically larger bodies. This paper will present a non-all-inclusive review of some of these techniques, including fast iterative methods, recursive methods, sparse matrix methods, and hybrid methods.

OVERVIEW OF THE METHOD OF MOMENTS

The method of moments (MM) is a numerical technique for solving a linear operator equation by transforming it into a system of simultaneous linear algebraic equations, i.e., a matrix equation (Harrington, 1982, 1987; Hansen, 1990; Miller, 1988; Miller et al., 1991; Wang, 1991). In electromagnetics, the linear operator equation is almost always a linear integral equation for the actual or equivalent current on or in a radiating or scattering body. Once these currents are known, most parameters of engineering interest, such as input impedance, efficiency and radiated (or scattered) fields, can be evaluated in a straightforward manner and with relatively small computer CPU time and storage (as opposed to that required to obtain the currents).

The first step in the MM solution of the integral equation is to approximate the unknown current $\mathbf{J}$ as

$$\mathbf{J} \approx \mathbf{J}^N = \sum_{n=1}^{N} I_n \mathbf{J}_n, \tag{10.1}$$

where $\mathbf{J}^N$ is an N term approximation to the true current $\mathbf{J}$, the $\mathbf{J}_n$ are a series of N known linearly independent expansion functions, and the I_n are a series of N unknown coefficients, $n = 1,2,3,\ldots,N$. The next step is to select a series of N linearly independent weighting functions $\mathbf{w}_m$, $m = 1,2,\ldots,N$, and enforce N weighted averages of the integral equation to be valid. This procedure will reduce the integral equation to a system of N simultaneous linear algebraic equations that can be compactly written as the order N matrix equation

$$[Z]I = V \tag{10.2}$$

In analogy with Ohm's Law, $[Z]$ is the $N \times N$ impedance matrix, V is the length N right-hand side or voltage vector, and I is the length N solution or current vector that contains the N unknown coefficients, I_n, from (10.1). Typical elements of the $[Z]$ matrix and V vector are given by

$$Z_{mn} = -\int_m \mathbf{E}_n \cdot \mathbf{w}_m dr \qquad m,n = 1,2,\ldots,N \tag{10.3}$$

$$V_m = \int_m \mathbf{E}^i \cdot \mathbf{w}_m dr \qquad m = 1, 2, \ldots, N, \qquad (10.4)$$

where $\mathbf{E}_n$ is the electric field of expansion function $\mathbf{J}_n$, $\mathbf{E}^i$ is the known incident electric field, and the integrals are over the region (line, surface, volume) of weighting function $\mathbf{w}_m$.

As it is usually applied in electromagnetics, there is no guarantee or proof that the MM solution will converge to the exact solution as N increases (Dudley, 1985). However, more than 30 years' experience has shown that it almost always does. In fact, except for classical shapes where exact convergent eigenfunction solutions are available (Bowman et al., 1987), the MM solution is often regarded as the most accurate and reliable method. In fact, it is often used as a reference solution to check other numerical methods and even measurements. Further, it is amenable to the analysis of very complex 2D and 3D geometries such as aircraft, buildings, biological bodies, and log periodic or spiral antennas.

In electromagnetics, the main limitation of the MM is that the number of unknowns, N, is proportional to the electrical size of the radiating or scattering body, i.e., its size in wavelengths. Assuming N_λ unknowns per wavelength (typically $N_\lambda = 4 \to 20$) and a body of characteristic dimension L_λ wavelengths, then Table 10.1 shows the computer resources for problems in which the unknowns are along a line (thin wires or perfectly conducting 2D cylinders) on a surface (perfectly conducting plate or penetrable 2D cylinder) and through a volume (3D penetrable body). The table also shows the matrix storage and the order of the CPU time assuming a direct solution of the matrix equation. For even modest-sized problems, the storage is dominated by the N^2 elements of the dense $[Z]$ matrix. For small or modest-sized problems (say, up to a few hundred unknowns), the CPU time is usually dominated by that to fill the N^2 elements of the $[Z]$ matrix. However, for large problems, the $O(N^3)$ CPU time to solve the matrix equation 2 by direct LU decomposition always dominates. The important point is that as the electric size of the body increases, the computer resources increase very rapidly. For example, assuming a problem in which the unknowns are distributed along a surface, and $N_\lambda = 10$ unknowns per wavelength, then $N = 100 L_\lambda^2$, storage $= 10^4 L_\lambda^4$, and CPU $\propto 10^6 L_\lambda^6$. Note that a factor of 10 increase in frequency will result in a factor of 100 increase in N, a factor of 10^4 increase in storage, and a factor of 10^6 increase in CPU time! Clearly, as the frequency increases, the computer resources for the MM increase dramatically, and for this reason the MM is limited to bodies that are not too large in terms of a wavelength.

Table 10.1 Computer Resources for a MM Solution

	Line	Surface	Volume
Unknowns = N	$N_\lambda L_\lambda$	$[N_\lambda L_\lambda]^2$	$[N_\lambda L_\lambda]^3$
$[Z]$ Storage = N^2	$[N_\lambda L_\lambda]^2$	$[N_\lambda L_\lambda]^4$	$[N_\lambda L_\lambda]^6$
CPU Fill $[Z] \propto N^2$	$O[N_\lambda L_\lambda]^2$	$O[N_\lambda L_\lambda]^4$	$O[N_\lambda L_\lambda]^6$
CPU Solve $I = [Z]^{-1} V \propto N^3$	$O[N_\lambda L_\lambda]^3$	$O[N_\lambda L_\lambda]^6$	$O[N_\lambda L_\lambda]^9$

The MM is often referred to as a "low-frequency" method; however, this is not strictly correct. The MM can be applied at optical frequencies if the bodies are of micron size. It is not the frequency that is important, but rather the electrical size of the bodies. The next section will present an overview of several techniques to increase the efficiency of MM solutions and thus allow them to be applicable to electrically larger bodies.

SUMMARY OF METHODS

Iterative Solutions

Without doubt, the main obstacle to the application of the MM to electrically large bodies is the $O(N^3)$ CPU time to solve the matrix equation 2 by direct methods such as LU decomposition. The most obvious solution to this problem is to employ an iterative solution to the matrix equation (Sarkar et al., 1981; Ferguson et al., 1976; Sarkar, 1991). For example, to generate a simple iterative solution of (10.2), add I to both sides of the equation to yield

$$[Z]I + I = V + I \quad \text{or} \quad I = [Z]I + I - V. \tag{10.5}$$

(10.5) suggests the iteration

$$I^{k+1} = [Z]I^k + I^k - V \quad k = 0,1,2,\ldots,K, \tag{10.6}$$

where I^0 is the initial guess for the I vector, I^k is the value of the I vector after k iterations of (10.6), and K is the number of iterations required to achieve some convergence criteria. The major computational effort in an iterative solution is the $O(N^2)$ $[Z]I^k$ matrix product required in each step of the iteration, and thus the total CPU time is $O(KN^2)$. Recent research has centered on reducing the number of iterations, K, required to achieve convergence, and also on reducing the $O(N^2)$ operation count per iteration.

Reducing the Number of Iterations

Iterative solutions of the dense matrix equations generated by the MM can be extremely slowly convergent, or even nonconvergent. If $K \propto N$, then the iterative solution is $O(N^3)$, and it is probably best to use a direct method. As illustrated in Figure 10.1, iterative solutions can exhibit *stalling* or *stagnation* in which the error initially drops rapidly, and then stalls for a number of iterations before again falling rapidly. As pointed out by Kalbasi and Demarest (1993), the stalling is due to the fact that the rapidly varying or high spatial frequency components in the solution typically converge rapidly, while the slowly varying or low-frequency spatial components are more slowly varying.

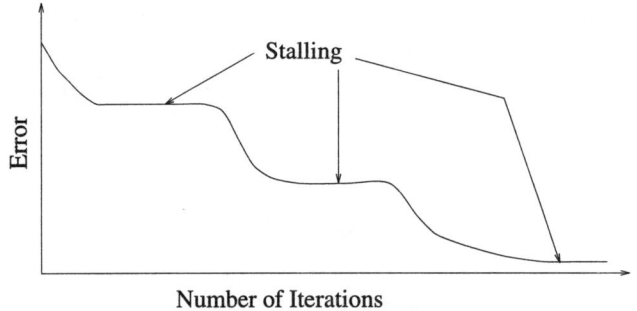

Figure 10.1 An illustration of the stalling or stagnation of an iterative solution.

As illustrated in Figure 10.2, one solution to the stalling problem is a multigrid or multilevel formulation of the MM (Kalbasi and Demarest, 1993). Figure 10.2(a) shows a strip split into N segments, where N is sufficiently large that the MM solution will yield acceptable results. This is referred to as level 1 or the fine level. Figure 10.2(a) also shows the strip split into $N/2$ and $N/4$ segments corresponding to the middle level 2 and coarse level 3. In the 3-level V formulation of Figure 10.2(b), the iterative solution is first solved at the fine level 1, then the coarser level 2, and finally the coarsest level 3. As the MM segment size is increased, low spatial frequency components are turned into high spatial frequency components, and stalling is hopefully avoided. The solution is continued by going back up to level 2, and then repeating the process. Results by Kalbasi and Demarest show that the multilevel MM can significantly reduce the number of iterations to obtain convergence, and even turn a nonconvergent iteration into one which does converge.

Murphy, Rokhlin, and Vassilou have developed a method that they term *complexification* to accelerate the convergence of iterative solutions (Murphy et al., 1993). When the MM is applied to electrically large bodies, the frequency is virtually always close to an interior resonance of the body. This results in a large condition number for the MM [Z] matrix, and slow convergence of the iterative solution. In the complexification method, one replaces the ambient medium, which is typically lossless

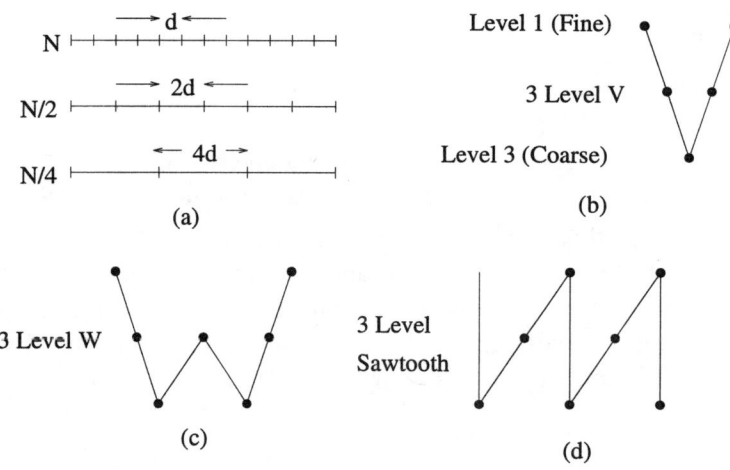

Figure 10.2 A three-level V-type multilevel formulation.

free space, by one with a small amount of loss. Adding loss to the problem reduces the matrix condition number, and thus accelerates convergence of the iterative solution. If the real wavenumber of the ambient medium is k, then one solves the problem in two lossy media of complex wavenumbers $k - jk_1$ and $k - jk_2$. One can then do a linear extrapolation of the currents from these two lossy media solutions, to obtain the currents for the lossless media with real wavenumber k. The same idea can be applied with solutions from three lossy media and quadratic extrapolation to real k. The results of Murphy et al. (1993) show that complexification does accelerate the convergence of the iterative solution, at times by an order of magnitude. An advantage of the complexification method is that it is relatively simple and straightforward to implement, since it typically requires relatively little additional effort to develop a computer code for a lossy, as opposed to lossless, medium.

The Fast Multipole Method

The fast multipole method (FMM), developed by V. Rokhlin and others, is a technique to perform a matrix vector product in $O(N^{3/2})$ or fewer operations (Murphy et al., 1993; Rokhlin, 1983, 1990; Engheta et al., 1992; Coifman et al., 1993). In an iterative solution, one must perform a matrix vector product of the form $V^k = [Z]I^k$, where I^k is the *known* solution vector after k iterations. Element m of V^k is the inner product between row m of $[Z]$ and I^k; i.e.,

$$V_m^k = \sum_{n=1}^{N} Z_{mn} I_n^k. \tag{10.7}$$

When performed in the straightforward manner of (10.7), each inner product requires $O(N)$ operations, and thus the matrix vector product requires $O(N^2)$ operations. The FMM provides a rapid method for performing those terms in the summation of (10.7), for which modes m and n are far removed.

As illustrated in Figure 10.3, in the FMM the basis functions are collected into P groups of $p = N/P$ basis functions each. If the groups containing the expansion and weighting functions are in the far zone of each other, then the contribution from *all* of the expansion functions in the expansion group can be evaluated through the use of a *multipole expansion* that requires far fewer operations than the straightforward superposition indicated in (10.7). In electromagnetics, roughly kD multipoles are required, where k is the wavenumber and D is the size of the expansion group. If $kD \ll p$, then considerable time can be saved using the multipole expansions.

As a simple example of a multipole, consider the problem of computing the gravitational field of a large collection of p small masses, $m_i, i = 1, 2, \ldots p$. If the observation point is far removed from the p masses, then the gravitational field will be close to that of a point mass

$$M_0 = \sum_{i=1}^{P} m_i \tag{10.8}$$

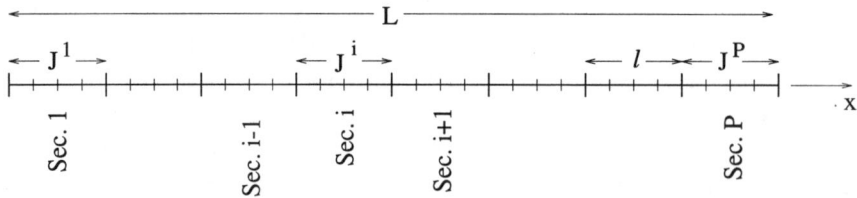

Figure 10.3 A scatterer is split into P sections of $p = N/P$ unknowns.

located somewhere near the center of the p masses. M_0 can be viewed as the 0th order multipole of the masses.

In an electrically large body, most of the groups will be in the far zone of each other, and thus most of the terms in the summation of (10.7) will be performed by the faster multipole expansions. Further, since the Z_{mn} corresponding to modes m and n in far zone groups are never explicitly used, they need not be computed or stored. Thus, the CPU time to compute the $[Z]$ matrix, and the memory to store it, can be dramatically reduced. Early results on problems with roughly $N = 1000$ unknowns show an order of magnitude reduction in CPU time and storage. More dramatic results may be possible for larger N.

Recursive Methods

Recursive methods are different from iterative methods. In an iterative method, one sets up the solution to the entire or global problem, and then attempts to generate in an iterative manner global solutions that approach the exact solution. By contrast, in a recursive solution the global problem is split into many smaller problems. The smaller problems are solved, and their solutions are combined in some recursive manner to build up to the solution to the global problem.

Exact Fixed Step Methods

Over the past several years W.C. Chew and others at the University of Illinois' Electromagnetics Laboratory (Chew, 1993) have developed a series of recursive algorithms for solving electromagnetic scattering problems that can be viewed as a large collection of N scattering bodies. In a recursive method, one begins by "analyzing" the individual scatterers. Typically, "analyzing" means determining the scatterer T-Matrix (Waterman, 1969), which defines the amplitude of the modal harmonics scattered by the body for a given incident mode. The N one body solutions are then combined to form $N/2$ two body solutions, which are then combined to form $N/4$ four body solutions, and the process is continued until the N body problem is solved. As opposed to the iterative methods described above, these recursive methods determine the exact (meaning the solution that would have been obtained if the N body problem were attacked directly) solution in a fixed number of steps. Several recursive algorithms have been developed by the group at the University of Illinois (Chew, 1989; Gurel and Chew, 1992a,b; Chew and

Yang, 1990; Chew et al., 1992, 1995; Lu and Chew, 1993; and Chew and Lu, 1993) and go under names such as "Recursive T-Matrix Algorithms" (RTMA). Depending upon the method, the complexity of the problem is reduced to $O(N^{1.5}) \rightarrow O(N^{2.5})$, as compared to $O(N^3)$ for a standard MM solution with LU decomposition.

Spatial Decomposition

As illustrated in Figure 10.3, in the spatial decomposition method, a large body is segmented into P smaller parts or sections, each having $p = N/P$ unknowns (Umashankar et al., 1992). Let us denote

$$\mathbf{J}^i = \sum_{n=1+(i-1)p}^{ip} I_n \mathbf{J}_n \qquad i = 1, 2, \ldots, P \tag{10.9}$$

as the current on section i, and thus the current on the entire strip $\mathbf{J} = \mathbf{J}^1 + \mathbf{J}^2 + \cdots + \mathbf{J}^P$. The method is begun by initializing $\mathbf{J}$ to the best estimate available. This might be the physical optics current $\mathbf{J} = 2\hat{\mathbf{n}} \times \mathbf{H}^i$, or it could simply be $\mathbf{J} = 0$. The first recursion is begun by determining the current, $\mathbf{J}^1$ on section 1, induced by the incident fields and by the best estimate currents on Sections $2 - P$. These Section 1 currents are determined by a standard MM solution with LU decomposition. This is an order $O(p^3) = O[(N/P)^3]$ process, which for large P is much less than the $O(N^3)$ direct MM solution of the entire problem. The next step is to determine the currents induced on Section 2 by the incident fields plus the best estimate currents on Sections 1 and $3 - P$ (i.e., the just-determined $\mathbf{J}^1$, plus the initial guess for the currents on Sections $3 - P$). This process is continued until the currents on all sections have been updated once, thus completing the first recursion. The second recursion is identical to the first, except the first recursion has provided an update of the best estimate current. The recursive process must then be continued until some convergence criterion has been met. Assuming that as N increases, P also increases to keep $p = N/P$ constant, the entire recursion is of order $O(KP) = O(KN/p)$, where K is the number of recursions necessary to obtain convergence. On modest-size problems of a few hundred unknowns, the method has been shown to produce essentially the same results as a standard MM solution, but with up to an order of magnitude savings in CPU time.

Sparse Matrix Methods

Method-of-moments (MM) impedance matrices are generally dense. Referring to (10.3), this is a result of the fact that $\mathbf{E}_n$ is an entire domain function that fills all space, and drops off only as $1/r$ from the expansion function $\mathbf{J}_n$. An MM formulation that did produce a sparse $[Z]$ matrix would save storage as well as CPU time by using sparse matrix methods for solving the matrix equation. Further, if the essentially zero elements could be determined a priori, then CPU time could also be saved in the matrix fill by simply not computing the essentially zero elements of $[Z]$. The methods presented below do not generate sparse matrices in the sense that most elements are exactly 0. Rather, they generate matrices in which most elements are much smaller than the largest or dominant elements. A threshold is set, and elements whose magnitude fall below that threshold are set to 0 and ignored. As the threshold level is raised, the $[Z]$ matrix becomes sparser, and the method becomes more efficient but less accurate.

The Impedance Matrix Localization Method

In the impedance matrix localization method (IML) developed by Canning (1990, 1993), it is helpful to think of Z_{mn} of (10.3) as the voltage induced in a receiving antenna with current $\mathbf{w}_m$ by a transmitting antenna with current $\mathbf{J}_n$. Thus, if modes m and n are not too close, Z_{mn} is proportional to the product

$$Z_{mn} \propto [\text{Transmit Pattern of } \mathbf{J}_n] \cdot [\text{Receive Pattern of } \mathbf{w}_m] / R_{mn}, \qquad (10.10)$$

where R_{mn} is the center to center distance between modes m and n and the patterns are evaluated at a look angle from the center of mode n to the center of mode m. The patterns of $\mathbf{J}_n$ or $\mathbf{w}_m$ are the free space electric fields of the currents, $\mathbf{J}_n$ or $\mathbf{w}_m$, respectively.

Typically MM solutions employ subsectional basis functions defined over an electrically small region of the body, not exceeding $\lambda/4$. These small currents have essentially omnidirectional radiation (and receive) patterns, and thus by (10.10) produce a dense $[Z]$ matrix. As illustrated in Figure 10.4 the idea of the IML is to use electrically large basis functions, and further to employ a traveling wave current distribution such that the basis functions have highly directional patterns. In Figure 10.4, it can be seen that Z_{mn} will only be significant if $\mathbf{w}_m$ falls within the main beam of $\mathbf{J}_n$ and vice versa. Otherwise Z_{mn} will be relatively small, and can be ignored or set to 0. In a sample computation with a threshold of 0.001 (i.e., ignore elements of $[Z]$ less than the threshold times the largest element), only 2 percent of the elements of $[Z]$ were above the threshold, and yet the method produced excellent results for the far-zone scattered fields.

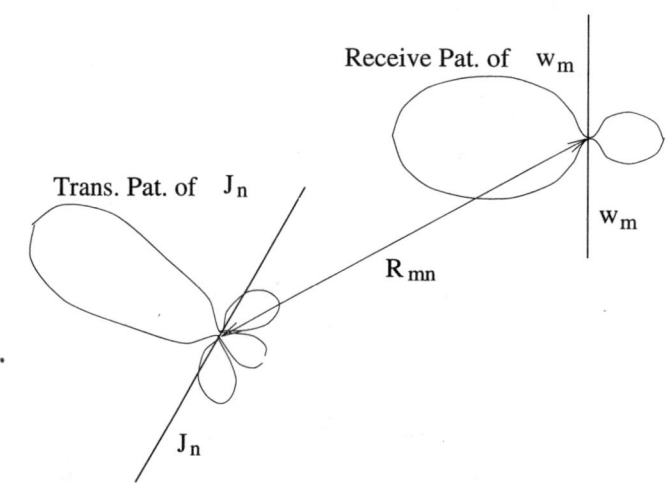

Figure 10.4 Z_{mn} is proportional to the product of the transmit pattern of $\mathbf{J}_n$ times the receive pattern of $\mathbf{w}_m$.

Pogorzelski developed a somewhat different method for obtaining a sparse MM [Z] matrix, which he termed near-field localization (Pogorzelski, 1993). Rather than use basis functions with highly directive far-field patterns as in the IML, Pogorzelski suggested the use of basis functions whose near-zone patterns are highly focused to a small region on the surface of the body. In this way there is only strong coupling between closed-spaced basis functions. As the electrical size of the body increases, and the total number of basis functions is increased, the total number of significant matrix elements remains approximately constant.

Wavelet Basis Functions

Although wavelets have many interesting properties that make them useful in signal processing applications and as basis functions for different numerical methods (Daubechies, 1992), the feature exploited by Steinberg and Leviatan (1993) to produce a sparse MM impedance matrix is that wavelets have 0 average value. Figure 10.5 illustrates the computation of the mutual impedance between expansion function $\mathbf{J}_n$ and Galerkin weighting function $\mathbf{w}_m$ for the cases of simple pulse and wavelet type basis functions. Referring to (10.3), Z_{mn} is the weighted average of the electric field of expansion n over the extent of weighting function m, and weighted by $\mathbf{w}_m$. For far separated modes, Z_{mn} can be approximated by

$$Z_{mn} \approx -\mathbf{E}_n(\mathbf{r}_m) \cdot \mathbf{P}_m, \tag{10.11}$$

where $\mathbf{E}_n(\mathbf{r}_m)$ is the electric field of $\mathbf{J}_n$ evaluated at the center of weighting function m, and

$$\mathbf{P}_m = \int_m \mathbf{w}_m dl \tag{10.12}$$

is the current moment of weighting function $\mathbf{w}_m$. Pulse basis functions (and virtually every other common subsectional basis function) have a finite current moment, and thus Z_{mn} is reduced only by the $1/r$ dependence of its far-zone fields. By contrast, wavelet basis functions have 0 average value, and thus 0 current moment. Although (10.11) predicts that for far separated wavelets $Z_{mn} = 0$, this is not exactly true. The reason is that (10.11) is based upon the approximation that $\mathbf{E}_n$ is constant over the extent of weighting function m, which is not true for the phase even in the limit as the separation goes to infinity. The wavelet far separated mutual impedances are further reduced by the fact that, since the wavelet expansion function has 0 current moment, its far-zone fields drop off as $1/r^2$, rather than as $1/r$ for conventional subsectional basis functions. On a rather small problem with only $N = 80$ unknowns, as the threshold was adjusted lower from 0.01 to 0.001, the number of elements of [Z] above the threshold increased from 7 percent to 25 percent.

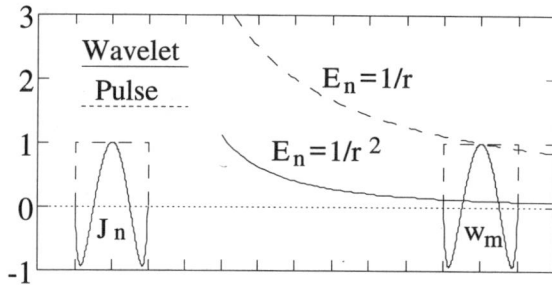

Figure 10.5 Mutual impedance computation for wavelet and pulse type basis functions.

Novel Expansion Methods

Traditionally the unknowns in integral equation methods have been the true electric currents flowing on perfectly conducting bodies, or rigorously obtained equivalent electric and/or magnetic currents flowing either on the surface of or through the volume of penetrable bodies (Balanis, 1989). In an attempt to reduce the number of unknowns in the MM solution, various authors have investigated the use of different unknowns.

Rather than use the true or rigorous equivalent currents to produce the fields scattered by the body, Leviatan and Boag have proposed the use of fictitious current filaments located inside (or outside) the scattering body (Leviatan and Boag, 1987). The use of these filaments avoids the time consuming numerical integrations typically needed to find the fields of the expansion functions. Also, since the filaments are removed from the surface on which the boundary conditions are being enforced, the usual singularity problem associated with computation of self-impedance terms is avoided.

Leviatan, Hudis, and Einziger have investigated the use of Gaussian beams as expansion functions in the MM (Leviatan et al., 1989). An advantage of the Gaussian beam expansion functions is that their fields can be expressed as a summation of analytic terms, thereby avoiding the necessity of a numerical integration. Boag and Mittra have investigated the use of multipole sources in complex space, and have found that the method not only reduces the number of unknowns in the MM solution, but also generates a banded (Z) matrix with a low condition number (Boag and Mittra, 1994).

Recently there has been work on using the electromagnetic fields rather than currents as the unknown. Ludwig developed the spherical-wave expansion technique (SPEX) (Ludwig, 1986, 1989), and Hafner developed the Generalized Multipole Technique (GMT) (Hafner 1990a,b) in which the unknowns are the electromagnetic fields, rather than currents. These fields are expanded in terms of a modal harmonics satisfying Maxwell's equations, and the unknown coefficients are determined by setting up a system of simultaneous linear equations that enforce boundary conditions at the surface of the scatterer. These modal expansion methods have the advantage that often the scattered field can be represented in terms of fewer unknowns than the current. These methods also totally avoid the sometimes numerically difficult problem of computing the fields of currents, since they deal directly with the fields.

Hybrid Methods

Despite all of the methods described above, the MM remains a "low-frequency" method, i.e., it is applicable only when the size of the radiating or scattering body is not too large in terms of a wavelength. By contrast, there are a variety of "high-frequency" or asymptotic methods that are applicable when the radiating or scattering body, and all of its important features, are electrically large. At X-band (10 GHz) there is no available single method that could analyze the radiation from a simple monopole antenna on an aircraft. The MM would fail because the aircraft is electrically too large, and thus would be impractical due to limited computer resources, while the asymptotic methods would fail because the monopole antenna is too small. In the authors' opinion, the only hope for analyzing complex bodies at high frequencies is through the use of *hybrid methods* (Burnside, 1980) that combine a high-frequency technique, such as the Geometrical Theory of Diffraction (GTD) (Hansen, 1981; Pathak, 1988), to analyze most of the electrically large body with a low-frequency technique, such as the MM, to analyze the remaining small parts that cannot be treated by the high-frequency method.

The Hybrid MM/GTD Method

In a standard MM solution, equivalence theorems are used to replace the entire radiating or scattering body by free space and by equivalent currents (Balanis, 1989, secs. 7.7, 7.8). The advantage of doing this is that all currents radiate in homogeneous free space, and thus the kernel of the integral equation contains the relatively simple free space Green's function. As a result, E_n and E^i in Equations 10.3 and 10.4 are the *free space* fields of expansion function $\mathbf{J}_n$ and the impressed currents, respectively.

By contrast, in an MM/Green's function solution only a portion of the body is replaced by free space and by equivalent currents (Newman, 1988). For example, Figure 10.6 shows a body split into Sections 1 and 2 (shown disjoint for clarity, but they may be touching). As suggested in Figure 10.6, Section 1 is considered as the small part of the body and requires N_1 expansion functions to approximate its current, while Section 2 is the large part and would require $N_2 \gg N_1$ expansion functions to approximate its current $(N_1 + N_2 = N)$. The advantage of the MM/Green's function method is that one must only place N_1 unknowns over that portion of the body that is being replaced by equivalent currents. Thus (10.1) is modified to

$$\mathbf{J}^1 \approx \mathbf{J}^{1N} = \sum_{n=1}^{N_1} I_n \mathbf{J}_n, \tag{10.13}$$

where $\mathbf{J}^1$ is that portion of $\mathbf{J}$ in Section 1. The MM matrix equation is still given by (10.2), however, now it is an order $N_1 \ll N$ matrix equation.

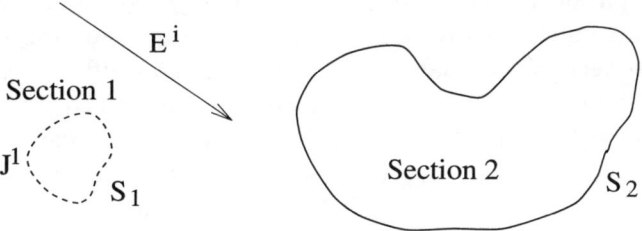

Figure 10.6 In the MM/Green's function method, Section 1, but not Section 2, of the body is replaced by free space and equivalent currents.

The disadvantage of the MM/Green's function method is that the kernel of the integral equation contains the Green's function for Section 2 of the body not replaced by equivalent currents, and thus $\mathbf{E}_n$ and $\mathbf{E}^i$ in (10.3) and (10.4) are the fields of expansion function $\mathbf{J}_n$ and the impressed currents, respectively, radiating in the presence of Section 2. A problem such as a dipole antenna (Section 1) radiating in the presence of a sphere (Section 2) would be ideal for the MM/Green's function method since the sphere is a classical shape whose Green's function is known, and also since many more unknowns would be required to model the current on the sphere as opposed to that on the dipole.

Of course, a limitation of the MM/Green's function method is that Section 2 must be of sufficiently simple shape that its Green's function is known, and there are only a few classical shapes with known Green's functions (Bowman et al., 1987). However, Thiele and Newhouse (1975) realized that the class of Section 2 geometries can be greatly extended if the GTD is used to approximate the Section 2 Green's function. The result is termed a hybrid MM/GTD method, and was applied by Thiele and Newhouse to analyze a small monopole antenna (Section 1) on a flat plate (Section 2). Today the GTD has been developed to the point where the asymptotic Green's function for a body as complex as an aircraft can be constructed.

The Hybrid GTD/MM Method

A virtual axiom in the MM, rooted in the Sampling Theorem, is that one needs at least 4 expansion functions per linear wavelength. Although one can develop "better" basis functions that can produce more accurate results with fewer unknowns per wavelength, in the end the number of unknowns is still proportional to the electrical size of the body (see Table 10.1) and the MM remains a "low-frequency" method. In an attempt to break this log jam, a technique known as a *physical basis* MM solution was developed by Richmond (1985). In a physical basis MM solution, one attempts to build the physics of the problem into the MM expansion functions so that their size can be increased to cover all or almost all of the body. For example, Richmond solved the problem of scattering by an electrically wide dielectric strip using only 3 expansion functions. Richmond's insight into the physics of the problem suggested that the fields or currents in an electrically wide strip are very close to the known currents in an infinitely wide strip, with the main difference coming from surface waves generated at either end of the strip. Thus, he expanded the current in terms of 3 expansion functions. The first had the form of the currents in an infinitely wide strip, while the second and third had the form of left and right propagating surface waves on the strip. The results for this 3-term physical basis expansion were in very close agreement with a standard pulse basis MM solution, except near the edges of the strip. However, the results for the far-zone fields were in essentially perfect agreement, since the far-zone fields are an integral of the current.

The problem with the physical basis MM solution is that, in general, one does not have enough physical insight into the problem to construct the physical basis expansion functions. However Burnside, Yu, and Marhefka (1975) recognized that, away from edges or other points of discontinuity on electrically large bodies, the form of the currents is closely approximated by the GTD. Thus, the GTD can be used to construct the physical basis expansion functions on a complex body, with the result being referred to as a GTD/MM solution. Near edges or other points of discontinuity they employed standard MM pulse functions. The result is a physical basis MM solution in which the number of unknowns is virtually independent of frequency.

Figure 10.7 illustrates the GTD/MM expansion for the current on a perfectly conducting strip. Note that the strip is divided into the GTD and MM regions. Away from the edges in the GTD region, the strip current will be closely approximated by the physical optics current $2\hat{\mathbf{y}} \times \mathbf{H}^i$ plus currents of the form

$$\frac{e^{-jk\rho_l}}{\sqrt{\rho_l}} \quad \text{and} \quad \frac{e^{jk\rho_r}}{\sqrt{\rho_r}}$$

corresponding to currents diffracted from the left and right edges. Here ρ_l and ρ_r denote distances from the left or right edges, respectively. In the MM regions near the edges, the shape of the current may be complex, and is represented in terms of standard MM expansion functions. Figure 10.7 illustrates a pulse expansion. Assuming N_e expansion functions are needed near each edge, the GTD/MM expansion for the current on the strip is

$$\mathbf{J} \approx \mathbf{J}^N = I_1 2\hat{\mathbf{y}} \times \mathbf{H}^i + I_2 \frac{e^{-jk\rho_l}}{\sqrt{\rho_l}} + I_3 \frac{e^{jk\rho_r}}{\sqrt{\rho_r}} \qquad \text{in the GTD region}$$

$$+ \sum_{n=4}^{N_e+3} I_n \mathbf{J}_n \qquad \text{in the left edge MM region}$$

$$\sum_{n=N_e+4}^{2N_e+3} I_n \mathbf{J}_n \qquad \text{in the right edge MM region.} \qquad (10.14)$$

Typically, the width of the MM region is about a wavelength, and thus N_e is typically about 10. Thus, the GTD/MM expansion for the strip requires $2N_e + 3 \approx 23$ unknowns, independent of the electrical width of the strip.

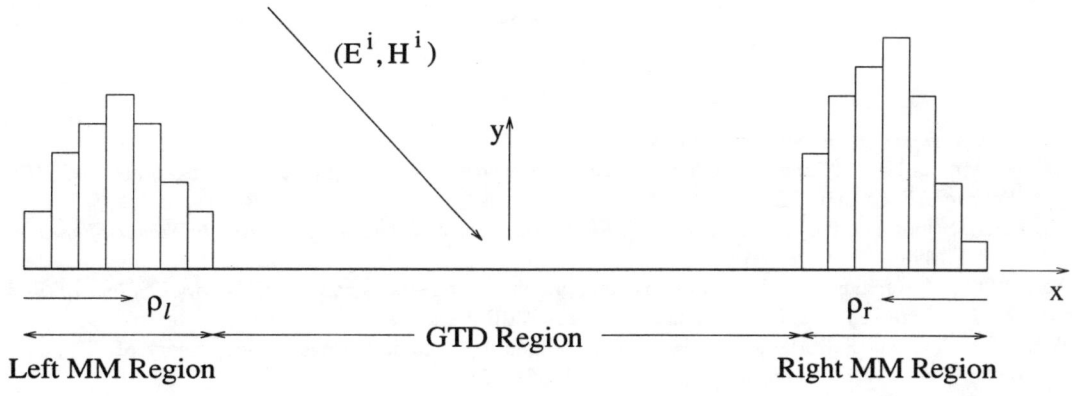

Figure 10.7 In the GTD/MM method, the strip is divided into GTD and MM regions for expanding the current.

CONCLUSIONS

This paper has presented a number of methods for improving the efficiency of the MM, in an attempt to permit it to be applicable to electrically large bodies. A number of interesting methods were described, each displaying the ability to reduce the CPU and/or storage requirements for MM solutions. However, in each case, the total CPU time and storage is still proportional to the electrical size of the body, and thus the MM remains a "low-frequency" technique applicable when the body is not electrically too large. Improvements in these techniques and the use of more and more powerful computers will permit the MM to be applied to higher and higher frequencies. However, at the same time, the operating frequency of radars and communication systems is being pushed to the millimeter or even terahertz band. Thus, the MM appears to be in a losing battle, and may never be able to cover the full range of desired operating frequencies.

Of the many methods presented, the authors feel that the fast multipole method (FMM) may be the best for arbitrary bodies and general purpose codes. The FMM not only reduces the matrix solution time to $O(N^{3/2})$ (or less), but it also reduces $[Z]$ matrix storage and fill time since only the "near-zone" elements need to be computed and stored. However, possibly the most attractive feature of the FMM is that it seems reasonably straightforward to apply it to arbitrary geometries, and it can even be implemented as a retrofit to an existing code.

The authors feel that the only hope of developing a computer code that is applicable at all frequencies is through the use of hybrid methods. In a hybrid solution a "high-frequency" method (such as the GTD) is used to treat most of the electrically large body. The remaining portions of the body that cannot be treated by the "high-frequency" method, because they are too small or are too close to discontinuities, are treated with a "low-frequency" method (such as the MM). Hybrid solutions offer the possibility of a solution in which, as the frequency is increased, the number of unknowns approaches a finite limit. The main problem with hybrid methods has been the difficulty in implementing them for arbitrary geometries.

REFERENCES

Balanis, C.A., 1989, *Advanced Engineering Electromagnetics*, New York: John Wiley and Sons.
Boag, A., and R. Mittra, 1994, "Complex multipole beam approach to electromagnetic scattering problems," *IEEE Trans. Antennas Propag.* **AP-42**, 366-372.
Bowman, J.J., T.B.A. Senior, and P.L.E. Uslenghi, 1987, *Electromagnetic and Acoustic Scattering by Simple Shapes*, New York: Hemisphere Publishing.
Burnside, W.D., 1980, "A summary of hybrid solutions involving moment methods and GTD." In: *Applications of the Method of Moments to Electromagnetic Fields*, St. Cloud, Fla.: The SCEEE Press.
Burnside, W.D., C.L. Yu, and R.J. Marhefka, 1975, "A technique to combine the geometrical theory of diffraction and the moment method," *IEEE Trans. Antennas Propag.* **AP-23**, 551-557.
Canning, F.X., 1990, "The impedance matrix localization (IML) method for moment calculations," *IEEE Trans. Antennas Propag.* **AP-32**, 18-30.
Canning, F.X., 1991, "Direct solution of the EFIE with half the computation," *IEEE Trans. Antennas Propag.* **AP-39**, 118-119.
Canning, F.X., 1993, "Improved impedance matrix localization method," *IEEE Trans. Antennas Propag.* **AP-41**, 659-667.
Chew, W.C., 1989, "An N^2 algorithm for the multiple solution of N scatterers," *Microwave Opt. Tech. Lett.* **2**, 380-383.
Chew, W.C., 1993, "Fast algorithms for wave scattering developed at the University of Illinois' Electromagnetics Laboratory," *IEEE Trans. Antennas Propag.* **AP-35**, 22-32.
Chew, W.C., and C.C. Lu, 1993, "NEPAL—An algorithm for solving the volume integral equation," *Microwave Opt. Tech. Lett.* **6**, 185-188.

Chew, W.C., and Y.M. Yang, 1990, "A fast algorithm for solution of a scattering problem using a recursive aggregate Tua matrix method," *Microwave. Opt. Tech. Lett.* **3**, 164-169.

Chew, W.C., L. Gurel, Y.M. Yang, G. Otto, R. Wagner, and Q.H. Liu, 1992, "A generalized recursive algorithm for wave-scattering solutions in two dimensions," *IEEE Trans. Microwave Theory Tech.* **40**, 716-723.

Chew, W.C., Y.M. Wang, and L. Gurel, 1995, "A recursive algorithm for wave-scattering using windowed addition theorem," *J. Electromagnetic Waves Appl.*, accepted for publication.

Cohoon, D.K., 1980, "Reduction of the cost of solving integral equations arising in electromagnetic scattering through the use of group theory," *IEEE Trans. Antennas Propag.* **AP-28**, 104-107.

Coifman, R., V. Rokhlin, and S. Wandzura, 1993, "The fast multipole method for the wave equation: a pedestrian prescription," *IEEE Trans. Antennas Propag.* **AP-35**, 7-12.

Daubechies, I., 1992, "Ten lectures on wavelets," *CBMS-NSF Series in Applied Mathematics*, SIAM, Philadelphia.

Dudley, D.G., 1985, "Error minimization and convergence in numerical methods," *Electromagnetics* **5**, 89-97.

Engheta, N., W.D. Murphy, and V. Rokhlin, 1992, "The fast multipole method (FMM) for electromagnetic scattering problems," *IEEE Trans. Antennas Propag.* **AP-40**, 634-641.

Ferguson, T.R., T.H. Lehman, and R.J. Balestri, 1976, "Efficient solutions of large moments problems," *IEEE Trans. Antennas Propag.* **AP-24**, 230-235.

Gurel, L., and W.C. Chew, 1992a, "A recursive T-matrix algorithm for strips and patches," *Radio Science* **27**, 387-401.

Gurel, L., and W.C. Chew, 1992b, "Scattering solution of three-dimensional array of patches using the recursive T-matrix algorithm," *Microwave Guided Wave Lett.* **2**.

Hafner, C., 1990a, *The Generalized Multipole Technique*, Boston: Artech House.

Hafner, C., 1990b, "On the relationship between the MoM and the GMT," *IEEE Trans. Antennas Propag.* **AP-34**, 12-19.

Hansen, R.C., 1981, *Geometric Theory of Diffraction*, New York: IEEE Press.

Hansen, R.C., 1990, *Moment Methods in Antennas and Scattering*, Boston: Artech House.

Harrington, R.F., 1982, *Field Computation by Moment Methods*, Malabar, Fla.: Krieger.

Harrington, R.F., 1987, "Matrix methods for field problems," *Proc. IEEE* **55**, 136-149.

Kalbasi, K., and K.R. Demarest, 1993, "A multilevel formulation of the method of moments," *IEEE Trans. Antennas Propag.* **AP-41**, 589-599.

Leviatan, Y., and A. Boag, 1987, "Analysis of electromagnetic scattering from dielectric cylinders using a multifilament current model," *IEEE Trans. Antennas Propag.* **AP-35**, 1119-1127.

Leviatan, Y., E. Hudis, and P.D. Einziger, 1989, "A method of moments analysis of electromagnetic coupling through slots using Gaussian beam expansions," *IEEE Trans. Antennas Propag.* **AP-37**, 1537-1544.

Lu, C.C., and W.C. Chew, 1993, "Electromagnetic scattering of finite strip array on a dielectric slab," *IEEE Trans. Microwave Theory Tech.* **41**, 97-100.

Ludwig, A.C., 1986, "A comparison of spherical wave boundary value matching versus integral equation scattering solutions for a perfectly conducting body," *IEEE Trans. Antennas Propag.* **AP-34**, 857-865.

Ludwig, A.C., 1989, "A new technique for numerical electromagnetics," *IEEE Trans. Antennas Propag.* **AP-37**, 40-41.

Miller, E.K., 1988, "A selective survey of computational electromagnetics," *IEEE Trans. Antennas Propag.* **AP-36**, 1281-1305.

Miller, E.K., L.M. Mitschang, and E.H. Newman, 1991, *Computational Electromagnetics-Frequency-Domain Method of Moments*, New York: IEEE Press.

Murphy, W.D., V. Rokhlin, and M.S. Vassilou, 1993, "Acceleration methods for the iterative solution of electromagnetic scattering problems," *Radio Science* **28**, 1-12.

Newman, E.H., 1988, "An overview of the hybrid MM/Green's function method in electromagnetics," *Proc. IEEE* **76**, 270-282.

Pathak, P.H., 1988, "Techniques for High-Frequency Problems." In: *Antenna Handbook: Theory, Applications, and Design*, Y.T. Lo and S.W. Lee (eds.), New York: Van Nostrand Reinhold Co.

Peterson, B., and S. Strom, 1973, "T matrix for electromagnetic scattering from an arbitrary number of scatterers and representations of E(3)," *Physical Rev. D* **8**, 3661-3678.

Pogorzelski, R.J., 1993, "Improved computational efficiency via near-field localization," *IEEE Trans. Antennas Propag.* **AP-41**, 1081-1087.

Richmond, J.H., 1985, "Scattering by thin dielectric strips," *IEEE Trans. Antennas Propag.* **AP-33**, 64-68.

Rokhlin, V., 1983, "Rapid solution of integral equations of classical potential theory," *J. Comput. Phys.* **60**, 187-207.

Rokhlin, V., 1990, "Rapid solution of integral equations of scattering theory in two dimensions," *J. Comput. Phys.* **86**, 414-439.

Sarkar, T.K., 1991, "Application of the conjugate gradient method to electromagnetics and signal analysis." In: *Progress in Electromagnetic Research* **5** (PIER 5), J.A. Kong (chief ed.), New York: Elsevier.

Sarkar, T.K., K.R. Siarkiewicz, and R.F. Stratton, 1981, "Survey of numerical methods for solutions of large systems of linear equations for electromagnetic field problems," *IEEE Trans. Antennas Propag.* **AP-29**, 847-856.

Steinberg, B.Z., and Y. Leviatan, 1993, "On the use of wavelet expansions in the method of moments," *IEEE Trans. Antennas Propag.* **AP-41**, 610-619.

Thiele, G.H., and T.M. Newhouse, 1975, "A hybrid technique for combining moment methods with the geometrical theory of diffraction," *IEEE Trans. Antennas Propag.* **AP-23**, 62-69.

Umashankar, K.R., S. Nimmagadda, and A. Taflove, 1992, "Numerical analysis of electromagnetic scattering by electrically large objects using spatial decomposition technique," *IEEE Trans. Antennas Propag.* **AP-40**, 867-877.

Wang, J.J.H., 1991, *Generalized Moment Methods in Electromagnetics*, New York: John Wiley and Sons.

Waterman, P.C., 1969, "New formulation of acoustic scattering," *J. Acoust. Soc. Am.* **45**, 1417-1429.

11

Discussion

JOHN TUCKER: It is my pleasure to introduce the individuals who will moderate the open discussion: Art Jordan from the Naval Research Laboratory and Tom Hughes from Stanford University.

THOMAS HUGHES: This moderated open discussion is intended as a period of fermentation where issues are put on the table, the most important of which will be summarized in the concluding panel discussion. In this spirit, we would like the audience to take the lead and raise issues. Some questions and comments have been proposed in writing, but we will introduce those after people express spontaneous perspectives.

OPEN DISCUSSION

RICHARD ZIOLKOWSKI: I have been struck these two days by how much overlap there appears to be between the acoustics community and the electromagnetics community on the type of modeling techniques applied to essentially the same kinds of wave propagation problems. There are a number of Helmholtz approaches at which both communities are looking. In the time domain, I believe there are similar approaches in both communities dealing strictly with waves.

For example, one of the outstanding electromagnetics problems is to get better absorbing boundary conditions. Antennas and propagation annual meetings have three or four parallel sessions each afternoon on nothing but absorbing boundary conditions and their similarities. The Enquist and Majda type of boundary conditions are well known. We started out using those but have developed other techniques, such as ion surface radiation conditions. There are new Berringer and absorbing loci electric materials that one can now put around the problem. From the acoustics standpoint, what kinds of techniques are being used for which there might be direct applicability to the electromagnetics regime, especially with regard to things such as absorbing boundary conditions?

I was also struck this morning by the idea of gridless methods. They would be wonderful. The biggest problem with three-dimensional modeling techniques right now is grid generation. Although the grids are sometimes good for the techniques being used, there are a number of unstructured grid capabilities for which, depending on the modeling technique being used, the generated grids sometimes cannot be used. These are merely a couple of representative issues of similarity between acoustics and electromagnetics that it would be useful to discuss.

HUGHES: The first question raised from the electromagnetics community concerns the status and thinking regarding absorbing boundary conditions in current use and being explored in acoustics, and what individuals in this allied field should consider: Where are the opportunities and current thinking?

LONNY THOMPSON: Some of the radiation boundary conditions that are used for electromagnetics and acoustics are actually the same for the lower-order boundary conditions. Some people have developed the same boundary conditions in both fields independently. If they were to look at each other's work, they would realize the conditions are very similar, if not the same.

I would stress that one should probably be using higher-order boundary conditions, instead of stopping at first- and second-order accurate boundary conditions as is typically done. The reason is that one would like to move the computational domain for exterior problems that arise in acoustic problems, as well as in electromagnetics. The radiation boundary should be brought as close to the object or the scatter as possible. If one of these lower-order boundary conditions is used, then the needed accuracy is unavailable when the boundary is brought very close. So one must go to a higher-order boundary condition. A number of papers have looked at and developed higher-order boundary conditions. The challenge is how to implement those in an efficient way either into finite difference, finite element, or other discrete techniques for basic computational methods.

LESZEK DEMKOWICZ: I have two comments. First, I believe those high-order absorbing boundary conditions are nothing other than a truncated form of an infinite element. In a theory of infinite elements, it turns out that all one really needs is an extra infinite expansion in the right direction. When second- or third-order truncated, nonreflecting boundary conditions are used, they actually correspond to a type of approximation in the real direction.

Second, when one looks at the classical separation-of-variables argument and the form of the exact solution, say, for the sphere problem under the Helmholtz equation, it becomes evident that the terms corresponding to a higher-order frequency on a sphere have simultaneously a larger exponent N in the denominator there. As one moves away from the body, the practical effect of those terms disappears, because having the denominator R raised to the power N for larger N makes the fraction converge to zero faster. For that reason and without any calculations, one can probably anticipate that there must be a trade-off between the number of finite elements put in between the scatter and that artificial boundary (on which are positioned the infinite elements or nonreflecting boundary condition), and the numbers of terms in the real direction for the infinite element. The closer one gets to the scatter, the smaller will be the number of elements in between; but the number of terms in the infinite element will grow, and there is nothing that can be done about that. It is just the nature of the problem. Analogously, of course, as one gets further away from the scatter, then the number of the terms in the expansion one can use will probably be smaller, but that is compensated for by the number of elements that are positioned in between the two.

THOMPSON: I wish to respond to that. The infinite element and local radiation boundary conditions are both very similar in that they approximate the exact impedance on the radiation boundary. However, when one uses infinite elements and goes closer to the scatter, the fact that more terms or more layers in the infinite element must be used is offset by the cost [of adding extra layers for the infinite element] being much less than it is for having to discretize with regular finite elements in the domain that has been eliminated.

Also, the alternative to going to local boundary conditions and local infinite elements is, of course, to look at nonlocal boundary conditions that are theoretically exact. I am referring here to the Direchlet-to-Neumann (DtN) boundary condition. It is nonlocal in this setting. So, if a direct solver were used, it would have disadvantages in storage costs. However, with some of the techniques for iterative solvers and some that Professor Pinsky presented—with matrix-free iterative solvers—one does not have to assemble a global matrix. There, some of the storage costs that arise from using nonlocal boundary conditions can be minimized or eliminated. So nonlocal boundary conditions require a very good approximation in order to go much higher up in the order of approximations that can be achieved.

HUGHES: I know that a number of acoustics techniques are already used in electromagnetics. Have infinite elements already been used also?

ZIOLKOWSKI: Yes, infinite elements have been used for a number of years.

HUGHES: It seems as though everything is used.

ZIOLKOWSKI: Yes, everything is used. What each technique is called for each community is going to be important with regard to interpretation. We in electromagnetics also call them infinite elements. My first project at Livermore was to do a global look-back scheme for the finite-difference time domain. It used Huygens' representation over a closed surface to provide the next point outside the surface in order to get an exact boundary condition. The problem with it is that, even though one can do it—in some sense—better and better with larger computers, one still has to back store for retarded time effects. It turned out it is not very efficient to do that. In a setting with extremely short pulses, one can get away with it. However, if one is working with very long pulses, then the back storage is a tremendous liability.

With regard to the techniques, something should be done to find out exactly what everyone knows and how well each approach works. For instance, we at the University of Arizona have difficulties with a lot of these conditions because we have evanescent wave problems. I do not know exactly what happens in applying boundary conditions for evanescent waves, but evanescent waves cancel out the use of many of the absorbing boundary conditions that people might try to use for certain problems.

PETER PINSKY: Concerning DtN versus infinite elements, the comment made by Leszek Demkowicz was a very sensible one in that there is an underlying relationship in the formulation between DtN and infinite elements. In fact, we have recently been performing numerical convergence studies at Stanford between these two types of boundary conditions, the local infinite element and the non-local DtN. We see numerically some very interesting asymptotic characteristics of these methods and, indeed, believe that fundamentally there are some similarities. In terms of the numerical implementation and the frequency domain for large-scale problems, as was pointed out by Lonny Thompson, maybe the differences between these things tend to diminish. I think they both represent very good opportunities.

I believe they can treat general complex waves, propagating waves, or evanescent waves. I see no difficulty in terms of absorbing complex wave numbers in these boundary conditions. The situation is very different, however, in the time domain. The issue of representation of boundary conditions in the time domain is an open question that deserves a lot of attention. To my knowledge, there is no notion of an infinite element in the time domain that is truly effective. One needs to create a class of methods and implement them in a way that provides a framework that will allow these boundary conditions to be placed close to the structure.

J. TINSLEY ODEN: In some work that I discussed yesterday on implementing Enquist-Majda-type local boundary conditions and nonabsorbing boundary conditions, these are locally applied. They work in the time domain, but a key feature in these considerations is the availability of an a posteriori error estimator that can be used to guide an adaptive process. If one has a robust error estimator, one can assess to what extent the boundary conditions are being correctly imposed. Such an estimation also allows decisions to be made on whether or not a very low order approximation at the boundary is sufficient, or whether one may go up to second-, third-, or in some cases even fourth-order approximations at the boundary.

HUGHES: Richard Ziolkowski raised the issue of gridless methods and their potential applicability in the electromagnetics area. Since Ted Belytschko is the most gridless person I know, I would like to ask him to make some remarks on the applicability of such methods.

TED BELYTSCHKO: The issue of gridless methods is something that still needs to be resolved. There are definitely computational penalties that are paid for using gridless methods in smooth problems, and consequently I am not sure that they will supplant methods that do have a grid. On the other hand, they do avoid the entire problem of meshing, which in three-dimensional problems or in problems with complex geometries is often an important consideration. Furthermore, one area where I see gridless methods as being particularly attractive is in adaptivity, because remeshing for adaptivity can generally be quite onerous; with gridless methods this burden is diminished considerably. However, I think the main application area of gridless methods is now in discontinuous problems. There, it is very easy to track fronts via gridless methods. That is exemplified to some degree in the dynamic fracture problems on which I have worked. There are also other problems that have clear fronts where gridless methods may be of considerable advantage.

LAKSHMAN TAMIL: Could somebody shed some light on the nonlinear problems encountered in acoustics?

IRA DYER: I would say the light to be shed is no light at all because acousticians always assume the problem to be linear, both in the fluid medium and in the structure. Some structures can behave nonlinearly in the structural acoustics problem, but we ignore that in favor of other things that seem currently to be of greater importance.

HUGHES: I am beginning to see some industry interest in nonlinear mechanisms in acoustics. One might model the acoustical field in a traditional way with a Helmholtz equation, but there are important sources of noise that are nonlinear, such as time-dependent turbulent phenomena and flows. There have been recent examples in which if you do not model that nonlinearity, you cannot possibly model the source of noise. Also, some nonlinear structural response is easy to incorporate in a general coupled acoustical structural problem. So some introduction of nonlinearity is beginning, although I do not believe it is yet widespread.

BELYTSCHKO: If people are going to come to grips with the interaction problem of structures, they will undoubtedly have to start looking at nonlinearities in structures. The more complex acoustical models may perhaps already exhibit some nonlinearities, and if one starts talking about a practical, real structure, one finds many joints and other, let us say, structure classes of that type that are inherently nonlinear. In order to truly assess their effect on the general behavior, nonlinear modeling will have to be included.

DYER: Indeed, nonlinearities ultimately will have to be included, but at this point they have not been. Many things in acoustics such as bells and gongs are basically structures that respond nonlinearly. They owe their musical qualities to the fact that they possess a nonlinear response. Although nobody bothered to find the differential equations in antiquity, nonetheless we are able to transmit the designs of bells and gongs. But modern-day acousticians seem to be focused on other problems. This is not to suggest that nonlinearities do not exist in structural systems, but acoustics, as it is usually defined, nowadays tends to ignore them.

TAMIL: Are inverse problems of interest to the acoustics community?

DYER: Half of the acoustics community solves inverse problems, and the other half solves direct problems. There are many inverse acousticians. For example, in a class of ocean acoustics called acoustical oceanography, sound waves in the ocean are used not to detect foreign objects, but rather to interact with the basic oceanic properties. Sound waves in the ocean have become a major oceanographic and geophysical measuring tool. For a long time, and still to this day, offshore exploration geophysicists have used acoustics to find oil. That is an old problem in acoustics, and there is an enormous field of activity in inversions.

ADRIANUS DE HOOP: One of the symposium goals that John Tucker put forward is to evaluate frequency domain versus time domain. As far as I know, one problem has been solved where the frequency-domain answer cannot predict the time-domain answer. This is the scalar wave scattering by an object of compact support. If one does the Neumann expansion of the relevant integral equation and looks for the convergence condition in the frequency domain, there is a combined convergence criterion in which the maximum contrast of the properties of the object with respect to its embedding occurs. One gets the square of the maximum diameter of the wavelength with another 2π, and so on; it must be less than one for convergence. Under that condition, the frequency-domain Neumann expansion of the integral equation converges. If you approach the same problem in the time domain, you end up with only a very simple restriction on maximum contrast. The maximum value of the absolute relative contrast must be less than one, and that is all. There is no condition on the size of the object or the convergence of the Neumann expansion of the time-domain integral equation.

I have not been able to generalize this to the basic, let us say, electromagnetic, electrodynamic problem because the bounds I need for the Green's function are not as simple as in the scalar wave scattering theory. On the other hand, someone here might know similar theorems, generalizations, or things of that kind.

ODEN: I do not know whether I can address the question in the particular framework you describe, but I made some very quick comments in my presentation that seem to be directly related to this feature. If one looks at certain frequency-domain formulations of the coupled problem, particularly with boundary element methods, finite element methods, where there is no damping—neither structural damping nor damping of any kind—then one can show that the governing operators are well behaved. They are strongly elliptic. Bilinear forms are strongly elliptic provided one is not close to resonant frequencies of the coupled system.

As the wave numbers get higher, these resonant frequencies stack up, and it becomes increasingly difficult to find a gap between the eigenvalues of the system and the natural frequencies in which a solution exists, and you experience this deterioration in stability and conditioning of the system numerically, as well as theoretically. A little damping changes the whole situation. If there is damping in the system, then at least from the theoretical point of view the associated sesquilinear forms are strongly elliptic, and one can infer that solutions exist.

If one takes an integral of these equations in time, and goes to the time-domain formulation, under suitable conditions on the initial data described in my presentation, one gets an abstract Cauchy problem with a self-adjoint operator on a complex Hilbert space that is completely well defined. There is no question of the existence of solutions, and then various methods of approximation essentially revert to ways of approximating the spectrum of that operator.

This recent observation would lend some weight to Professor de Hoop's comments.

ARTHUR JORDAN: A written question asks: "In what ways are methods that alternate implicit and explicit schemes to improve stability while maintaining efficiency analogous to those that alternate time-domain and frequency-domain methods for the same purposes? Can these parallels be used to further beneficial ends here?"

HUGHES: I would say, "In what ways, if any?" I am not sure they are related at all.

JORDAN: If the author of the question is here, maybe he could clarify it.

TUCKER: When I was speaking with Hermann Haus the other day, one thing he described was a way in which the time-domain and frequency-domain methods were alternated as a means of maintaining stability of the computational scheme, and it struck me that this parallels what was done about 15 years earlier to achieve stability without sacrificing efficiency for the solution of stiff systems of ordinary differential equations. Are there any other parallels that could beneficially improve what this symposium addresses?

THOMPSON: Implicit-explicit methods partition the domain into different regions where one uses different techniques depending on the character of the subdomains. I am not sure there is any parallel, but that brings up an interesting idea. Can one somehow use a time-dependent or time-domain technique for a certain domain and then use a frequency-domain technique for a different domain, and then use some transformation between the two to hook them up between different domains? I am dreaming, but perhaps something like that could be done.

ZIOLKOWSKI: At the University of Arizona, we actually do something like that; people are now looking at domain decomposition techniques to decompose the region in which the problem is to be solved so that in one particular region, say, one frequency-domain-type technique can be used, and one takes another domain in the same region and solves the problem there with some other frequency-domain technique, and provides the means of joining the regions together. That is something people have been working on. I do not know of anyone who has done it in one region in the time domain and in another region in the frequency domain, but if you can do one, you could probably do the other.

HUGHES: I am gratified that you did not say, "We have already done that. So, it is not Thompson's method." I would also like to mention that there are lots of techniques of domain decomposition in the structural and fluid area where one uses very different algorithms in different subdomains. That is commonly done nowadays in multiphysics problems.

HERMANN HAUS: To elaborate on that conversation with Dr. Tucker, my students solved a nonlinear Schrödinger equation. It turns out that if you stay solely in the time domain, the solution becomes unstable. So, if you go to a frequency-domain approach, there is a smoothing for the d^2/dt^2 operator in the nonlinear Schrödinger equation, and the instability is avoided.

DEMKOWICZ: I want to say a little more on the argument Tinsley Oden put forward. The essential difference between the transient- and frequency-domain formulations is that the solution in the transient domain is regular. It is very nice; it has no singularities in it. The original physical problem is stable. If energy is not pumped in, the energy actually stays constant. If stable schemes are used, both explicit and implicit, in principle it should be no problem to solve the transient formulation except, of course, for higher-order theories. With no explicit damping mechanism in the model then, from a purely formal mathematical point of view, when one takes the Laplace transform of the fully coupled problem, one ends up with a solution that has at least a single pulse at resonance frequencies.

Now, if one continues with an infinite time signal, those single pulses become double pulses. Furthermore, when it comes to the Fourier inverse transform of such a function involving double pulses, it is no surprise that some difficulties arise because one cannot numerically integrate a function even with a single pulse; actually, one can do it by using the Cauchy principal value integral sum, but it is not simple, and it is almost hopeless with a double-order pulse. That double pulse shows up in practical computations. Put simply, if one hits or gets too close to one of the numerical resonance frequencies, a double pulse shows up in that Fourier inverse transformed solution of the transient problem. However, this does not happen when one simulates the problem by treating it as a transient problem from the very beginning.

DYER: Those who work in structural acoustics never have any problems with damping because at a minimum, even if we are perhaps ignorant of what the energy loss mechanism is in the structure, we always have the energy carried to, let us say, infinity through the coupling to the external medium. So the practical answer to the question is that frequency-domain methods ought not to be difficult because there is always a way in which energy is pumped out of the system, and so out of the computation, that is, at the boundary conditions or at infinity. To the extent one does a computation that couples to an external fluid, there should be no problem. However, I do not perform these numerical calculations, and so I cannot be certain. Every problem that I know of has a coupling to an external fluid.

DEMKOWICZ: This was exactly the problem I was puzzled by when I started looking at dependence of Ladyzhenskaya-Babuška-Brezzi (LBB) constants on the wave number and, in particular, on the radiation damping. Of course, we really have only one problem in three dimensions with the exact solution (and please correct me if I am wrong): the sphere problem. For that particular problem I managed to evaluate the LBB constant that governs the stability. It turns out that, except for the first two or three eigenmodes, the LBB constant drops below 10 to roughly minus 8 for typical data for steel shelves sitting in water. For a computer, that means this is a full resonance. The computer does not see merely the radiation density. It simply cannot solve those problems.

DYER: Then one should treat a more realistic structure.

DEMKOWICZ: That is correct; stop playing with just the sphere.

PINSKY: Using domain-based methods to solve frequency-domain problems does not seem to introduce any difficulties provided that there is appropriate resolution of the wavelengths. We have experienced no difficulties for any kind of coupled problem. Indeed, the systems do experience damping in the form of radiation damping as was suggested. I am wondering if some of the problems that have been alluded to here do not really pertain to resonance. Are they simply manifestations of the lack of uniqueness in the integral representation that has been used in the boundary element method as opposed to real resonances that are somehow causing difficulty in the numerical solution?

EDWARD NEWMAN: In electromagnetics, the resonance corresponds to physical cavity modes. If the setting is scattering by a sphere, there is a cavity mode. When one is at a frequency where one of those cavity modes can exist, then depending on the integral formulation, the solution is nonunique.

DEMKOWICZ: First, I want to defend boundary integral formulations. As you know, some of them do have the defect of forbidden fictitious frequencies; some do not. The boundary obviously does not, and it is uniformly equivalent to the original formulation with respect to wave number. So, it is not a problem with the formulation. The entire problem of the existence of eigenvalues for the coupled problem of the scattering frequencies, as some people prefer to call them (people differ in the way they name these frequencies), is very much dependent on the geometry of the problem.

If the domain is concave, and you experience what I call a bay phenomenon, by which I mean a cavity, then even when solving the Helmholtz problem you do end up with a resonance. The corresponding eigenvalue, the scattering frequency, is going to have a small negative real contribution resulting in decay in time, some damping in time, but it is going to be very small; and the more concave the domain is, the more that particular resonance phenomenon may show up in the simulation.

JORDAN: The following written question was prompted by Tom Hughes' talk: "Does the poor estimation observed for the Galerkin least squares with adaptivity indicate a need for straightforward and more accurate upper bounds?"

HUGHES: First, what we observed was that the Galerkin least-squares (GLS) method on a mesh of an equal number of nodes compared with Galerkin was more accurate whether the solution is computed using a uniform approach or an adaptive one. So GLS is a more accurate method. On the other hand, the particular a posteriori error estimator that we derived from Galerkin least squares was indicating a larger error. So, my answer would be: Absolutely! One should endeavor to use a better error estimation to account for the improvement that is inherent in that method. Such a method would automatically fall into the category of implicit techniques that solve a local problem, because this explicit procedure simply has some shortcomings. It is an error estimation procedure based on very crude functional analysis in which inequalities start to concatenate, and every additional term that is actually doing something beneficial ends up appearing on the right-hand side. As error is currently estimated, the right-hand side, which is your bound, just piles up. Something that is more implicit, where the value of the method also appears in the left-hand side in the local problem, would definitely improve the error estimation.

PINSKY: To add a comment, we have been looking at that implicit error estimation through residual-based approaches for the scalar advection diffusion equations, steady state, in which case we have been using GLS to stabilize the basic discretization methodology. Indeed, the residual-based formulation does lend itself very naturally to stabilization of GLS-type methods even in the local problem for the estimation of the error. Therefore, I believe that one can, in fact, apply implicit-type error estimation methods directly to GLS-type formulations.

JORDAN: The next written question is: "What are the effects of stretched grid on transmissivity of acoustic waves through a medium? How does the grid affect amplitude and frequency of the signal? In predictions for finite difference methods, what are the effects?"

BELYTSCHKO: I do not believe we have studied the problem for stretched grids, but the transmissivity and the wave propagation characteristics through grids of both finite element and finite difference types have been studied for plane waves. I wrote a paper on that with Bob Mullin, and several other dispersion analyses have appeared. There are analogous results in the finite difference literature, too.

In general, these results seem to indicate that waves that are not traveling along directions of the grid lines suffer much more severe dispersion than waves that go along the grid lines. In fact, it is interesting that in several recent papers the authors have in essence tried to get dispersion-minimizing schemes by combining several finite difference schemes or finite element schemes. I believe someone has combined Galerkin least squares in order to minimize dispersion. There is considerable research in this area, and it is a fruitful area that does need further study.

PINSKY: We have been using the techniques of complex dispersion analysis to understand the dispersion that is introduced, for example, for a plane wave propagating obliquely to mesh lines. We have sought to enhance the accuracy of basic Galerkin methods by appending Galerkin least-squares terms and then designing these terms in such a way that the dispersion error is minimized in some rather general way to optimize the accuracy of the method of arbitrary wave directions. In fact, we have extended this to higher-order elements as well.

ZIOLKOWSKI: In the finite-difference time-domain community in electromagnetics, people have been trying to use stretched grids for many years and essentially the same phenomenon occurs. One has to be very careful about not proceeding too quickly so that artificial numerical reflections within the grid do not arise, and be careful of the phase front propagation errors that can result from dispersion errors. These have been similar experiences in the electromagnetics community.

THOMPSON: In a paper by, I believe, Zaunt, who did a series of papers looking at stretched elements, he had the domain spread into small elements and then he abruptly changed to a domain that had large elements, and thus addressed the effect of any numerical transmission or reflection that occurs there. He did find additional discretization there due to big changes in mesh sizing. So it is preferable to keep the mesh sizing relatively uniform or at least graded from a small to a large mesh.

ZIOLKOWSKI: A number of people on the electromagnetic side are interested in so-called hybrid techniques, in which discretized methods are coupled with other things. An example we saw was Newman's presentation of the method of moments coupled with the Geometrical Theory of Diffraction. I described approaches with FDTD, with integral formulations to do near-field, far-field types of things, and attempts to combine FDTD with ray asymptotics. There are a number of hybrid approaches, and for the last couple of years using hybrid approaches has seemed to be the big push in electromagnetics modeling. Yet in all the talks from the structural acoustics side, I did not hear about anyone trying hybrid approaches. Is that not done, or is it done in just this community, or is it something of a new interest?

PINSKY: First, I would suggest that the DtN formulation itself, the Direchlet-to-Neumann boundary condition, is in a sense an example, at least, of combining an exact analytic boundary condition within the framework of a numerical approximation. So DtN can be viewed as something of a step in a hybrid direction. We have thought about trying to extend some of the analytic solutions, asymptotic solutions that describe near-field phenomena, around discontinuities. Essentially, such an approach creates a kind of bubble around these fine features in which analytical solutions are suitably embedded in the bubble boundary. That extension is then coupled with more global techniques where the solution is varying in a smooth way. Some of the work done at Stanford by Joe Keller in the geometric theory of diffraction for underwater acoustics would certainly be applicable here. It again goes back to solving certain canonical problems for certain kinds of features.

We certainly do have a good framework now for pursuing this approach, but for some reason it has, to my knowledge, never really been pursued in a vigorous way in the context of underwater acoustics. It probably should be.

DYER: From a perspective in structural acoustics—underwater acoustics—there is an interest in hybrid methods, but I believe of a rather different kind than Peter has mentioned. One motivation for a kind of experimental research that MIT has been involved with is identifying processes rather than domains within which analytical techniques then might be coupled domain to domain. It was initially hoped that some people in the numerical computation community and the mathematics community would be motivated to find a way to combine processes that cut across domains—spatial domains, say, or time domains. To some degree, I am not discontented that I see no activity in that direction. However, there is a distinction between an analytical solution and a spatial domain that might be coupled to a set of processes.

LOUISE COUCHMAN: A number of areas in which hybrid solutions are being attempted happen not to be represented today. The problem is somewhat more difficult in acoustoelasticity than it is in electrodynamics because the elasticity has to be dealt with. Nevertheless, hybrid solutions are being attempted in which rays are traced on shelves and the elasticity due to the shelf is included in the formulation of the ray tracing; that is, coefficients of coupling of energy onto and off the shelf are computed from canonical solutions.

There are also examples of solutions in which discontinuities in structures are being treated using canonical solutions and coupled to a more global solution. Those are at least two instances of hybrid coupling-type approaches being studied.

I.C. MATHEWS (Imperial College): Every U.S. submarine for the last, say, 20 years has been designed for shock loading using, effectively, asymptotic-type numerical techniques and the USA codes. Almost everyone here must have come across that work, and it should also be represented. John DeRhams pushed it into the frequency domain, and I think that approach could be used for higher frequencies. Boundary integral techniques were very good because they were in the sort of midfrequency range with which people had trouble. To go to higher frequency, one would use some asymptotic technique; the point is that the scattering response at the higher frequencies is generally quite smooth. It is in the midfrequency range, where it is terribly peaky, that one wants the accuracy.

THOMPSON: Regarding the USA code and the use of the doubly asymptotic approximation, it too is a combining of numerical techniques. For the radiation boundary, an asymptotic-type boundary condition is used that is basically exact for very high frequencies. So at infinite frequency, where that boundary condition would be exact, the ultimate goal at the low end is also exact. However, the doubly asymptotic approximation presents difficulties for structural acoustics problems because it has a large gap in the midfrequency range, which Professor Mathews just suggested was where you really want to capture the accuracy. This is an area where one has to be careful when using asymptotic methods. One needs to be aware in what part of the frequency range is the accuracy valid for the methods.

MATHEWS: We have been using integral techniques, the Burton-Miller approach, for the last 10 years, and they do work. Within this frequency regime they give extremely accurate results for full three-dimensional structures and for realistic sorts of wave numbers.

DE HOOP: Let me return to the question of whether there are any similarities of combined techniques—hybrid techniques used in elastodynamics—that more or less parallel the ones in electromagnetics. In work done roughly 10 years ago, S.K. Datta of the University of Colorado performed matched asymptotic experiments. For a part of the regime he used analytic techniques of an asymptotic nature, and in the remaining part of the structure he used separation-of-variable techniques and other techniques. Everything went fine for a couple of situations.

HUGHES: Another example, in shell analysis rather than in acoustics, is work that Charles R. Steele at Stanford has done over the years involving asymptotic solutions around boundary conditions that are highly oscillatory in shell response, typically with fixed boundary conditions, and also around junctures in shells. He has, further, used asymptotic methods in dynamic wave propagation shell problems and combined some of this technology into what he calls a big-element finite element approach that is based on the asymptotic solution. He can model a complex intersecting shell such as that associated with the North Shore problem with a very small number of elements, because all of the critical asymptotic features are built into the basis functions of the elements. These codes are actually used quite extensively by designers in industry. He has made the comment that, due to the fine scale features of shells that appear when one does an exact or an asymptotic analysis, he has never seen a finite element mesh for a shells problem that was sufficiently resolved to actually calculate a good solution.

JORDAN: With regard to using various asymptotic methods for inverse scattering, in inverse scattering the ground rule is that one wants all the scattering information that can be obtained. People will start with volume scattering to get the rough volume of the unknown object, but then, to get finer detail, go to higher frequencies, physical optics. Those give all the fine details, all the cracks and corners and such, but the combination of the two gives the whole picture for the unknown object. This is all done in the frequency domain and so is an example of a hybrid combining technique in the frequency domain for inverse scattering.

Philip Abraham of the Office of Naval Research has submitted a list of fairly general questions that I have held off addressing until last because it provides a type of summary for this discussion. I would like Phil to pose these questions in some prioritized order that he thinks is most effective.

PHILIP ABRAHAM: Of the topics listed [see box], some have already been covered in the preceding discussion. One that has not been discussed is the physics underlying these problems, although some people have referred to it and stated differential equations for a particular system. Similarly the question regarding the desired end product has not been formally addressed. What do you want from a solution? Do you want the pressure or the radiation? Do you want the displacement of a structure, response of the structure, response to the fluid if it is a structure acoustic? What I mean is that the calculations or computations to be done depend entirely on what information is sought about a particular system.

Other issues to be addressed include how computational methods apply to each category of linear problems and nonlinear problems. There are common obstacles to accuracy and efficiency in the various methods. There is a challenge in dealing with real-time applications. If the context is surveillance problems, sonar, radar, and control problems for the Navy and Air Force, one needs to obtain a solution in real time. One cannot wait for the computational method to work in 6,000 hours. We just heard comments that there is damping in structures, but it may not influence the results.

In systems that are built up from many subsystems, there are various components to be considered. For example, in a microstructure, we have seen that one must faithfully model what is happening at the junction of the subsystems, and do so not just computationally. There is a prior setting in which that faithful modeling must be done; it has to be done experimentally and perhaps physically. One must attempt to model what is happening at the physical level, and then give this result to those who perform the computations. This experimental and fundamental modeling issue has not been discussed here. While it may not necessarily be part of this symposium's discussions, it is an important aspect.

What are the limitations of the computational tools? In the computational methods described at this symposium, are the present limitations due to hardware or software? Can these methods be adapted from one area to another? These are some of the topics on which further comments could be useful.

TAMIL: To address the first question, with regard to optics, the time-domain technique has the advantage of bringing out some of the physics. For example, when a nonlinear medium with a short pulse and with Raman scattering is analyzed, these aspects of the physics can be obtained out of the computation. However, extreme care should be taken not to mix up the physics with erroneous outcomes that are due to inexact computations. By doing nice filtering, the physics can definitely be brought out.

Concerning the last question, the methods can definitely be adapted from one area to another. Any equation that is normalized can be used irrespective of the originating area of science. As we have seen, most of the equations of acoustics and electromagnetics are of the same type so that the same methodology can be used. I am sure this generalization extends to other areas.

> **PHILIP ABRAHAM'S TOPICS FOR DISCUSSION**
>
> 1. Delineate *domains of applicability* of frequency and time domain, in terms of the
>
>   a. *Underlying Physics* (what is the desired end product),
>   b. *Linear vs. Nonlinear* problem under consideration,
>   c. Common problems of *Accuracy and Efficiency (cost)*, and
>   d. *Real-time applications*, such as surveillance and control, which impose different demands on the two computational methods (in all of their variants).
>
> 2. The *need for good, physically sound, descriptive models* of large-scale built-up systems; e.g., the actual modeling of damping resulting from connections between subsystems, and of prestresses and residual stresses present in a system, is a vitally needed input to any computational effort.
>
> 3. Present *limitations of computational tools:*
>
>   a. Hardware, and
>   b. Software, i.e., the computational methods presented.
>
> 4. *Cross-fertilization:* Can methods from one area of application be adapted to other areas, or can different variants of the same method be somehow joined (e.g., frequency-domain or time-domain method).

DE HOOP: Let me comment on the problem of built-up systems composed of subsystems for which, for instance, we have determined scattering properties and so on. One of the most fruitful methods is the no-field or T-matrix method in which, within a certain set of basis functions that are selected for a particular three-dimensional object of compact support, you essentially make a signature of scattering properties in terms of what I call the T matrix. In combining these with an adjacent part, in building up the structure—again, with its own signature—if you have used basic functions that for the most part have a common domain exterior to what is being built up, easy combinations are possible, because if there is an orthogonal basis one uses the orthogonality properties. The combined signature of two parts can be easily constructed by matrix multiplication, and then one can go on with the third.

None of these computations has singular integrals to evaluate as in standard boundary element techniques simply because the integrals that are evaluated are easily done. One either can take, in the no-field point-source case, solutions for which the source points are not in the domain of interest, or can take wave function expansions in the frequency domain, such as spherical waves or cylindrical waves or any other type of waves convenient to the problem. Based on what I have seen, this signature-combination approach is the only technique involving a specific strategy for combining properties in the acoustics or electrodynamics problems, and also electromagnetic problems, so that once the signatures of the separate elements have been built up, one can combine them in a systematic mathematical manner.

JORDAN: I have had a general question in mind during these two days. We have discussed acoustics, electromagnetics, and electrodynamics, and, let us say, the differential equations for each of these three disciplines. These differential equations have certain symmetry properties. Are these properties the same for each of the three types of equations? Are they all different? Are there some areas of common symmetries, and if there are common symmetries, can they be used to generate common solutions in these three areas? This is a very general mathematical question, but I hope it addresses the basic question that we have considered in this symposium.

DE HOOP: I pointed out in my presentation the general structure of all the differential equations we have considered [see equation 3.12]. It is a first-order coupled system of equations where the spatial differentiation appears as a square array of spatial differential operators. The time differentiation is a first-order convolution of the material properties of the wave function, with losses present. An interesting property, which was very surprising to me, is that this differential operator enfolds all the types of wave motions in a common structure. For the acoustic case, I have shown that its shape is that of a square array where, at the main diagonal, it is block zero. The elastodynamic case and the electromagnetic case have exactly the same form. In that respect all the wave motions that we know of in physics have this structure. This differential operator satisfies one additional property that leads automatically to what in my presentation I referred to as the reciprocity theorem. It surprised me that all these wave motions had this common structure, apart from reciprocity. One other thing: the acoustic waves in porous media also have the same structure. It is a more extensive array, but again, it has this same structure. So if one seeks either general mathematical theorems or particular numerical applications, this is one thing on which to focus.

HUGHES: To make an addendum to those observations, that structure is the structure of a symmetric hyperbolic system. Thus, all existence and uniqueness are governed by Friedrich's theory, for example, and in addition there are theorems due to Mach and Gudanov that point out the equivalence of a symmetric hyperbolic set of equations and the existence of an entropy function. So there is a convex function associated with each of these theories that in the linear and nonlinear cases takes on the interpretation of an entropy. It is the basic quantity one would look at not only when examining the nonlinear stability of nonlinear systems such as these, but for particular linear systems as well. They share many features, but the possibilities in structural mechanics are a bit greater. First, the tensorial nature of the theory is one level higher. There is the possibility of anisotropy that is somewhat precluded by the scalar nature of the acoustical problem.

DE HOOP: For the elastodynamic formulation of these equations, inhomogeneity and anisotropy are present.

HUGHES: Yes, one maintains the symmetry, but as to techniques to solve problems, there are decomposition theorems for at least the isotropic case where one can do some classical analysis by breaking up the elastodynamic problem into a set of scalar wave equations. We do not know whether one can do that for the anisotropic case.

DE HOOP: It is not only isotropy that counts, because the breaking up into waves applies only to homogeneous domains. On homogeneous and isotropic media, of course, one again goes to the eigenvalue. I was puzzled a little by the fact that the hyperbolic system generally does not require only the symmetry of that spatial differentiation to operate. It does not require the block of zeroes at the diagonal. But for any wave propagation problem, starting from the physics, it puzzled me why those blocks of zeroes occur on the diagonal. In a general hyperbolic system one would expect this matrix to be symmetrical and the like, but that is not indicated for wave propagation in general. Only for these differential equations for physical wave propagation examples in acoustics, elastodynamics, and electromagnetics do I get these block zero diagonal terms. That is not necessarily so for general wave point systems. So not every hyperbolic system is a wave motion. Each wave motion is a hyperbolic system, but it is not a one-to-one mapping, because there is an extra feature in wave propagation that, let us say, is a restriction on the total amount of operators possible in general hyperbolic systems. With a more restrictive operating context, one can prove more than can be proved for general hyperbolic systems. Of course, one can borrow the results from Cauchy problem stability, and so on, but it is even more puzzling then, knowing that there is this main diagonal block zero structure.

HUGHES: If you look, though, at a theory such as compressible flows, say for a perfect gas, and put it in a quasi-linear form where these matrix operators could be examined, I do not believe they would be diagonal.

DE HOOP: It still leaves open the mathematical problem: Is a subclass of hyperbolic systems included in this framework? When there are zeroes on the main diagonal, does it offer nicer, better, more extensive, or whatever properties than the general hyperbolic systems have?

HUGHES: I am unsure what the significance of that is.

DE HOOP: I do not know either. But the appearance of this common form suggests underlying structure and relationships.

PANEL DISCUSSION

TUCKER: There have been many thought-provoking observations in the open discussion. To now close the symposium, Ted Belytschko, Adrian de Hoop, Tinsley Oden, Peter Pinsky, Lakshman Tamil, and Richard Ziolkowski have kindly agreed to form a discussion panel. They will attempt to synthesize the presentations and the discussions into a summary that captures the most important points, issues, and ideas and indicates what are the best directions in which to focus research, and what approaches or ideas are the ones that merit emphasis for the future.

ODEN: No matter how objective one attempts to be, in the final analysis one sees the world through one's own, perhaps tinted, glasses. Consequently, when asked to identify areas that are important for future research, one cannot help but identify some of those areas that are specifically of interest to oneself. With that admitted bias, I hope the rest of the panel will help add balance to what I say.

In the lectures, I saw a number of intriguing and interesting activities under way that will ultimately have a significant impact on how large-scale structures are addressed in acoustics and electromagnetics. One area was that of adaptivity. I lump adaptive methods into a very large collection of techniques I call "smart" algorithms. What that means is that one attempts to endow the algorithm with some decision-making capability, based on some future aspect of the solution that is calculable, and which has some bearing on the quality of the solution. Generally that aspect is the numerical error.

Only a decade ago it was a farfetched idea that one could actually calculate with any degree of confidence some estimate of error. Now, methods are emerging that lead one to believe otherwise, that one can in fact estimate the error, and on the basis of that estimated error, control the computational process. We have seen several examples in this symposium indicating that. Indeed, this is a viable approach to large-scale computation. I will warn those who are novitiates to adaptivity in error estimation that this is not necessarily a straightforward area of technology. Error estimates, at least the good ones, are expensive. If one develops a successful adaptive scheme based on parallel computing, multiprocess computing, and domain decomposition techniques, be prepared to include the error estimation package as part of the package that must be parallelized.

Hand in hand with error estimation and adaptive methods is the ability to control data at the boundary. If nonabsorbing, nonreflecting boundary conditions are going to be imposed, there are viable techniques described in these presentations. More work needs to be done on these methods, but the availability of error estimation and adaptivity will influence how well those boundary conditions are imposed.

During the open discussion the word "damping" was mentioned. Every physical system has damping. I believe that good *models* of acoustics and electromagnetism should also include damping, and that the models of damping should arise from the physics of the problem at hand, and not be included merely as a numerical artifact to control the stability of the method. That means research needs to be done on physical mechanisms that give rise to disipation in a system, that is, material damping, structural damping, and so on. Structural damping is a very complex phenomenon. It is manifested by frictional effects, for example, and contact conditions in structural systems. Viscoelastic properties of materials give rise to material damping. I believe that a systematic look at physical mechanisms that produce damping is needed, and that these damping mechanisms need to be incorporated in successful models.

Finally, the behavior of all of these numerical techniques must be better understood, for frequency-domain approaches on mesh parameters and on frequencies and wave numbers. The interplay needs to be understood between mesh parameters and frequency wave number content and solution. These behaviors can have a dramatic influence not only on the rates of convergence of methods but also on their stability.

BELYTSCHKO: I agree wholeheartedly with Tinsley Oden's comments. The only area that seems rather underrepresented is that of experimental acoustics. Professor Dyer stated fairly categorically that acousticians do not believe in anything that is not linear. Here is an area where one could perhaps find significant payoffs if one were trying to match experiments. In structural dynamics, one finds many small components that have nonlinear behavior. Friction and the associated damping, to which Tinsley alluded, are a consequence of that. In joints, found in real structures, and in connections, one quickly finds nonlinear behavior. If one is trying to model experiments on the basis of first principles, this nonlinear behavior must be treated from those first principles.

This issue will open a tremendous amount of needed research because the entire area of acoustics has in a sense been built on a technology that is basically linear in character. How to study scattering when one has nonlinear components is an open question. However, if one is going to deal with real structures, experiments will be very difficult to match. This was already indicated in Professor Dyer's comment that nobody wants to deal with a submarine model that has some internal structure. When one deals with a real submarine, one has very complex internal structure. That problem should be investigated.

PINSKY: Echoing Tinsley Oden, we tend to concentrate on those parts of the problem with which we are immediately concerned. In that way, my comments pertain to computation. One outcome of this symposium is recognition of the potential significance of a confluence of ideas. Concerning the numerical treatment of the acoustic fluid and the electromagnetic problem, this is true, for example, in discretization techniques and in radiation boundary conditions. One essential difference that needs to be addressed, with regard to the acoustics problem versus the electromagnetics problem, is the feedback mechanism, the fluid loading on the structure.

Many of us would probably agree that the numerical approximation in the acoustic medium is quite good. We have heard reports in this meeting, and indeed in many other meetings, of the lack of ability of computational techniques to treat truly complex structures in the midfrequency range. Some blame for this has been placed on the representation not of the fluid, but of the structure. Though many of us have been working in the area of structural dynamics and have considerable experience in that, it may still be worthwhile to revisit the issue of thoroughly understanding the nature of the structural representation for the coupled acoustics problem, and completely understanding the role of the coupling.

At Stanford, we have attempted to do some of this when looking at the effects of point-loaded cylindrical shells in the context of finite elements, and making an exact representation of the pressure loading on a cylindrical shell. We are analyzing the ability of the finite element discretization to represent propagating waves emanating from the point drive as well as the decaying evanescent waves. We found that the finite element representation for that particular class of problems worked surprisingly well.

However, for the structural acoustics problem where interaction is a major consideration, attention may need be paid again to how one captures the dynamic properties of the shell as it is coupled to the fluid, and it may be necessary to truly understand the role of the near fields and the effects of discontinuities in the complete coupled problem.

The other area that is certainly going to be fruitful, and in which enormous progress is already visible, is error estimation. For example, the work of Tinsley Oden and others in this field displays a clear trend that should be exploited to the greatest possible degree, both in electromagnetics and acoustics. I concur with Tinsley that this needs to be done in the context of parallel methods. There is additional cost to be borne in the cost of error estimation, and indeed the adaptive procedure itself creates additional difficulties in the solution of the problem. Therefore, one should be taking a systematic approach to the problem of developing error estimation for the fully coupled situation. However, the approach should also take into account the nature of the matrix problem that arises from this adaptive procedure, and should aim for tailored global solution methods that are somehow optimized for this class of formulations.

As a final remark, this meeting has been fruitful in confirming my suspicion that much of the work in electromagnetics is, in fact, directly applicable to computational structural acoustics. I would certainly encourage these two communities to continue to exchange ideas.

ZIOLKOWSKI: There are a number of areas that I believe would benefit from more work. In electromagnetics, an obvious need, considering the direction in which computing is going, is to move our co-development efforts into the parallel arena. That move is beginning, and it is also clearly now occurring in the acoustics area, as we observe so many overlaps in algorithms, strategies, and so on. There are many areas where electromagnetics could benefit, from both ends of the spectrum, especially with regard to parallel processing because of the difficulties involved with moving codes into parallel environments. If there are common numerical approaches in acoustics and electromagnetics, we should definitely be working together to minimize the amount of effort devoted to converting codes, as well as to rediscovering the various wheels that we travel on in the parallel environment.

From the physics standpoint, and particularly as the physics is driven to shorter time scales and smaller distance scales, one of the major concerns for electromagnetics is to develop better material models that are more completely integrated into Maxwell's equations. My presentation showed some examples of what we at Clemson are trying to do. The nature of coupling microscopic effects with macroscopic effects will come much more radically into play in commercialization, for example. Everybody would like to have things in smaller packages and still be able to behave in the fully macroscopic manner. To model those integrated systems effectively, we have to couple together and integrate better material models with the numerical approaches.

One direction that the electromagnetics community has decided to pursue is that of hybrid approaches. A number of different hybrid approaches need to be addressed further, including coupling the discrete techniques with integral techniques and with asymptotic techniques, and doing so in an effective manner for the particular types of problems of interest as well as for the different computer architectures that are coming on line.

Last but not least, we saw in this symposium efforts to couple basic device modeling with basic system modeling. As we move into larger integrated packages of different systems, we will have to address not merely how to model devices, but also how to scale up—how to put devices into basic system packages, and put those systems together into a larger complex system that works. As one moves up the scale from devices to integrated packages, there will be a number of issues; it is a very different kind of hybridization. It is not simply a wave propagation. A serious systems integration issue will be how to take modeling from one regime and couple it effectively with modeling in other regimes. That is another important issue on which future work will be needed.

TAMIL: Complex systems will be made up of subsystems of components. Once the components and subsystems are characterized and put together to form the complex system, there will be a perturbation to the overall specification of the desired complex system. So the systematic study of such perturbations will be a part of putting subsystems or components together. This is very close to what Professor Ziolkowski just mentioned. Methods will also need to be found for isolating the perturbations between subsystems. If that can be done, then building complex systems out of individual components will be easy.

In electronics, there are very sophisticated packages for very large scale integration; it is time that the integrated optics community developed one also. People may be doing that, or maybe a package for optics integration exists but is not on the open market. Nevertheless, it is very important to have a means whereby one can design and synthesize an integrated component, and then get it fabricated. If this can be done for electronics, then surely the field is mature enough to do at least a small type of integration in optics.

We saw commonality between acoustics and electromagnetics. I am sure that a similar commonality exists also among other branches of science. We need somebody like the Russian who first compiled the integration tables to compile all these equations—in a normalized form—that, say, identifies what the important parameters are. If one could also learn what are the most efficient codes available for what kinds of structures, that would be very helpful. It is another means to avoid energy being wasted in repeating the same inquiries. It would help inform people in acoustics about what the folks in electromagnetics have already done. And if tomorrow the people in biology need such knowledge, they need not redo it. For example, the chaos studied in biology has also been studied by many people in other areas. Whatever commonality there is from the equations should be compiled.

As to modeling for nonlinear systems, if the system is integrable, then deriving conservation laws to check the accuracy of a numerical computation is easy. However, for nonintegrable systems, which includes most of the realistic systems, how would one check the results of numerical computation? In certain cases, one closes one's eyes and extrapolates from the integrable systems. Is there a mathematically justifiable method of doing it?

Lastly, many people are pursuing the study of chaos and instability. To increase the rate at which we communicate, the rate of transmitting information optically, may require looking into chaos and instabilities. One of the necessary conditions for chaos is nonlinearity, but I do not know whether that is a sufficient condition.

DE HOOP: I have very little to add to all these magnificent remarks and suggestions and will merely focus on what I call quasi-analytical issues that are promising subjects of further research. The first, mentioned by Professor Oden, is the element of damping. Damping is present in any physical system, and it needs to be modeled in the computational scheme in accordance with the physics. In the frequency domain, this means essentially that the real and imaginary parts of the constitutive coefficient should obey the Kramers-Konig causality relations. In the time domain, they should be of the convolution type, at least within the linear framework of the Boltzmann type.

This may seem a completely superfluous remark for anybody who has studied elementary physics. However, my experience is with the inclusion of damping phenomena in seismic prospecting. Roughly 90 percent of all seismic prospecting papers interpret particular experimental results and try to match the results to a frequency dependence that can be easily shown to be noncausal. The better an author succeeds in matching the experimental results to that kind of frequency behavior, the better the damping phenomenon is believed to have been covered. It is essential that damping be included in the model, but it has to be done carefully.

Another promising subject is that of no-field methods as a possibility for modeling acoustic, elastodynamic, and electromagnetic wave fields. This was a very popular approach some 10 years ago, but since then, it has completely died out. Almost all of the papers that have been published, mainly in the *Journal of the Acoustical Society of America* and on related areas such as wave motion, deal exclusively with the application of frequency-domain methods.

I believe, but am not 100 percent sure, that one can also set up the no-fields approach in the time domain. As I mentioned in earlier discussion, it is an ideal method for putting things together from elementary building blocks. You can determine the T matrix, the relevant signature matrix of that object, in a fairly elementary way and then put things together in a manner that is not a perturbation technique but is exact. Being able to combine things exactly would be very important, and so developing this no-fields time-domain method presents a very interesting research opportunity.

The third subject with promise involves looking at the complex frequency domain for positive real values of the complex frequency. One takes the time Laplace transform for any causal function and works with the positive real values of S to solve the problem. On the one hand, one wants to return to the time domain, for which one uses specific algorithms that have been developed in certain aspects of physics (mostly of the acoustic diffusive type used in rock fracturing for, say, geophysical reservoir enhancement). On the other hand, some particular aspect may be sought that can be considered independent of S. An example is in inverse problems where reconstruction of a particular profile of a constitutive parameter that is independent of frequency is desired, for instance, the mass density in the interior of the earth. If damping is neglected in this setting, then of course, the whole analysis and signal processing can be done for positive real values of S, because the quantity you are seeking has nothing to do with S. So, one can approach things in the frequency domain, in the time domain, and in the S domain and ask what is the cheapest, and best, way to do it. There are indications at the moment that using the positive real values of S gives the cheapest way, and if done in a clever way, one can easily reconstruct those profiles.

The first thing you have to specify is what are you after. First tell me: What do you really want? In doing university research for industrial companies, the difficulty one usually encounters is that the company tries to think for the researcher. The only role the company should play is providing support; let the researcher do the thinking. What may happen is that the company formulates the problem, and the researcher says, "I can do that." A contract is set, and the researcher obtains the required result and writes an invoice. Then the company says, "Oh, now we see we actually would have preferred . . . ," and it really wants something else. Thus the first thing to dig out is what the customer is after. It can take much time and involve considerable difficulty to fully get the flavor of what they really want. Once that is identified as something that can be determined, then accuracy becomes an issue. The usual university perspective is that if one has accuracy to three decimal places, six would be much nicer but one must then beware of round-off errors, and so on. The company's message might be that a factor of two is not important; that is, if the result is correct to within a factor of two, it is good enough.

Given that, one has to rethink everything one has ever learned about computational modeling. One's thinking must be recalibrated so that only being within a factor of two is important, and nothing else. That means the speed of the computational scheme for that particular application can be tremendously increased, because there is no interest in obtaining so many decimal places. One still has to be very accurate, but only under the constraint that the result is accurate up to a factor of two. So the first thing one must find out in all these computational techniques is what one is after.

That brings me to my final remark. One goes to a customer and says, "Do you want to compute something?" If the customer answers, "Yes," one then says, "Okay, tell me your accuracy requirements; specify the domain in space where you want accurate results, and specify how accurate they should be." Then one devises an error criterion that satisfies certain conditions (for instance, positive definiteness). The ideal situation occurs when a computational scheme automatically comes out from the criterion and conditions, an iterative scheme that the computer executes by minimizing the errors after each step in an iterative procedure. I believe this would be a very nice future direction to pursue. First, an error criterion is specified, usually in L^2 norm, but more general ones are possible. Then one shows that each iterative step decreases the error. One has no guarantee of the decrease to zero in a general system, but in an algebraic system of 10,000 equations with 10,000 unknowns, we have had accuracy to three decimal points after three iterations, whereas in an ordinary method that would be out of the question. This was done on a PC; the program was begun on a Friday afternoon, and on Monday morning with the last cup of coffee at nine o'clock, everything was finished.

ABRAHAM: When calculating wave propagation via time domain, one can focus on the specific end of the structure of concern in the calculation. The rest of the structure does not need to be addressed except up to the point where the wave is propagating. In the frequency domain, on the other hand, when one frequency is sought, it must be calculated for the whole structure. One cannot do just one part of the structure. These two things are separate. People say that they want to calculate in the frequency domain because they need to know about one particular frequency. That is fine, but when one has to calculate something in time, this is where the time dimension is useful and should be the preferred method. When computing in the time domain, one determines how long one must compute in time. It is not necessary to go to infinity. Perhaps when computing, for instance, in the frequency domain, one may want to go back to the time domain. It would be necessary to take a Fourier transform and then the time domain is an infinite domain. There are thus advantages in the time domain that do not exist in the frequency domain, yet some areas of frequency domain may be more useful.

DE HOOP: Perhaps I can comment on that from experience with geophysical prospecting. Relating it to one of your remarks, first tell me what you are after. In geophysical prospecting, for example, one always has an inverse problem. One is trying to characterize the structure of the earth by doing acoustic electromagnetic or elastodynamic measurements. It turns out that oil is always beneath a discontinuity in surface material properties. So the first thing one must determine is where the discontinuities are. Those you detect by arrival times of reflected waves, which means that part of the analysis can be done better in the time domain. You are interested only in whether the oil is 7 kilometers deep or 5 kilometers deep. It turns out that once one has decided to drill a hole and has taken fairly detailed measurements in it, one wants to look a little bit sideways. There one runs into trouble because the travel times across, say, a layer 5 millimeters thick with an observational time window of a few milliseconds produce almost an infinity of wiggles. If one looks only at the wiggles, the interpretation of what is just a little outside that hole will miss quite a lot of detail. The fine details in the wiggles tell, for instance, if something is rather flat or has another bend or is perhaps a small but sharp aspect. So the standard practice is to take a Fourier transform, after which, fortunately, doing many of these fairly detailed things becomes much easier. In the picture, details show up more easily in the frequency spectrum than in the time domain.

So, again, the major question is what is one after, and the kind of approach one intends to use. If the object of interest is an arrival time of a wave, in the time domain one can see when it is coming. It can be traced; the philosophy of how deep it is is known. But if details get washed out because of high oscillations when it is the tiny details that are sought, then in many instances the frequency domain is the way to go. Then, of course, the computational scheme has to be tailored to what is being sought.

ZIOLKOWSKI: In electromagnetics, people have been working for many years in the frequency domain. In fact, computational electromagnetics started in the frequency domain, particularly because people were first interested in antennas. Once that interest was satisfied, then the methods-of-moments calculations started in the 1960s, with the numerical electromagnetics, called NEC, as the first large code for calculating. As we have moved up in frequency into the resonance regime, there are areas with multiple resonances present, and so people have had to move into using the time domain. If one is doing nonlinear problems, there is no question. As one moves into different regimes, as was pointed out, most measurements are done with pulses. They are finite windowed sinusoids, but one never really has a true continuous wave situation. One always has some pulse effects present in a system. If the intent is to compare with experiments—something we all should eventually be trying to do—then calculating solutions in the time domain can actually provide some information that is very useful.

ODEN: I do not see any other comments. On behalf of the audience and the panelists, I want to give a word of thanks first to Phil Abraham for having conceived the idea of holding this symposium, to Louise Couchman for providing the resources of her office toward making this event happen, and to John Tucker of the National Research Council for his excellent hospitality and organization.

TUCKER: And I want to express my thanks to all of you for coming. Without you, this event could not happen and the benefit that it provides would be lost. I feel sure the research directions described in these presentations and discussions will bear fruit for these scientific areas, and that this information will be viewed by others as very interesting and provocative with regard to research opportunities.

Appendices

Appendix A
Symposium Agenda

MONDAY, SEPTEMBER 26, 1994

8:30 AM   **WELCOME AND OPENING REMARKS**
Fred E. Saalfeld
Deputy Chief of Naval Research and Technical Director
Office of Naval Research

8:45 AM   **INTRODUCTION**
John R. Tucker
Director, Board on Mathematical Sciences, National Research Council

8:55 AM   **HIGH-ORDER, MULTILEVEL, ADAPTIVE TIME-DOMAIN METHODS FOR STRUCTURAL ACOUSTICS SIMULATIONS**
J. Tinsley Oden
Texas Institute for Computational and Applied Mathematics
University of Texas at Austin

9:45 AM   **DISTRIBUTED FEEDBACK RESONATORS**
Hermann A. Haus
Department of Electrical Engineering
Massachusetts Institute of Technology

10:35 AM   **ACOUSTICS, ELASTODYNAMICS, AND ELECTROMAGNETIC WAVEFIELD COMPUTATION—A STRUCTURED APPROACH BASED ON RECIPROCITY**
Adrianus T. de Hoop
Faculty of Electrical Engineering
Delft University of Technology

11:25 AM   **NUMERICAL MODELING OF THE INTERACTIONS OF ULTRAFAST OPTICAL PULSES WITH NONRESONANT AND RESONANT MATERIALS AND STRUCTURES**
Richard W. Ziolkowski
Department of Electrical and Computer Engineering
University of Arizona

1:45 PM   **ACOUSTIC SCATTERING FROM COMPLICATED SUBMERGED STRUCTURES**
Ira Dyer
Department of Ocean Engineering
Massachusetts Institute of Technology

2:35 PM   **ADVANCES IN TIME-DOMAIN COMPUTATIONAL ELECTROMAGNETICS USING STRUCTURED/UNSTRUCTURED FORMULATIONS AND MASSIVELY PARALLEL ARCHITECTURES**
Vijaya Shankar
Director, Computational Sciences
Rockwell International Science Center

3:25 PM   **ADAPTIVE FINITE ELEMENT METHODS FOR THE HELMHOLTZ EQUATION IN EXTERIOR DOMAINS**
Thomas J.R. Hughes and James R. Stewart
Division of Applied Mechanics
Stanford University

4:15 PM   **MODELING OF OPTICALLY "ASSISTED" PHASED ARRAY RADAR**
Alan Rolf Mickelson
Department of Electrical Engineering
University of Colorado at Boulder

5:05 PM   **ADJOURN**

TUESDAY, SEPTEMBER 27, 1994

8:30 AM   **MULTI-TIME STEP EXPLICIT SCHEMES FOR TIME INTEGRATION IN STRUCTURAL DYNAMICS**
Ted Belytschko
Department of Mechanical Engineering
Northwestern University

9:20 AM   **SYNTHESIS AND ANALYSIS OF LARGE-SCALE INTEGRATED PHOTONIC DEVICES AND CIRCUITS**
Lakshman S. Tamil
Electrical Engineering Program and Center for Applied Optics
University of Texas at Dallas

10:10 AM   **DESIGN AND ANALYSIS OF FINITE ELEMENT METHODS FOR TRANSIENT AND TIME-HARMONIC STRUCTURAL ACOUSTICS**
Peter M. Pinsky and Lonny L. Thompson
Department of Civil Engineering
Stanford University

11:00 AM   **AN OVERVIEW OF THE APPLICATION OF THE METHOD OF MOMENTS TO LARGE BODIES IN ELECTROMAGNETICS**
Edward H. Newman
Department of Electrical Engineering
Ohio State University

1:15 PM   **MODERATED OPEN DISCUSSION**
Moderators: **Arthur Jordan**
Naval Research Laboratory

Thomas J.R. Hughes
Stanford University

3:00 PM   **PANEL DISCUSSION**
Ted Belytschko, Adrianus T. de Hoop, J. Tinsley Oden, Peter M. Pinsky, Lakshman S. Tamil, Richard W. Ziolkowski

5:00 PM   **ADJOURN**

Appendix B
Speakers

Ted Belytschko, Walter P. Murphy Professor of Computational Mechanics at Northwestern University, received a Ph.D. from I.I.T. in 1968 and taught at the University of Illinois at Chicago before joining Northwestern University in 1977. He is the author of more than 300 works on a wide variety of applied mechanics problems, with emphasis on the application of numerical techniques and finite element methods. He is the editor of seven books, including *Computational Methods for Transient Analysis* (with T.J.R. Hughes). He is the editor of the journals *Nuclear Engineering and Design* and *Engineering with Computers*. He has received the Pi Tau Sigma Gold Medal from the American Society of Mechanical Engineers, the Walter L. Huber Research Prize from the American Society of Civil Engineers, the Thomas Jaeger Prize from the International Association for Structural Mechanics in Reactor Technology, the Computational Mechanics Prize of the Japanese Society of Mechanical Engineers, and the ASCE Aerospace Structures and Materials Award. He is a fellow of the American Society of Mechanical Engineers and the American Academy of Mechanics, past chairman of the Engineering Mechanics Division of the American Society of Civil Engineers, and past chairman of the Applied Mechanics Division of the American Society of Mechanical Engineers. He was elected to the National Academy of Engineering in 1992.

Ira Dyer, Weber-Shaughness Professor and professor of ocean engineering in the Department of Ocean Engineering at the Massachusetts Institute of Technology, specializes in ocean acoustics. He has focused much of his acoustics research on ambient noise, fluctuations in the transmission of sound, scattering from topographic features in the ocean, scattering from submerged shells with internal complexities, and dynamics of truss-like structures. He has participated in and/or led acoustic tests for nine field experiments, and many more laboratory studies. Dr. Dyer has sought understanding of and applications for acoustic topographic mapping, acoustic radiation from fracture of oceanic ice, and acoustic scattering as related to submarine design. His findings have been published in numerous technical publications. Professor Dyer earned a S.B. (1949), S.M.(1951), and Ph.D. (1954) from MIT, and is a member of the National Academy of Engineering as well as a fellow of several professional societies.

Adrianus Teunis de Hoop was born in Rotterdam, the Netherlands. He received a M.Sc. degree in electrical engineering in 1950 and a Ph.D. in 1958 from Delft University of Technology, Delft, the Netherlands. He was a research assistant at Delft University of Technology from 1950 to 1952, assistant professor of electromagnetic theory from 1953 to 1957, associate professor of electromagnetic theory from 1957 to 1960, and professor of electromagnetic theory and applied mathematics from 1960 to 1992. Since 1992 he has held a position as emeritus professor. He has spent sabbatical leaves with the Institute of Geophysics, University of California at Los Angeles, and with the Philips Research Laboratories, Eindhoven, the Netherlands. On a regular basis he is a visiting scientist with Schlumberger-Doll Research, Ridgefield, Connecticut, and with Schlumberger Cambridge Research, Cambridge, England. He received an honorary doctoral degree in applied sciences from the State University of Ghent, Belgium, and the Gold Research Medal from the Royal Institution of Engineers in the Netherlands, and he was the recipient of awards from the Stichting Fund for Science, Technology and Research (a companion organization to the Schlumberger Foundation in the United States) in 1986, 1989, 1990, 1993, and 1994. He is a member of the Royal Netherlands Academy of Arts and Sciences. His main research area is the interdisciplinary approach to the theory of the excitation, propagation, and scattering of acoustic, elastic and electromagnetic waves, with emphasis on their technological applications. He is a member of the Acoustical Society of America, the Society of Exploration Geophysicists, the European Association of Exploration Geophysicists, and the Royal Institution of Engineers in the Netherlands.

Hermann A. Haus was born in Ljubljana, Slovenia. He attended the Technische Hochschule, Graz, and the Technische Hochschule, Vienna, Austria. He received a B.Sc. degree from Union College, Schenectady, New York, a M.S. degree from Rensselaer Polytechnic Institute, and a Sc.D. degree from the Massachusetts Institute of Technology. He joined the faculty of electrical engineering at MIT in 1954, where he is an institute professor. He is engaged in research in electromagnetic theory and lasers. He is the author or co-author of five books and more than 200 journal articles. Dr. Haus is a member of Sigma Xi, Eta Kappa Nu, Tau Beta Pi, the American Physical Society, the National Academy of Engineering, and the National Academy of Sciences, and a fellow of the American Academy of Arts and Sciences. He is the recipient of the 1984 Award of the IEEE Quantum Electronics and Applications Society, the 1987 Charles Hard Townes Prize of the Optical Society of America, and the 1991 IEEE Education Medal of the Institute of Electrical and Electronics Engineers. He holds honorary doctoral degrees from Union College, the Technical University of Vienna, Austria, and the University of Ghent, Belgium.

Thomas J.R. Hughes holds B.S. and M.S. degrees in mechanical engineering from the Pratt Institute and a M.S. in mathematics and a Ph.D. in engineering science from the University of California at Berkeley. He began his career as a mechanical design engineer at Grumman Aerospace, subsequently joining General Dynamics as a research and development engineer. Upon graduation from U.C.-Berkeley, he received the Bernard Friedman Memorial Award in Applied Mechanics. After receiving his Ph.D., he joined the Berkeley faculty, eventually moving to the California Institute of Technology and then to Stanford University. At Stanford, he has served as chairman of the Division of Applied Mechanics and chairman of the Department of Mechanical Engineering. He is currently professor of mechanical engineering at Stanford. He is a fellow of the American Academy of Mechanics and the American Society of Mechanical Engineers, coeditor of the international journal *Computer Methods in Applied Mechanics and Engineering*, and past president of the U.S. Association for Computational Mechanics. He is a founder and executive council member of the International Association of Computational Mechanics, and founder and chairman of Centric Engineering Systems, Inc. Dr. Hughes has been a leading figure in the development of the field of computational mechanics. He has published more than 200 works on computational methods in solid, structural, and fluid mechanics and is one of the most widely cited authors in the field. He is the author or editor of 14 books, including the frequently cited text *The Finite Element Method*. His research has included many pioneering studies of basic theory as well as diverse applications to practical problems. He received the Walter L. Huber Civil Engineering Research Prize in 1978, the Melville Medal in 1979, and the 1993 Computational Mechanics Award of the Japan Society of Mechanical Engineers for his research contributions and service to the field of computational mechanics. His seminal studies on contact-impact, plate and shell elements, time integration procedures, incompressible media, algorithms for inelastic materials, nonlinear solution strategies, iterative equation solvers, parallel computing, and finite elements for fluids have had a major impact on the development of software used throughout the world today.

Alan Rolf Mickelson received a B.S.E.E. degree from the University of Texas, El Paso in 1973, and M.S. and Ph.D. degrees from the California Institute of Technology in 1974 and 1978, respectively. Dr. Mickelson spent a postdoctoral year at Caltech in 1978-1979 before going to the Byurakan Astrophysical

Observatory, Byurakan, Armenia, U.S.S.R. (currently Armenia) for 1970-1980. Following this period, he joined the Elektronikklaboratoriet (Electronics Research Laboratory) of the Norwegian Institute of Technology, at first as an NTNF postdoctoral fellow and later as a staff scientist. His research in Norway primarily concerned characterization of optical fibers and fiber-compatible components and devices. In 1984, he joined the Department of Electrical and Computer Engineering of the University of Colorado, Boulder, where he became associate professor in 1986. His present research concerns integrated optical device fabrication and characterization, microwave measurement and device characterization, and application of optical devices and techniques in high-speed systems. Professor Mickelson has published more than 60 articles in refereed publications and has graduated more than 10 Ph.D. students.

Edward H. Newman received B.S.E.E., M.S., and Ph.D. degrees in electrical engineering from the Ohio State University in 1969, 1970, and 1974, respectively. Since 1974 he has been a member of the Ohio State University Department of Electrical Engineering, ElectroScience Laboratory, where he is currently a professor. His primary research interest is in the development of method-of-moments techniques for the analysis of general antenna or scattering problems, and he is the primary author of the *Electromagnetic Surface Patch Code* (ESP). Other research interests include printed circuit antennas, antennas in inhomogeneous media, scattering from material coated edges, artificial dielectrics, and chiral media. He has published more than 45 journal articles in these areas, and is a co-author of the IEEE Press book *Computational Electromagnetics (Frequency Domain Method of Moments)*. Dr. Newman is a fellow of the IEEE and is a member of Commission B of URSI and the Electromagnetics Institute. He is a recipient of the 1986 and 1992 College of Engineering Research Award and is a past chairman of the Columbus sections of the IEEE Antennas and Propagation and Microwave Theory and Techniques Societies.

J. Tinsley Oden received a B.S. from Louisiana State University in 1959 and a M.S. and a Ph.D. in structural mechanics from the Oklahoma State University in 1962. He is director of the Texas Institute for Computational and Applied Mathematics at the University of Texas-Austin, and holds the Ernest and Virginia Cockrell Chair in engineering as professor of aerospace engineering and engineering mechanics there. He is a member of the National Academy of Engineering and has received from the National Academy of Sciences the Billy and Claude R. Hocott award for distinguished engineering research. He has also received the Worcester Reed Warner medal, the Chevalier dans l'Ordre des Palmes Academie from the French government, and the Eringen Medal from the Society for Engineering Science. He is the author or co-author of 17 books and monographs. His research includes nonlinear continuum mechanics, approximation theory, and numerical analysis of nonlinear problems in continuum mechanics.

Peter M. Pinsky received a Ph.D. in civil engineering from the University of California, Berkeley, in 1981. He joined Brown University as an assistant professor in the Division of Engineering in 1982. In 1984 he moved to Stanford University where he is currently an associate professor of civil and mechanical engineering. His research is in the areas of structural, solid, and computational mechanics.

Vijaya Shankar received M.S. and Ph.D. degrees in aerospace engineering, from Iowa State University. He joined the Rockwell International Science Center in 1976. He is responsible for developing multidisciplinary computing technologies involving fluid dynamics, electromagnetics, structures, and other related disciplines that play a critical role in many defense and defense-conversion (commercial) projects. For his contributions to advancing the state of the art in supercomputing, he has received numerous awards. Dr. Shankar, a fellow of the American Institute of Aeronautics and Astronautics

(AIAA) and the Institute for the Advancement of Engineering (IAE), is a 1984 Rockwell Engineer of the Year, and a recipient of the 1985 Lawrence Sperry Award and the 1991 Dryden Research Lectureship Award from AIAA, the 1985 Outstanding Young Alumnus Award and the 1991 Professional Achievement Citation in Engineering Award from Iowa State University, the 1986 NASA Public Service Medal for exceptional scientific achievement, the 1990 CRAY Gigaflop Performance Award, and the 1992 Distinguished Engineering Achievements Award from IAE. His work in computational electromagnetics (CEM) received the 1993 Computerworld Smithsonian Award in Science. He is an associate editor of the *AIAA Journal* and a member of the editorial board of the *Journal on Computing Systems in Engineering*, and he has authored more than 50 publications in the technical literature. Dr. Shankar is also an adjunct professor at the University of California, Davis.

Lakshman S. Tamil received a B.E. degree in electronics and communication engineering from the Madurai Kamaraj University, Madurai, India, in 1981, and a M.Tech. degree in microwave and optical communication engineering from the Indian Institute of Technology, Kharagpur, India, in 1983. He also received a M.S. degree in mathematics and a Ph.D. degree in electrical engineering from the University of Rhode Island, Kingston, in 1988. Dr. Tamil joined the University of Texas at Dallas in 1988 and is currently an associate professor in electrical engineering there. He is also a member of the Center for Applied Optics at the same university. His research interests include fiber optics, photonic integrated devices and circuits, nonlinear guided wave optics, semiconductor lasers, inverse scattering theory, numerical methods applied to electromagnetic problems and wireless communication. Dr. Tamil is a member of Sigma Xi, the Optical Society of America, and the Electromagnetic Academy, and he is an elected member of Commissions B and D of the International Union of Radio Science.

Richard W. Ziolkowski received a Sc.B. degree in physics magna cum laude with honors from Brown University in 1974 and M.S. and Ph.D. degrees in physics from the University of Illinois at Urbana-Champaign in 1975 and 1980, respectively. He was a member of the Engineering Research Division at the Lawrence Livermore National Laboratory from 1981 to 1990 and served as the leader of the Computational Electronics and Electromagnetics Thrust Area for the Engineering Directorate from 1984 to 1990. Dr. Ziolkowski joined the Department of Electrical and Computer Engineering at the University of Arizona as an associate professor in 1990. His research interests include the application of new mathematical and numerical methods to linear and nonlinear problems dealing with the interaction of acoustic and electromagnetic waves with realistic materials and structures. Dr. Ziolkowski is a member of Tau Beta Pi, Sigma Xi, Phi Kappa Phi, the Institute of Electrical and Electronics Engineers (IEEE), the American Physical Society, the Optical Society of America, the Acoustical Society of America, and Commission B (Fields and Waves) of the International Union of Radio Science (URSI). He is an IEEE fellow and an associate editor for the IEEE *Transactions on Antennas and Propagation*. He served as the vice chairman of the 1989 IEEE/AP-S and URSI symposium in San Jose. He is currently serving as the secretary of the United States' URSI Commission B and as a member of that commission's Technical Activities Committee. He was awarded the Tau Beta Pi Professor of the Year Award and the IEEE and Eta Kappa Nu Outstanding Teaching Award in 1993.

Appendix C
Symposium Participants

Philip B. Abraham
Office of Naval Research
Arlington, Virginia

Dare Afolabi
Indiana University-Purdue University
Indianapolis, Indiana

Roshdy S. Barsoum
Office of Naval Research
Arlington, Virginia

Ted Belytschko
Northwestern University
Evanston, Illinois

Jacobo Bielak
Carnegie Mellon University
Pittsburgh, Pennsylvania

Michael Butler
General Dynamics Electrical Boat Division
Groton, Connecticut

Raymond S. Cheng
David Taylor Model Basin
Bethesda, Maryland

Louise Couchman
Office of Naval Research
Arlington, Virginia

John D'Angelo
GE Corporate Research and Development
Schenectady, New York

Adrianus T. de Hoop
Delft University of Technology
Delft, The Netherlands

Leszek Demkowicz
University of Texas at Austin
Austin, Texas

David C. Dobson
Texas A&M University
College Station, Texas

Ira Dyer
Massachusetts Institute of Technology
Cambridge, Massachusetts

Gordon C. Everstine
David Taylor Model Basin
Bethesda, Maryland

David Feit
Bethesda, Maryland

Po Geng
University of Texas at Austin
Austin, Texas

Jonathan Goodman
New York University
New York, New York

Fernando F. Grinstein
Naval Research Laboratory
Washington, D.C.

Gary Guthart
SRI International
Menlo Park, California

Hermann A. Haus
Massachusetts Institute of Technology
Cambridge, Massachusetts

Thomas J.R. Hughes
Stanford University
Stanford, California

Myles Hurwitz
Naval Surface Weapons Center
Bethesda, Maryland

T. Igusa
Northwestern University
Evanston, Illinois

Arthur Jordan
Office of Naval Research
Arlington, Virginia

K. Kailasanath
Naval Research Laboratory
Washington, D.C.

Loukas Kallivokas
Carnegie Mellon University
Pittsburgh, Pennsylvania

David Karchov
Fairfax, Virginia

Naum Khutoryansky
Drexel University
Philadelphia, Pennsylvania

Ralph Kleinman
University of Delaware
Newark, Delaware

Ron Kolbe
Naval Research Laboratory
Washington, D.C.

Timothy D. Leigh
University of Florida
Gainesville, Florida

R. Lohner
George Mason University
Fairfax, Virginia

I.C. Mathews
Imperial College
London, United Kingdom

John J. McCoy
Catholic University of America
Washington, D.C.

Alan Rolf Mickelson
University of Colorado at Boulder
Boulder, Colorado

Peter Monk
University of Delaware
Newark, Delaware

Richard Morrow
GE Corporate Research and Development
Schenectady, New York

Arje Nachman
Air Force Office of Scientific Research
Bolling AFB, D.C.

Edward H. Newman
Ohio State University
Columbus, Ohio

William Nowlin
SRI International
Menlo Park, California

Ruth E. O'Brien
National Research Council
Washington, D.C.

J. Tinsley Oden
University of Texas at Austin
Austin, Texas

Peter M. Pinsky
Stanford University
Stanford, California

Fred E. Saalfeld
Office of Naval Research
Arlington, Virginia

Andrzej Sasjan
University of Texas at Austin
Austin, Texas

Vijaya Shankar
Rockwell International Science Center
Thousand Oaks, California

Horacio Sosa
Drexel University
Philadelphia, Pennsylvania

Frederic Stasker
University of Maryland, Baltimore County
Baltimore, Maryland

James R. Stewart
Stanford University
Stanford, California

Lakshman S. Tamil
University of Texas at Dallas
Richardson, Texas

Lonny L. Thompson
Clemson University
Clemson, South Carolina

John R. Tucker
National Research Council
Washington, D.C.

Paul Wesson
University of Delaware
Newark, Delaware

John Wilder
General Dynamics Electrical Boat Division
Groton, Connecticut

Barbara W. Wright
National Research Council
Washington, D.C.

Richard W. Ziolkowski
University of Arizona
Tucson, Arizona